DATE DUE

| | | | |
|---|---|---|---|
| | | | |
| | | | |
| | | | |
| | | | |
| | | | |
| | | | |
| | | | |
| | | | |
| | | | |
| | | | |
| | | | |
| | | | |
| | | | |
| | | | |
| | | | |
| | | | |
| | | | |
| | | | |
| | | | |
| | | | |
| GAYLORD | | | PRINTED IN U.S.A. |

# Modeling Marine Systems

## Volume I

Editor

**Alan M. Davies**
Principal Scientific Officer
Proudman Oceanographic Laboratory
Bidston Observatory
Birkenhead, Merseyside, England

CRC Press, Inc.
Boca Raton, Florida

**Library of Congress Cataloging-in-Publication Data**

Modeling marine systems / editor, Alan M. Davies.
p. cm.
Bibliography: v. 1, p.
Includes index.
ISBN 0-8493-4221-X (v. 1)
1. Oceanography—Mathematical models. I. Davies, Alan M.
GC10.4.M36M6 1989
551.46'001'5118—dc20 89-9803
CIP

This book represents information obtained from authentic and highly regarded sources. Reprinted material is quoted with permission, and sources are indicated. A wide variety of references are listed. Every reasonable effort has been made to give reliable data and information, but the author and the publisher cannot assume responsibility for the validity of all materials or for the consequences of their use.



Direct all inquiries to CRC Press, Inc., 2000 Corporate Blvd., N.W., Boca Raton, Florida, 33431.



International Standard Book Number 0-8493-4221-X (Volume I)
Library of Congress Card Number 89-9803
Printed in the United States

# NORMAN HEAPS

## IN MEMORIAM

Sadly, after a long illness, Norman Heaps died in June 1986. Norman's scientific career was truly exceptional in both its productivity, diversity of ideas, and originality.

He was born in Prescot, Lancashire, and before entering a career in oceanography was employed in the aircraft industry and subsequently as a lecturer in Mathematics. An account of Norman's life and work is presented as the first paper in this volume written by Eric Jones.

Norman's research papers in oceanography covered a wide range of topics, with classic papers on storm surges, lake circulation, shelf sea problems, and topics such as tidal resonance and tidal power. He developed a range of both analytical and numerical techniques for solving a broad class of oceanography problems; being regarded as a world expert on shallow sea modeling.

In view of his worldwide reputation in both analytical and numerical modeling of oceanography problems it is appropriate to begin this volume with a dedication to Norman.

In the course of inviting people to contribute articles to such a volume it was clear that Norman's work and influence had brought him into contact with an extremely large number of people actively researching on numerical modeling. Regretably not everyone could be invited, those who were, wrote very substantial articles which will appear in this volume and a subsequent volume both of which are dedicated to Norman Heaps.

Norman was a true friend who will be greatly missed by all who knew him, personally or through his work.

**Alan M. Davies**

# THE EDITOR

**Alan M. Davies, B.Sc., Ph.D.,** is a senior scientist at the Proudman Oceanographic Laboratory, which he joined in 1971.

The early part of his career was concerned with preparative chemistry in particular organic chemistry, leading to the synthesis of several new compounds and the subsequent publication of a number of papers describing this research in chemistry journals. In 1968, Dr. Davies received a B.Sc. degree in Chemistry from Sheffield University.

Future research was concerned with theoretical and quantum chemistry, in particular the calculation of electro-magnetic properties of small molecules. This work involved significant numerical calculations, and during this period he developed an interest in computational science and numerical methods. This research lead to the development of a number of new methods for computing electro-magnetic properties which were subsequently published in the scientific literature. Based upon this research he was awarded a Ph.D. degree in 1971.

His interest in computers led him to accept a position as Systems Analyst at the Proudman Oceanographic Laboratory, initially as a computer specialist, although subsequently he worked on a tidal prediction program. Further work involved the development of an operational storm surge prediction model leading onto his interest in three dimensional models.

In addition to his work at the Proudman Laboratory, Dr. Davies has spent 1- to 3-month periods as a visiting scientist and guest lecturer at various universities and institutes in Europe, Scandinavia, the U.S., Canada, and Australia. He has also been an invited speaker at a number of international meetings, summer schools, and workshops, at which he has presented papers on oceanography, numerical modeling, and computing.

His major interest in the field of oceanography is in the development and validation of three dimensional shallow sea tidal and circulation models. Dr. Davies has been a member of the ICES Flushing Time Group on pollution problems and is presently a member of GESAMP Working Group 25, concerned with water quality models of shallow seas. He is also secretary of JONSMOD (a shallow sea modeling group) and has published over 60 scientific papers on a range of topics.

## ADVISORY BOARD

## CONTRIBUTORS

**Jan O. Backhaus, Dr.rer.nat.**
Professor
Institut für Meereskunde
Universität Hamburg
Hamburg, West Germany

**Anthony J. Bowen, Ph.D.**
Professor
Department of Oceanography
Dalhousie University
Halifax, Nova Scotia, Canada

**Byung Ho Choi, Ph.D.**
Professor
Department of Civil Engineering
Sung Kyun Kwan University
Suwon, Korea

**Roger A. Flather, Ph.D.**
Principal Scientific Officer
Proudman Oceanographic Laboratory
Bidston Observatory
Birkenhead, Merseyside, England

**Gunnar K. Furnes, Dr. Philos.**
Senior Scientist
Bergen Scientific Center
Bergen, Norway

**Bjorn Gjevik, Dr. Philos.**
Professor
Department of Mathematics
University of Oslo
Oslo, Norway

**Kevin P. Hubbert, Ph.D.**
Higher Scientific Officer
Proudman Oceanographic Laboratory
Bidston Observatory
Birkenhead, Merseyside, England

**David A. Huntley, Ph.D.**
Professor and Reader
Institute of Marine Studies
Plymouth Polytechnic
Plymouth, Devon, England

**Kolumban Hutter, Ph.D.**
Professor
Department of Mechanics
Technische Hochschule
Darmstadt, West Germany

**John E. Jones**
Proudman Oceanographic Laboratory
Bidston Observatory
Birkenhead, Merseyside, England

**G. W. Lennon**
Professor and Dean
School of Earth Sciences
Flinders University of South Australia
Bedford Park, South Australia

**James J. O'Brien**
Secretary of the Navy Professor
Department of Meteorology and
Oceanography
Florida State University
Tallahassee, Florida

**Vijay G. Panchang, Ph.D.**
Assistant Professor
Department of Civil Engineering
University of Maine
Orono, Maine

**Roger Proctor, Ph.D.**
Proudman Oceanographic Laboratory
Bidston Observatory
Birkenhead, Merseyside, England

**R. B. Schahinger**
School of Earth Sciences
Flinders University of South Australia
Bedford Park, South Australia

**Thomas F. Stocker, Dr. sc. nat.**
Laboratory of Hydraulics, Hydrology
and Glaciology
ETH Zurich
Zurich, Switzerland

**Cai Wenju**
School of Earth Sciences
Flinders University of South Australia
Bedford Park, South Australia

**Judith Wolf, Ph.D.**
Proudman Oceanographic Laboratory
Bidston Observatory
Birkenhead, Merseyside, England

# TABLE OF CONTENTS

Chapter 1

# THE SCIENTIFIC CAREER OF DR. NORMAN STUART HEAPS, D.Sc.

**John E. Jones**

By any standards the scientific career of Norman Stuart Heaps was positively outstanding. He is perhaps best remembered for his pioneering work on the numerical modeling of tides and storm surges which eventually led to advanced flood-prediction models in use today. He is also remembered as a pioneer in other areas, such as the effects of tidal barriers, long-term circulation due to density currents, and the dynamical consequences of stratification in lakes, shelf seas, and the continental shelf slope.

Norman Heaps was born in 1928 in Prescot, just outside Liverpool, and received his first scientific education at Prescot Grammar School. It is startling to discover in his school reports that he never received more than a ''fair'' grade for mathematics. In 1945 he entered the University of Liverpool and after 4 years obtained an upper second class honors degree in mathematics, with physics and oceanography as subsidiary topics. He then went to work at the Royal Aircraft Establishment, Farnborough, on the theory of wing structures. This early period was a proving ground for some of the basic techniques that Norman developed later. Here, also, Norman gave an early sign of his promise by solving a problem which had kept other mathematicians puzzled for weeks. After 5 years however, Norman began to feel restricted by his work and also wished to return to the North of England. As he put it, he preferred his landscapes ''more rugged''.

He spent a further 5 years in the aircraft industry as a mathematician in the Aerodynamics Department of the English Electric Company's Aircraft Division at Warton in Lancashire. From there he moved into the academic world by becoming Senior Lecturer at the Royal College of Advanced Technology, Salford (which is now the University of Salford). Apart from anything else these two posts enabled him to travel home to Prescot at the weekends for his mother's cooking.

The post at Salford brought with it long summer vacations. Norman took advantage of these to spend some time at Ferry House, the laboratory of the Freshwater Biological Association on Windermere in the Lake District. During these visits, he carried out a theoretical study of water movements in Windermere under stratified conditions. This became a part-time M.Sc. thesis accepted by the University of Liverpool in 1960.

His time at Ferry House proved crucial. By sheer chance during one visit, he happened to hear of a vacancy at Bidston. So it was in 1962 that Norman began his long and very fruitful association with Bidston Observatory.

During his first decade at Bidston, Norman started on three main themes of work which he developed throughout the rest of his life, i.e., storm surges, tidal barriers, and residual drifts. However, from time to time he continued to look at stratified systems, eventually returning more fully to the topic in later years. In those early days at Bidston, computing facilities were limited and so the work followed much the tradition of Proudman and Doodson by being analytic or semi-analytic.

The storm-surge problem was of national importance following the disastrous 1953 storm surge. Norman investigated several approaches, using elaborate analytical techniques at first, and gained several valuable insights. However, as better computing facilities became available he converted to a numerical approach which culminated in his Royal Society paper of 1969 which demonstrated a two-dimensional model of the North Sea. From this basis followed the advanced development work by Norman and his colleagues leading to the fully operational schemes that the United Kingdom Meterological Office is running at the moment.

In the field of tidal power, Norman's early analytical work led eventually to the numerical approach adopted by two of Norman's Ph.D. students with respect to the Bay of Fundy and the Bristol Channel.

A similar pathway was followed when considering long-term residual drifts. Norman's early analytical work on density currents in Liverpool Bay once again led to a conversion to numerical techniques with a wider-scale Irish Sea investigation.

We can see here a common pattern, first the analytic study, then the numerical approach followed up by Norman himself or others. Norman did not, however, abandon analytic techniques completely. He would love to cover page after page of carefully saved scrap computer paper with acres and acres of formulae and would positively relish using the Laplace Transform. One consequence of his methods was a paper published in 1969 where he developed a modification to a tidal theorem by Proudman regarding the estimation of tidal conditions in the interior of a sea from coastal measurements. This is now known as the Proudman-Heaps tidal theorem.

Norman's work spanned the period from the earliest computers right up to the present-day super-computer era. He used to talk about the early days where he had to run computer programs three times just to check that the computer was producing the same answer each time. Norman eagerly took advantage of every technical advance and some of his most recent models were very large indeed.

It was interesting to see the way in which Norman dealt with computers. He had great attention to detail, and although as time went on, there were more and more data analysis methods such as graphical output, Norman did insist wherever possible on seeing every number that the models produced. He would spend hours examining great quantities of computer printout, carefully annotating in pencil various points of interest until in some way he had built up in his head an overall picture of what the model was doing. This labor meant that he could build up a clear interpretation of the physical significance of any model run. It also meant that he was able to tell very quickly if there was something amiss, even if in a subtle way. This apparently laborious approach therefore actually helped to save time.

It was in Norman's computing work that he pioneered the representation of current structure through depth by means of orthogonal functions. This technique enabled the full three-dimensional current structure of a sea area to be determined quickly and efficiently. He first tried this scheme on a simple model of the North Sea which he called IBASIC. It was only years later that Norman revealed that IBASIC was based on an advertising slogan of a certain vacuum cleaner company—"It Beats As it Sweeps as It Cleans". IBASIC was followed by the Irish Sea model which was used as a testbed for many of Norman's ideas. It is a pitifully small model by modern standards, but over a period of 7 years Norman threw every possible physical process he could imagine at this model, some of which he came to apply elsewhere.

Norman's M.Sc. thesis of 1960 dealt with stratification in lakes. In recent years the wheel turned full circle as he took up the topic of stratification in various shallow-sea areas and at the continental shelf slope. Even during the last year of his life as lung cancer and the associated painful treatments made working very difficult for him, he produced a new model of the North Sea to include fronts.

Apart from the actual scientific content of Norman's career, mention must be made of the characteristic way that he dealt with the presentation of his results and with those about him.

He would take great care in the writing of his scientific papers and developed a concise yet easily readable style. It is not surprising, therefore, that many of his papers have become classics. Furthermore, this style of writing became perceptible in the work of his colleagues as well. This is one of the ways in which his influence will be felt for many years.

Attending a lecture by Norman Heaps at a conference was always a treat. He had acquired a wealth of lecturing experience over the years and as his distinctive style developed, a classic Heaps lecture was always eagerly anticipated. Sometimes, like at the opera, the audience would tend to neglect the content of the talk, which was nevertheless usually excellent, to concentrate on the performance, which was breathtaking. His talks would start quietly with Norman checking his pocket watch to note the time available for his talk and also to let the Chairman know that he was fully aware that he had a limited time. Then the talk itself would start quietly until some crucial point was reached when his voice would get louder, his face would flush with enthusiasm, and he would bubble over with the information he had to convey. As the time left for his talk ran out he would signal to the chairman that he was aware of this by consulting his pocket watch more and more frequently and thus by this ruse he would manage to sneak an extra 5 or 10 min until the talk was ended with a rousing conclusion.

Apart from these public performances, Norman was a genuinely quiet person. He would give himself entirely to his work and preferred to quietly pursue his studies. Whereas some might seek personal glory, Norman was genuinely modest and selfless. When he came to lead the numerical modeling section at Bidston some of his other qualities became apparent. He had the ability to gently motivate others to give of their best and he was always ready to encourage and give advice. He would always try to promote the careers of promising scientists, not only in his own group at Bidston but workers at other institutes around the world.

Outside work, Norman also pursued quiet interests. He enjoyed fell walking, especially in the Lake District, and reached a high standard in Crown Green bowling, winning several trophies. However, his work was always of prime importance to him.

Norman Heaps never married but he fully appreciated the fact that his mother, by continuing to take care of his domestic arrangements, gave him 22 years of freedom to pursue his studies without distraction. Norman leaves a tangible legacy of some 70 published papers but he also leaves an influence on many other people's lives which will be felt for a very long time into the future.

Chapter 2

# ON THE DETERMINATION OF HYDRAULIC MODEL PARAMETERS USING THE ADJOINT STATE FORMULATION

**Vijay G. Panchang and James J. O'Brien**

## TABLE OF CONTENTS

# I. ABSTRACT

The estimation of various parameters involved in numerical modeling of hydraulic systems is discussed. The method described here uses the adjoint equation approach, which has also been used in ground-water hydrology, with the strong constraint formulation in the relevant functional. The parameter-correction equation is used to modify parameter estimates, with the philosophy that there is no modification when the data misfit vanishes. The application to a set of equations (e.g., equations of motion) is illustrated. In particular, this data-assimilation method is used to determine the friction factor for tidal rivers. The method is thus an alternative to those suggested by Dronkers,[17] Yeh and Becker,[18] or Chiu and Isu.[19]

# II. INTRODUCTION

The mathematical modeling of almost all natural dynamical systems involves some physical aspects that defy convenient quantification, or render the equations too difficult to solve. Examples of these in the equations of motion are boundary stresses (bottom friction, wind shear), eddy viscosity, etc. Such physical processes are therefore included in the model in some parameterized form, but the associated unknown parameters usually cannot be estimated accurately from experiments. Values are assigned to them based on empirical relationships, experiments (where the circumstances are, in all likelihood, different from the situation under consideration), individual judgment, or precedent. If no data are available for the situation of interest, it is difficult to estimate the parameters.

We are concerned in this article with those cases where at least some data is available. This availability enables the model to then be "tuned", i.e., the model is run repeatedly for different values of the parameter until the model results match the data. This is often an expensive process, since the parameterized process may not bear a linear relationship to the data, or, for different values of the parameter, the departure of the model solutions from data may be small for different segments of the $(\bar{x},t)$ domain for which the data is available.

In the field of ground-water hydrology, a more systematic attempt has been made in recent years at estimating the model parameters (hydraulic transmissivities). Yeh[1] gives a review of the various methods used in the inverse ground-water problem. Of particular interest is the adjoint-state method used by Neuman,[2] Neuman and Carrera,[3] and Carrera and Neuman[4-6] to determine the transmissivities from data, and by Sykes et al.[7] to study the sensitivity of ground-water model results to the input parameters. In these studies the ground-water flow is described by a single (diffusion-type) governing equation.

A method similar to the ground-water hydrologic case is described in this paper for estimating parameters of other flow models, such as those governed by the equations of momentum and continuity. The adjoint equation approach (described in Section IV) is combined with the strong constraint formulation of Sasaki,[8] giving a general framework that is readily applicable to a set of steady or unsteady state equations. The philosophy adopted is that, if an initial estimate of the parameter yields a mismatch of the model results and data, this mismatch (hereafter referred to as "data misfit") should be minimized (reduced to noise level, if possible). One will then have obtained the desired value of the parameter that produced the data. As such, the data play a role in providing successively better values of the parameter. The method may briefly be described as follows:

Let $D(x,t)$ be the variable for which data $D'(x_i,T)$ exists at some locations $x_i$ for duration $T$. Let $p(x,t)$ be some parameter (such as eddy viscosity, etc.) of the model which produces the data $D'$. If the governing differential equations are written as:

$$L_j(D) = 0, \qquad (j = 1(1)N) \tag{1}$$

then we will try to minimize the square of the data misfit,

$$\iint \Gamma_i (D - D')^2 dxdt, \qquad (\Gamma_i = 1 \text{ for } x = x_i$$
$$= 0 \text{ otherwise}) \tag{2}$$

subject to the constraints (Equation 1). Since minimizing Equation 2 will not necessarily yield the true value of p (unless, of course, this minimum is exactly zero), we will obtain only an estimate p′ of p, and we require that

$$\iint (p(x,t) - p'(x,t))^2 dxdt \tag{3}$$

also be a minimum. The above expression is the "plausibility criterion" of Neumann and Carrera.[3]

The variational formulation of the problem with the constraint (Equation 1) implies minimizing the functional F, given by[9]

$$F = \iint K_D \Gamma_i (D - D')^2 dxdt + \iint K_p (p - p')^2 dxdt + \sum_{j=1}^{N} \iint \lambda_j L_j(D) dxdt \tag{4}$$

where $\lambda_j$ is the Lagrange multiplier associated with the jth equation, and the Gauss precision moduli[10] $K_D$ and $K_p$ are introduced for scaling purposes (these are discussed later). The above is a "strong constraint" formalism, following the classification of Sasaki.[8] A "weak constraint" formalism could be used by replacing L(D) in Equation 4 by $[L(D)]^2$, and this formulation can be physically interpreted by saying that the governing equation is only approximately satisfied. This formalism has been used by Provost and Salmon.[11] If necessary, filter terms can be added in Equation 4 to eliminate small-scale variability. They are desirable when the data are noisy or sparse,[8] e.g., Provost and Salmon[11] have included a filter term that describes spatial roughness, since the sparsely sampled data provide no information about length scales smaller than the data separation. The functional F can be minimized by differentiating, resulting in extra "adjoint" equations for the Lagrange multipliers and an algebraic equation for systematic correction of the initial estimate p′.

The foregoing ideas are illustrated in this paper by application to the problem of estimating friction factors for a tidal river. The flow situation is described by one-dimensional time-dependent equations of motion and flow data from laboratory simulations that were taken from Harlemann and Lee.[12] In Section III some details of the laboratory study and the governing equations are given. The formulation of the functional F and its minimization, together with a solution algorithm, are discussed in Section IV. The results of this study and concluding remarks are given in Section V.

## III. ILLUSTRATIVE EXAMPLE

We consider a one-dimensional tidal-flow model, with the objective of determining the bottom friction coefficient. Harlemann and Lee[12] have reported data pertaining to tidal-flow-simulation studies performed at the U.S. Waterways Experiment Station. The model consists of a rectangular flume, closed at one end, and connected at the other end to a large basin where sinusoidal tidal motion is produced. Harlemann and Lee[12] have reported making a number of runs of their finite-difference model for various values of Manning's n to obtain agreement of their computer model results with measurements along the flume. The data pertaining to Test 1 described in Harlemann and Lee[12] are shown in Figure 1.

The phenomenon may be described by the following equations of continuity and momentum:

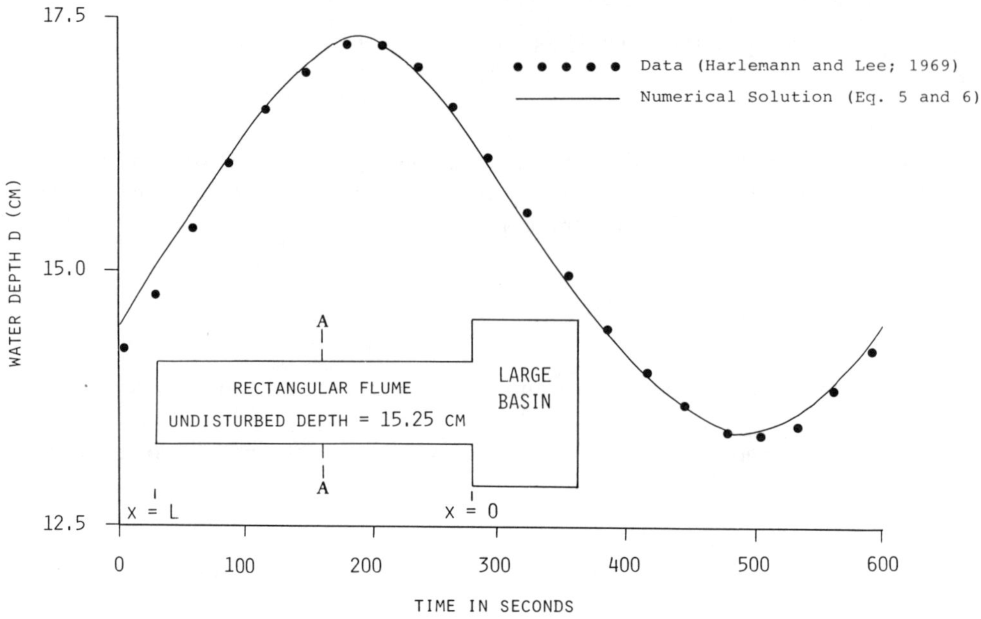


FIGURE 1. Variation of surface elevation with time, at $x \approx 0.5L$. Dots refer to hydraulic model data, and curve to numerical solutions obtained with $f = 0.189$.

$$D_t + q_x = 0 \tag{5}$$

$$q_t + gDD_x + \frac{f|q|q}{8D^2} = 0 \tag{6}$$

where D(x,t) is the distance between the water surface and channel bottom, q(x,t) is the flow per unit width, g is the gravitational acceleration, and f is the friction factor. The boundary conditions are

$$D \text{ (in cm)} = 15.25 + 1.525 \sin\omega t \qquad \text{at} \quad x = 0$$

$$(\omega = 2\pi/600)$$

$$\frac{\partial q}{\partial x} = 0 \quad \text{at} \quad x = L$$

A finite-difference scheme staggered in space and time is used, and when the friction factor f is assigned the value of 0.189 (determined by Harlemann and Lee[12] with hydraulic experiments and model tuning), the resulting surface elevations at one point in the flume (at $x = L/2$, approximately) are shown in Figure 1.

We will now assume that this data was available, and consider the inverse problem of determining f.

## IV. THE ADJOINT EQUATIONS

Corresponding to Equation 4, the functional F for the problem described in the preceding section is

$$F(D,q,\lambda_D,\lambda_q,f) = \iint \frac{K_D}{2}\Gamma_i(D - D_i')^2 dxdt + LT\frac{K_f}{2}(f - f')^2$$

$$\iint \lambda_D(D_t + q_x)dxdt + \iint \lambda_q\left(q_t + gDD_x + \frac{fq|q|}{8D^2}\right)dxdt \tag{7}$$

where the third and fourth integrals contain the strong constraints (viz. the governing Equations 5 and 6), and $\lambda_q(x,t)$ and $\lambda_D(x,t)$ are the Lagrangian multipliers corresponding to these. L and T are characteristic length and time scales of the problem, introduced for proper scaling since the parameter f is not considered to be a function of x and t, as in Equation 4. If, for instance, $f = f(x)$, then the second term in Equation 7 would read

$$T\int K_f(f - f')^2 dx$$

Here we choose $L = 100$ m (length of the channel) and $T = 600$ s (length of the data). The integrals in Equation 7 are in the domains $0 < x < L$ and $0 < t < T$.

The functional F is minimized by differentiating:

$$\frac{\partial F}{\partial \lambda_D} = 0, \quad \text{resulting in Equation 5}$$

$$\frac{\partial F}{\partial \lambda_q} = 0, \quad \text{resulting in Equation 6}$$

$$\frac{\partial F}{\partial f} = 0, \quad \text{resulting in } f = f' - \frac{1}{LT}\iint_{x\,t} \frac{\lambda_q|q|q}{K_f 8D^2}dxdt \tag{8}$$

$$\frac{\partial F}{\partial D} = 0, \quad \text{resulting in } \frac{\partial \lambda_D}{\partial t} + gD\frac{\partial \lambda_q}{\partial x} + \frac{\lambda_q f|q|q}{4D^3} - \Gamma_i K_D(D - D_i) = 0 \tag{9}$$

$$\frac{\partial F}{\partial q} = 0, \quad \text{resulting in } \frac{\partial \lambda_q}{\partial t} + \frac{\partial \lambda_D}{\partial x} - \frac{2\lambda_q f|q|}{8D^2} = 0 \tag{10}$$

While the first three derivatives are easily derived, the derivation of the last two is described in Appendix A. It is therefore seen that by minimizing the functional, the original governing equations are recovered, along with Equation 8 for estimating a corrected value f from the original estimate f′, and two adjoint Equations 9 and 10 which allow the calculation of the Lagrangian multiplier needed in Equation 8. A more formal minimization of the functional, resulting in Equations 8 to 10, is presented in Appendix B.

Some simplification of Equations 8, 9, and 10 is wrought by letting $\lambda_q/K_f = \Lambda_q$ and $\lambda_D/K_f = \Lambda_D$ and $K_D/K_f = K$, so that we have only one Gauss precision modulus K instead of $K_i$ and $K_f$. Equations 8, 9, and 10 then take the form:

$$f = f' - \frac{1}{LT}\iint \frac{|q|q}{8D^2}\Lambda_q dxdt$$

or,

$$f = f' - \Delta f \tag{8a}$$

$$\frac{\partial \Lambda_D}{\partial t} + gD\frac{\partial \Lambda_q}{\partial x} + \frac{\Lambda_q f|q|q}{4D^3} - \Gamma_i K(D - D_i') = 0 \tag{9a}$$

$$\frac{\partial \Lambda_q}{\partial t} + \frac{\partial \Lambda_D}{\partial x} - \frac{2\Lambda_q f|q|}{8D^2} = 0 \tag{10a}$$

The adjoint equations have two important features. We note that the friction terms in the governing Equation 6 and the corresponding adjoint Equation 10a have opposite signs. This implies that the adjoint equations have to be integrated backwards in time. Secondly, if the initial and boundary conditions for this backward integration are chosen to be

$$\Lambda_q(x,T) = 0 \quad \text{and} \quad \Lambda_D(x,T) = 0$$

$$\Lambda_q(L,t) = 0 \quad \text{(since q is specified at L,t)}$$

$$\Lambda_D(0,t) = 0 \quad \text{(since D is specified at 0,t)} \tag{11}$$

then the adjoint equations are forced only by the data misfit term in Equation 9a. The data misfit thus plays the dominant role in determining the corrected friction factor (Equation 8a), since the Λ's depend on this forcing function. The procedure for determining the parameter f can be outlined as follows:

1. Choose an initial estimate $f = f_i$.
2. Run the model for the governing Equations 5 and 6 with this value of f. Obtain q(x,t) and D(x,t) for $0 < x < L$ and $0 < t < T$.
3. Calculate the resulting data misfits $D(x_i,t) - D'(x_i,t)$.
4. Run the adjoint model equations 9 and 10 backwards in time, using the q's and D's calculated in 2. The numerical scheme for this will be very similar to that of the original model, as the Λ-equations are very similar.
5. Use the Λ's calculated in 4 and the q's and D's calculated in 2 to determine $f_{i+1} = f_i - \Delta f$ (Equation 8a). If $|\Delta f| <$ some prescribed tolerance, the solution is reached, STOP.
6. Set $f = f_{i+1}$, and go to 2.

The use of the parameter-correction equation in step 5 above is a direct realization of the philosophy set out in Section II, since Δf is dependent on $\lambda_q(x,t)$, which is forced only by the data misfit in the adjoint equations, as mentioned earlier.

The parameter correction Equation 8 can also be discussed in the context of an optimization procedure. Consider the minimization of the data misfit:

$$F' = \frac{1}{LT} \iint \frac{K_D}{2} [D(x,t) - D'(x,t)]^2 \Gamma_i dxdt$$

We can deduce (from Equation B7) that

$$\frac{1}{LT} \iint_{xt} \frac{\lambda_q |q| q}{K_f 8D^2} dxdt = \Delta f = \frac{\partial F'}{\partial f}$$

and Equation 8 may be rewritten as

$$f_{i+1} = f_i - \left(\frac{\partial F'}{\partial f}\right)_i$$

Thus, using Equation 8 is equivalent to minimizing F′ by the method of steepest descent, which may be written more generally as

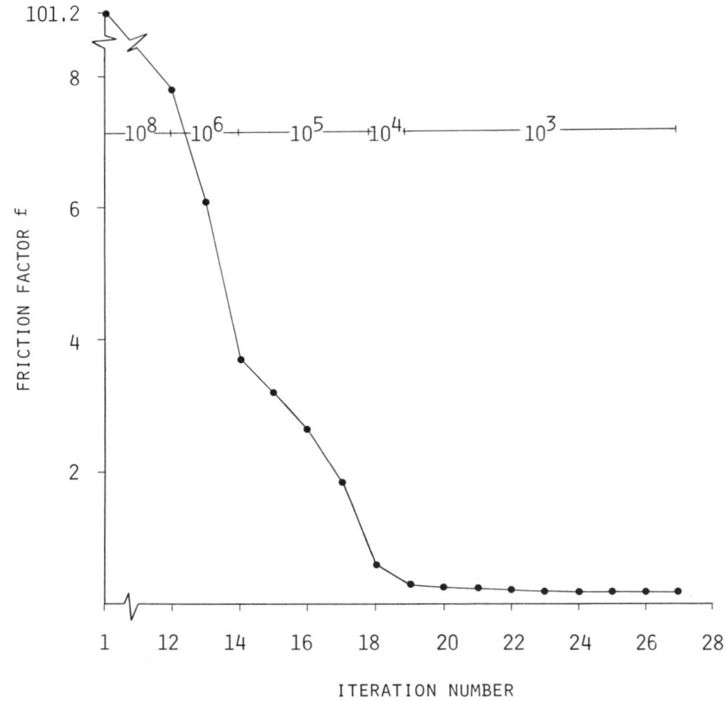


FIGURE 2. Convergence of f when initial estimate $f_1 = 101.2$. Changes made in K during the convergence process are indicated.

$$f_{i+1} = f_i - \alpha\left(\frac{\partial F'}{\partial f}\right)_i$$

where $\alpha$ is a variable weighting factor (implicitly represented here by $1/K_f$) used to ensure that $F'$ is reduced. It is also possible to use other minimization schemes, such as the conjugate gradient method, which may be written generally as:

$$f_{i+1} = f_i + \gamma_i d_i$$

where $d_i = -(\Delta f)_i + \beta_{i-1}\, d_{i-1}$.

The $\gamma$ and $\beta$ are determined by appropriate formulae (e.g., Walsh[13]). Steepest descent and conjugate gradient algorithms, or some variations thereof, have been commonly used in adjoint ground-water modeling studies (e.g., Chen et al.,[14] Chavent et al.,[15] Neumann,[2] Carrera and Neumann,[5] Yziquel and Bernard[15]) and although several comparisons have been made, there appears to be no consensus as to which algorithm is superior.

## V. RESULTS AND DISCUSSION

It is seen from the algorithm given in the preceding section that $\Delta f \rightarrow 0$ when the data misfits $(D - D_i') \rightarrow 0$. The number of iterations required to obtain the solution will therefore depend on the magnitude of the departure of the first solution from the data. If a good initial estimate $f_1$ can be made, the data misfit will be small to start with, and the process will converge quickly. Otherwise more iterations are required. Figures 2 and 3 show the convergence of f when the initial estimates $f_1 = 101.2$ (an inordinately high value which damps out almost all motion, deliberately chosen) and $f_1 = 0.067$, respectively. In both cases, f

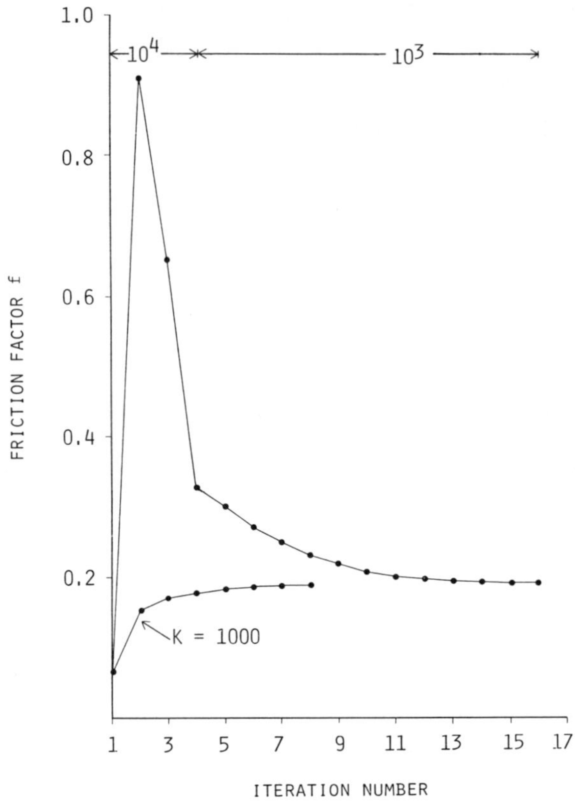


FIGURE 3. Convergence of f when initial estimate $f_1$ = 0.067. Two cases are shown, K being constant (= 1000) for the lower curve, and varied as indicated for the upper curve.

converges to a value of 0.189, which corresponds to a Manning's n = 0.02 (determined by Harlemann and Lee[12] by hydraulic experiments and model tuning). Figures 4 and 5 show how the data misfit is reduced during successive iterations. (In this study, no effort was made to accelerate convergence.)

Some remarks are appropriate regarding the choice of K. It is best to choose a large value for K first, and if, during the succeeding iteration, some measure of the data misfit (like $\int (D - D_i')^2\, dt$) is found to have increased, then K can be reduced. It is easy to see from Equations 8a to 10a that $\Lambda_q(x,t)$, and hence $\Delta f$ are directly proportional to K. In the case where $f_1$ was chosen to be 101.2, a value of K larger than $10^8$ in the initial stages would give more rapid convergence. However, Figure 3 shows one case where the second iterate is actually farther from the solution than the first iterate, due to K being too large (10,000). This, however, only increases the number of iterations required to obtain the true solution. If K= 1000 this does not happen and convergence is rapid. Also, if during the convergence process, $\Delta f$ is seen to reverse sign after a few iterations, then this is an indication that the solution region has been reached, but a large K causes f to overshoot. Oscillations may be then observed if K is not reduced, since the overshoot reverses the sign of the data misfit. In this study, K is reduced by a factor of 10 (see Figure 2) whenever $\Delta f$ changed sign. (This reduction factor could also be allowed to vary for greater efficiency.) Similar treatment of K is necessary if a large K yields a negative f, or, alternatively, the constraint that $f > 0$ can be included in the functional. In any case, K can be easily treated as a self-adjusting parameter based on $\Delta f$, and the above criteria or similar ones can be built into the

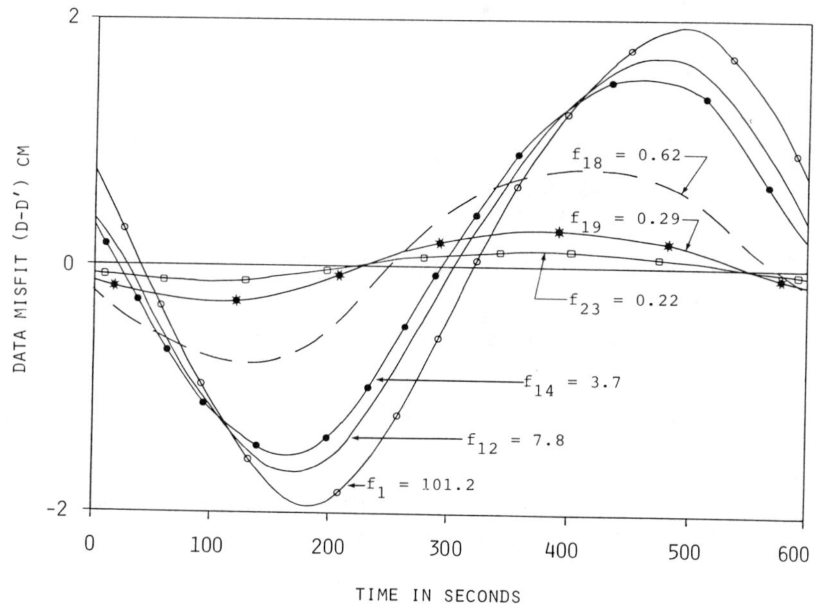


FIGURE 4. Data misfit for various iterates ($f_1$ = 101.2).

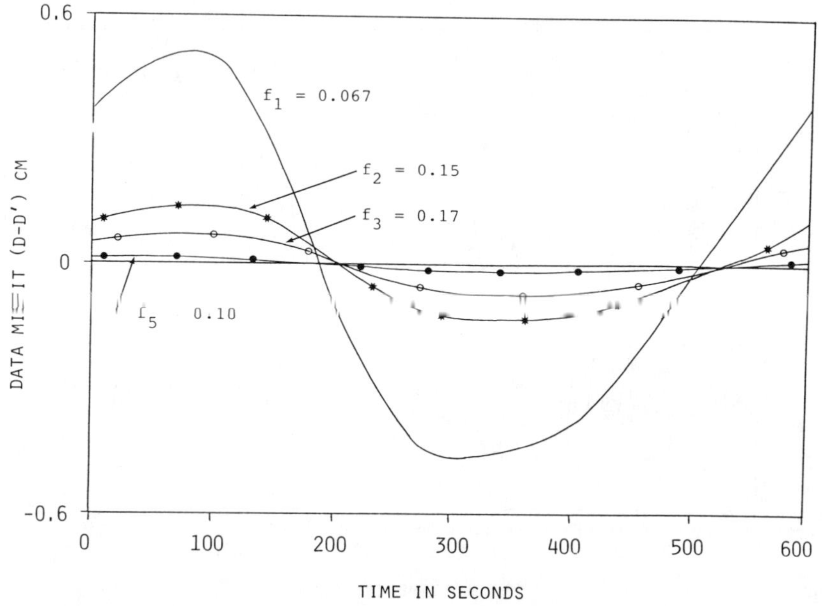


FIGURE 5. Data misfit for various iterates ($f_1$ = 0.067).

program, so that the iterations can proceed unmonitored. The most important observation is that, regardless of the magnitude of K, the method always modifies a given iterate in the direction of the true friction factor, and never away from it.

Unlike the laboratory data used in the above calculations, most field data are usually contaminated by noise. In order to assess the effect of noisy data on the performance of the adjoint model, Gaussian noise was superimposed (Figure 6) on the data used previously. Figure 7 shows the data misfit shrinking to noise level when the true value of the friction

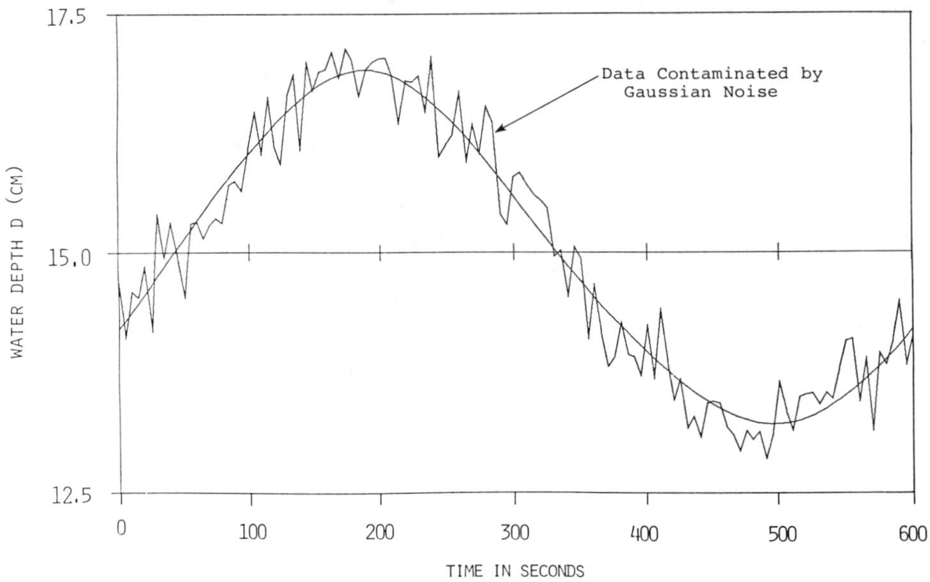


FIGURE 6. Variation of surface elevation with time, at $x \approx 0.5L$ (as in Figure 1), with noise superimposed on model output.

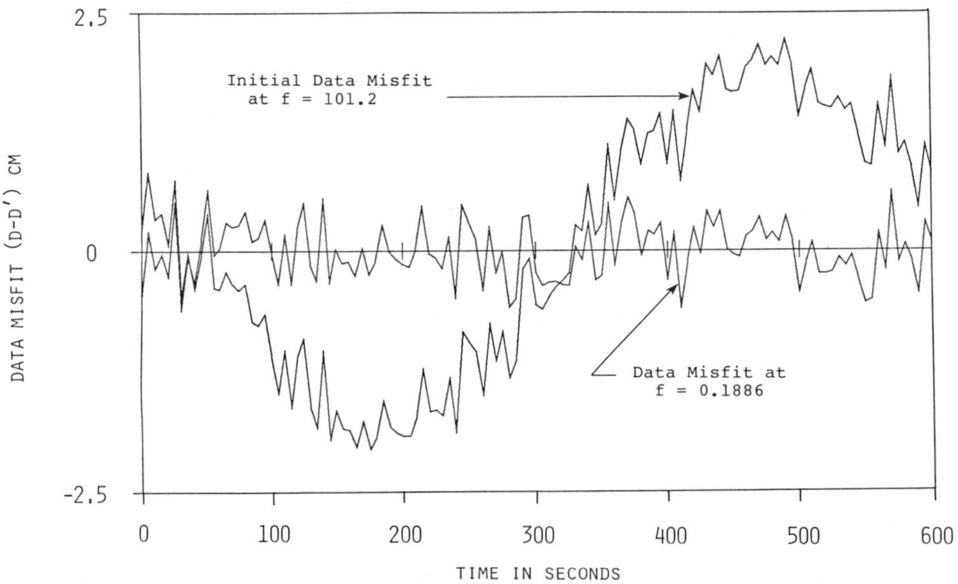


FIGURE 7. Reduction of data misfit ($f_1 = 101.2$) for data shown in Figure 6.

factor is obtained from the algorithm. At least in the case of this well-posed problem, the model appears to be unaffected by the noise in the data.

The above example demonstrates that the strong constraint variational formulation can be used successfully to determine friction factors for tidal rivers. Dronkers[17] has suggested a rather elementary method for estimating the Chezy coefficient for short river sections, by algebraically manipulating governing equations similar to equation 5 and 6, and assuming a linear variation of $D_x$ and other quantities along the section. The data required for this method consists of the tidal elevations at the ends of the channel section and discharge measurements at some point within. The method proposed in this paper also requires boundary conditions (for the original forward model) but can utilize either surface elevations or discharge measurements as data, and makes no simplifying assumptions. It may be considered an alternative to other more sophisticated methods of estimating the friction factor in open channel flow, e.g., Yeh and Becker,[18] who have devised a linear programming formulation to minimize the data misfit; or Chiu and Isu,[19] who have used Kalman filtering in conjuction with the standard-step method to compute the friction factor from steady-state flow data in the Rio Grande.

This paper has discussed another adaptation of a data assimilation technique (e.g., Harlan and O'Brien,[20] Talagrand and Courtier,[21] Courtier and Talagrand[22]) to water-resources engineering. The technique determines the friction factor for river flows, by utilizing the parameter-correction equation in conjunction with the adjoint model. The approach can be used for other hydraulic models as well, in the context of model tuning, as a viable alternative to the laborious process of varying the parameters (probably arbitrarily) and comparing results for many runs. The iterative scheme suggested here automatically stops when the data misfit is reduced, essentially, to noise level. As such, the data is used as a guide to obtain better estimates of the parameter.

## ACKNOWLEDGMENTS

This research was supported by NASA, Oceanic Processes, Contract NAGW-985. We wish that Norman Heaps was with us today. He would be very interested in using this technique in his storm-surge models.

## APPENDIX A

We first rewrite Equation 7 after integrating once the last two of the four terms in the right hand side

$$
\begin{aligned}
F(D,q,\lambda_D,\lambda_q,f) = & \iint \frac{K_D}{2} \Gamma_i (D - D_i')^2 dxdt + LT \frac{K_f}{2} (f - f')^2 \\
& + \int_x \lambda_D D\Big|_{t=0}^{t=T} dx - \iint D \frac{\partial \lambda_D}{\partial t} dxdt + \int_t \lambda_D q\Big|_{x=0}^{x=L} dt \\
& - \iint \frac{\partial \lambda_D}{\partial x} \cdot q dxdt + \int_x \lambda_q \cdot q\Big|_{t=0}^{t=T} dx - \iint q \frac{\partial \lambda_q}{\partial t} dx \\
& + \int_t \frac{gD^2}{2} \lambda_q\Big|_{x=0}^{x=L} dt - \iint \frac{gD^2}{2} \frac{\partial \lambda_q}{\partial x} dxdt + \iint \lambda_q \cdot \frac{fq|q|}{8D^2} dxdt
\end{aligned}
$$

The boundary integrals (viz. the third, fifth, seventh, and ninth terms on the right-hand side) are all equal to zero, as explained in Appendix B. Differentiation of the remaining terms in F with respect to D and q yields equations 9 and 10.

## APPENDIX B

In the following, we assume that $f = f(x)$, since this case has been briefly alluded to in the text. The functional F is formulated slightly differently, including only the least squares error and the plausibility criteria (as in Neuman[2]), and the governing equations are subsequently accounted for.

$$\text{Let } F = \iint\limits_{tx} \frac{K_D}{2} [D(x,t) - D'(x,t)]^2 \Gamma_i dxdt + T \int_{x=0}^{L} \frac{K_f}{2} [f(x) - f'(x)]^2 dx \quad \text{(B1)}$$

To minimize F, $\delta F = 0$

$$\delta F = \iint K_D [D(x,t) - D'(x,t)] \delta D dxdt + T \int_{x=0}^{L} K_f [f(x) - f'(x)] \delta f dx = 0 \quad \text{(B2)}$$

Taking the first variation of Equation 5 gives:

$$\delta(D_t + q_x) = 0$$

$$\frac{\partial}{\partial t} (\delta D) + \frac{\partial}{\partial x} (\delta q) = 0 \quad \text{(B3)}$$

Multiplying Equation B3 by an arbitrary function $\lambda_D(x,t)$, and integrating over all x and t,

$$\iint \lambda_D \frac{\partial}{\partial t} (\delta D) dxdt + \iint \lambda_D \frac{\partial}{\partial x} (\delta q) dxdt = 0$$

Integrating by parts once

$$\int_x \lambda_D \delta D \Big|_{t=0}^{t=T} dx - \int_x \int_t \frac{\partial \lambda_D}{\partial t} \delta D dxdt + \int_t \lambda_D \delta q \Big|_{x=0}^{L} dt - \iint \frac{\partial \lambda_D}{\partial x} \delta q dxdt = 0 \quad \text{(B4)}$$

We then take the first variation of Equation 6,

$$\delta\left(q_t + \frac{g}{2} (D^2)_x + \frac{f|q|q}{8D^2}\right) = 0$$

$$\frac{\partial}{\partial t} (\delta q) + \frac{g}{2} \frac{\partial}{\partial x} (\delta D^2) + \frac{|q|q}{8D^2} \delta f - \frac{fq|q|}{4D^3} \delta D + \frac{f}{8D^2} 2|q| \delta q = 0 \quad \text{(B5)}$$

Multiplying Equation B5 by an arbitrary function $\lambda_q(x,t)$ and integrating,

$$\iint\limits_{xt} \lambda_q \frac{\partial}{\partial t} (\delta q) dxdt + \frac{g}{2} \iint\limits_{xt} \lambda_q \frac{\partial}{\partial x} (\delta D^2) dxdt + \iint\limits_{xt} \frac{\lambda_q |q| q}{8D^2} \delta f dxdt - \iint\limits_{xt} \lambda_q \frac{fq|q|}{4D^3} \delta D dxdt + \iint \frac{f|q|}{4D^2} \delta q dxdt = 0$$

Integrating by parts once gives

$$
\begin{aligned}
&\int \lambda_q \delta q\Big|_{t=0}^{T} dx - \iint \frac{\partial \lambda_q}{\partial t}\, \delta q dx dt \\
&\quad + \frac{g}{2} \int_t \lambda_q \delta D^2\Big|_{x=0}^{L} dt - \frac{g}{2} \iint \frac{\partial \lambda_q}{\partial x}\, \delta D^2 dx dt \\
&\quad + \iint \frac{\lambda_q |q| q}{8D^2}\, \delta f dx dt - \iint \frac{\lambda_q f |q| q}{4D^3}\, \delta D dx dt \\
&\quad + \iint \frac{\lambda_q f |q|}{4D^2}\, \delta q dx dt = 0 \qquad \text{(B6)}
\end{aligned}
$$

Adding Equations B2, B4, and B6 (collecting terms in $\delta f$, $\delta q$, $\delta D$), yields

$$
\begin{aligned}
\delta F = &\int_{x=0}^{L} \left[ \int_t \frac{\lambda_q |q| q}{8D^2}\, dt + TK_f(f(x) - f'(x)) \right] \delta f dx \\
&- \iint \left( \frac{\partial \lambda_D}{\partial t} + \frac{\partial \lambda_q}{\partial x} \cdot gD + \frac{f|q|q}{4D^3}\, \lambda_q - K_D[D(x,t) - D'(x,t)] \right) \delta D \cdot dx dt \\
&- \iint \left( \frac{\partial \lambda_D}{\partial x} + \frac{\partial \lambda_q}{\partial t} - \frac{f|q|}{4D^2}\, \lambda_q \right) \delta q dx dt \\
&+ \int_{x=0}^{L} (\lambda_D \delta D + \lambda_q \delta q)\Big|_{t=0}^{T} dx + \int_{t=0}^{T} (\lambda_D \delta q + gD\lambda_q \delta D)\Big|_{x=0}^{L} dt = 0 \qquad \text{(B7)}
\end{aligned}
$$

For the forward model, initial conditions must be specified. This implies $\delta D(x,0) = \delta q(x,0) = 0$, so that the lower limit (i.e., $t = 0$) of the integrand in the first boundary integral in Equation B7 vanishes. The upper limit (i.e., $t = T$) vanishes if $\lambda_D(x,T) = \lambda_q(x,T) = 0$, and these are the conditions imposed on the adjoint model, as described in the text. Thus, the first boundary integral vanishes. Similarly the second boundary integral in Equation B7 vanishes, since $\delta q(L,t) = \delta D(0,t) = 0$, q and D being given at $x = L$ and $x = 0$, respectively, and due to the boundary conditions used for the adjoint model, viz. $\lambda_q(L,t) = \lambda_D(0,t) = 0$.

Having dispensed with the boundary terms, the arbitrary quantities $\lambda_q$ and $\lambda_D$ can be so chosen that the coefficients of $\delta q$ and $\delta D$ in Equation B7 (i.e., the second and third terms) vanish. This yields the adjoint Equations 9 and 10, respectively. Consequently $\delta F = 0$ implies that the coefficient of $\delta f$ must also vanish, resulting in the parameter-correction equation (similar to Equation 8).

## REFERENCES


1. **Yeh, W. W.-G.,** Review of parameter identification procedures in groundwater hydrology: the inverse problem, *Water Resour. Res.* 22(2), 95, 1986.
2. **Neuman, S. P.,** A statistical approach to the inverse problem of aquifer hydrology. III. Improved solution method and added perspective, *Water Resour. Res.,* 16(2), 331, 1980.
3. **Neuman, S. P. and Carrera, J.,** Maximum likelihood adjoint-state finite-element estimation of groundwater parameters under steady- and nonsteady-state conditions, *Appl. Math. Comp.,* 17, 405, 1985.
4. **Carrera, J. and Neuman, S. P.,** Estimation of aquifer parameters under transient and steady state conditions. I. Maximum likelihood method incorporating prior information, *Water Resour. Res.,* 22(2), 199, 1986.

5. **Carrera, J. and Neuman, S. P.,** Estimation of aquifer parameters under transient and steady state conditions. II. Uniqueness, stability and solution algorithms, *Water Resour. Res.*, 22(2), 211, 1986.
6. **Carrera, J. and Neumann, S. P.,** Estimation of aquifer parameters under transient and steady state conditions. III. Application to synthetic and field data, *Water Resour. Res.*, 22(2), 228, 1986.
7. **Sykes, J. F., Wilson, J. L., and Andrews, R. W.,** Sensitivity analysis for steady state groundwater flow using adjoint operations, *Water Resour. Res.*, 21(3), 359, 1985.
8. **Sasaki, Y.,** Some basic formations in numerical variational analysis, *Mon. Weather Rev.*, 98(12), 875, 1970.
9. **Hildebrand, F. B.,** *Methods of Applied Mathematics,* Prentice-Hall, New York, 1952, 139.
10. **Whitaker, E. and Robinson, G.,** *The Calculus of Observations,* Blackie and Son, Ltd., London, 1944, 397.
11. **Provost, C. and Salmon, R.,** A variational method for inverting hydrographic data, *J. Mar. Res.*, 44, 1, 1986.
12. **Harlemann, D. R. F. and Lee, C. H.,** The computation of tides and currents in estuaries and canals, Tech. Bull. No. 16, Committee on Tidal Hydraulics, U. S. Army Corps of Engineers, 1969, 121.
13. **Walsh, G. R.,** *Methods of Optimization,* John Wiley & Sons, New York, 1975.
14. **Chen, W. H., Gavalas, G. R., Seinfeld, J. H., and Wasserman, M. L.,** A new algorithm for automatic history matching, *Soc. Pet. Eng. J.*, 14(6), 593, 1974.
15. **Chavent, G., Dupuy, M., and Lemonnier, P.,** History matching by use of optimal theory, *Soc. Pet. Eng. J.*, 15(1), 74, 1975.
16. **Yziquel, A. and Bernard, J. C.,** Automatic computing of a transmissivity distribution using only piezometric heads, in *Finite Elements in Water Resources,* Brebbia, C. A., Gray, W. G., and Pinder, G. F., Eds., Pentech, London, 1978, 1.
17. **Dronkers, J. J.,** *Tidal Computations in Rivers and Coastal Waters,* North-Holland, Amsterdam, 1964, 423.
18. **Yeh, W. W.-G. and Becker, L.,** Linear programming and channel flow identification, *J. Hydraul. Div. Am. Soc. Civ. Eng.*, 99(HYII), 2013, 1973.
19. **Chiu, C. L. and Isu, E. O.,** Application of Kalman filter to open channel flow profile estimation, in *Applications of Kalman Filter to Hydrology, Hydraulics, and Water Resources,* Chiu, C. L., Ed., University of Pittsburgh, Pittsburgh, 1978, 15261.
20. **Harlan, J. and O'Brien, J. J.,** Assimilation of scatterometer winds into surface pressure fields using a variational method, *J. Geophys. Res.*, 91(D7), 7816, 1986.
21. **Talagrand, O. and Courtier, P.,** Variational assimilation of meteorological observations with the adjoint vorticity equation. I. Theory, *Q. J. R. Meteorol. Soc.*, 113, 1311, 1987.
22. **Courtier, P. and Talagrand, O.,** Variational assimilation of meteorological observations with the adjoint vorticity equations. II. Numerical results, *Q. J. R. Meteorol. Soc.*, 113, 1329, 1987.

Chapter 3

# ON BOTTOM TRAPPED WAVES OVER THE WESTERN SLOPE OF THE NORWEGIAN TRENCH

**Gunnar K. Furnes**

## TABLE OF CONTENTS

# I. ABSTRACT

Current measurements in the proximity of the western slope of the Norwegian Trench have revealed significant low-frequency (subdiurnal) wave activity. The most prominent observed signal has a period of about 36 h, and is occasionally found to exceed the dominating $M_2$ tidal constituent.

The purpose of this paper is to give an overview of some of the existing literature concerning this matter. In addition, a model is set up to study wave trapping over a topography approximating a section between Norway and Scotland at a latitude of 60.75°N. In particular this study focuses on the effects of stratification on the frequency of waves trapped over the western slope. A two-layered model is considered in which the upper layer is continuously stratified and the lower layer is homogeneous. The motions are set up by a prescribed wind field. It is apparent from the model studies that the 36 h period is that of the gravest mode wave of zero group speed trapped over the western slope. For a shallow stratification as employed in the numerical experiments, it turns out that the effect of stratification on the frequency is rather small, but depends upon the wavelength. For the zero group speed wave, having a wavelength of about 360 km, the increase in frequency amounts to only about 2%.

# II. INTODUCTION AND REVIEW

Over the last 10 to 15 years, a large number of papers and reports have been published elucidating more and more details of the circulation and mass distribution in the North Sea. This research has been stimulated by an extensive cooperation among scientists in the countries surrounding the North Sea, both with respect to field programs (e.g., JONSDAP '76) and development of models (e.g., JONSMOD). It is also clear that the demand from the offshore oil industry for oceanographic data, in connection with design and marine operation, has been an important motivation for this research.

The scientific interest has mainly been in connection with studies of wind and tidal phenomena for description of sea-surface displacement and current fields. Less attention has been devoted to the study of low-frequency (subdiurnal) oscillations. In this paper we address these motions for the Norwegian Trench area, northern North Sea, by means of a short literature review and by considering a simple model.

The Norwegian Trench is a conspicuous topographic feature of the North Sea, following the coast of Norway from the Skagerrak into the Norwegian Sea (Figure 1). The depth of the Skagerrak reaches more than 700 m, decreasing to a sill depth of 275 m off the southwest corner of Norway. From there the depth gradually increases into the Norwegian Sea. The width of the trench at the 250 m depth contour varies from 80 to 35 km at the sill depth.

It is now well established that the Norwegian Trench has a strong influence on the current system of the North Sea. Practically all the water entering the North Sea is steered into the trench and flows out along the Norwegian coast (for more details, see Furnes et al.[1] and references therein). It is also well known from the scientific literature and textbooks that pronounced bottom topography, like for example, the western slope of the Norwegian Trench, acts as a wave guide for the propagation of subinertial fluctuations.

## A. OBSERVED FLUCTUATIONS

The Norwegian Trench and adjacent sea areas have been the site of a great number of current-meter moorings, some of them being deployed continuously over periods of over 1 year. The major part of these data has been collected for oil companies and has a rather limited distribution. Available analyses of these current-meter series, as well as those obtained from the open literature, have revealed the existence of significant variance contribution on time scales ranging from 1 d up to several days. Furnes and Sælen[2] found from analysis of

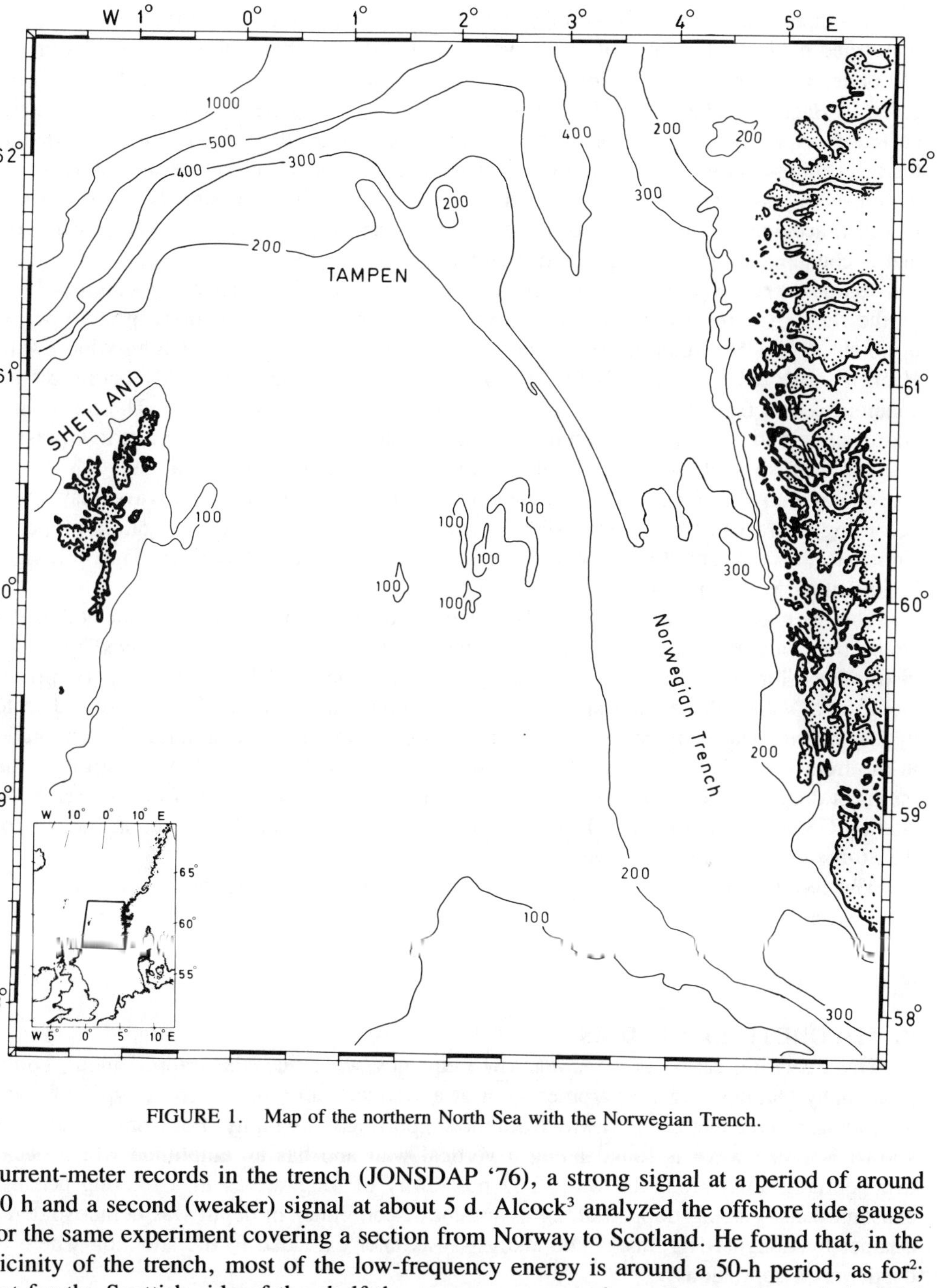


FIGURE 1. Map of the northern North Sea with the Norwegian Trench.

current-meter records in the trench (JONSDAP '76), a strong signal at a period of around 50 h and a second (weaker) signal at about 5 d. Alcock[3] analyzed the offshore tide gauges for the same experiment covering a section from Norway to Scotland. He found that, in the vicinity of the trench, most of the low-frequency energy is around a 50-h period, as for[2]; but for the Scottish side of the shelf the energy was spread over bands of a period of 5 d and longer, similar to the spectrum of atmospheric pressure variations at Kirkwall.

An important analysis has been carried out by Koltermann,[4] who also reported on the JONSDAP '76 current and tide-gauge data. He found signals in the currents in the trench with periods of about 10 d, 3.3 to 5 d and 2.5 to 3 d, and suggested barotropic topographic Rossby waves as a governing mechanism. From analysis of the same data, Mork[5] reported a dominant signal at 5.5 d, which he identified as a frontal wave related to the front between the Norwegian Coastal water and the denser Atlantic water.

A comprehensive analysis with respect to low-frequency motions in the trench has been performed by Hackett[6]. His data base was obtained during the NORSEX '79 experiment, and is described more fully in Johannessen et al.[7] A detailed presentation of this data set is given in Hackett.[8] The study of Hackett[6] indicated that, in general, the subdiurnal energy decreases as we move down through the water column at all positions. In the horizontal there are high-energy areas near the Norwegian coast and near the western slope of the trench, with less-energetic areas between. As a rule, he found that signals in the range 25 to 120 h account for 70 to 90% of the subdiurnal energy and exhibit a rotary polarization that is counterclockwise over the trench and clockwise westwards on the shelf. In the very-low-frequency range (> 120 h), the motion is clockwise over the trench and counterclockwise on the shelf. Among several signals recovered from the autospectra, there are two periods of about 36 and 58 h that are particularly persistent and connected to strong wind events. Hackett[6] analyzed these signals further and concluded that they are highly barotropic and coherent across the western slope and denoted them as "trench waves". The 58 h signal was attributed to atmospheric forcing in the northeast Atlantic, suggesting that low-pressure systems force a shelf wave along the shelf break west of the British Isles which then propagates into the trench as a free, gravest mode trench wave. This wave may, again according to Hackett,[6] excite its "sibling" trench wave, which has propagation properties similar to those observed for frontal waves in the Coastal Current regime. The 36 h signal he identified as the gravest mode zero-group-speed trench wave.

The 36 h period came also clearly out of analysis of current-meter data obtained in connection with Oseberg Pipeline studies.[9] The data from this study were collected by moored instruments along a section from the Norwegian coast to the North Sea Plateau, approximately along 60.5°N latitude. Strong signals, with amplitudes up to about 50 cm, were detected by the pressure gauges in the proximity of the western slope, with a counterclockwise rotation in the trench, clockwise on the shelf. In some situations these signals were stronger than the most dominant tidal constituent $M_2$. Gordon et al.[10] presented data from the same area (station OS at the shelf break). It is evident also from these data that the contribution from the 36-h signal is quite significant.

The results summarized above are from a few of the many reports on low-frequency motions in the Norwegian Trench area. In recent years, more attention has been paid to baroclinic processes, mesoscale eddies, and meanders. (See, for example, Johannessen et al.[11] and references therein.)

## B. THEORETICAL STUDIES

On a rotating earth, propagating low-frequency waves can be trapped along bottom topography features (e.g., escarpments) or at a vertical wall (coast). These types of waves have an amplitude that decays away from the supporting topography. For instance, the well-known Kelvin[12] wave is found along a vertical wall and has an amplitude which decays exponentially away from the wall. The possibility of long waves in a rotating ocean of constant density being trapped along a straight discontinuity in depth was demonstrated in a paper by Longuet-Higgins.[13] The analysis was later extended to the situation where the depth varies continuously.[14]

Concerning the Norwegian Trench and adjacent sea areas, the first simplifying geometrical representation can, for example, be two lateral walls, one representing the Norwegian coast and the other the Scottish coast, and a "partial wall" for the western slope. The "partial wall" can be described either by a single step or by a smooth transition from the trench deep to the shallower North Sea plateau. If the walls are assumed to be straight and infinitely long we can expect trapped waves in three regions. These waves propagate with shallow water (or coast) to the right and have a decaying nature on both sides of the slope (or from the coasts). Their influence in the sea outside the supporting topography

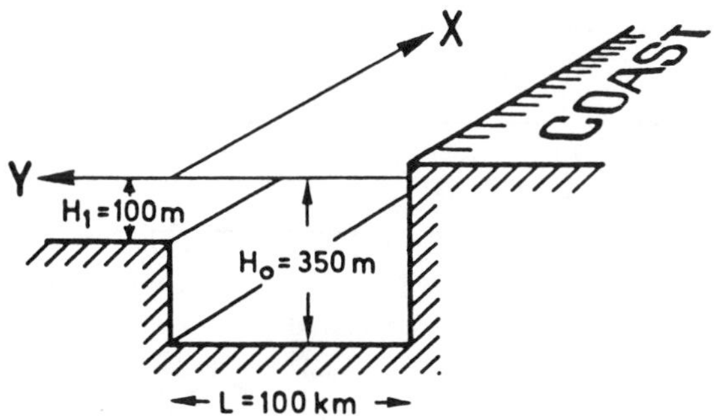


FIGURE 2. Trench model geometry applied by Martinsen.[15]

depends upon two important parameters, namely the barotropic and the baroclinic Rossby radii of deformation. The barotropic radius of deformation reaches far away from the trapping area, depending upon the sea depth. For a 100-m sea depth, the barotropic radius at the relevant latitude is about 250 km, and consequently one can expect barotrophic Kelvin waves to affect areas almost across the whole section from Norway to Scotland (approximately 400 km).

The baroclinic radius of deformation, however, is less than 40 km[8] and is therefore much smaller than its barotropic counterpart. Internal Kelvin waves are thus expected to be trapped in a narrow band near the supporting topography and to have little influence in the other above-mentioned regions.

The first model study of wave trapping in the Norwegian Trench was carried out by Martinsen.[15] In his study the western slope was modeled by a depth discontinuity between the trench (360 m deep) and the plateau (100 m). Martinsen excluded the western boundary and hence eliminated any effects of wave trapping at the Scottish Coast (see Figure 2).

The model equations used by Martinsen were the linear depth-integrated hydrodynamical equations for a homogeneous sea. The motions in the model were driven by applying a wind stress parallel to the coastline. Two different wave phenomena could clearly be recovered from this experiment: a coastal right-bounded wave and a wave trapped at the depth discontinuity. The coastal trapped wave is denoted "modified Kelvin wave" as it is an ordinary Kelvin wave being modified by the shallow sea. The other wave system is denoted topographic planetary wave, also termed trench waves (adapted from Mysak et al.[16]), and propagates at a lower speed with the shallow sea to the right. Other important properties of the trench waves deduced from the model are that the maximum surface elevation and a strong shear occur at the edge. Another interesting feature caused by the trench wave is that the volume transport in the trench is opposite to the applied wind stress. Finally, it is evident from Martinsen's work that the modified Kelvin wave has a cutoff and exists only for long waves. The cutoff occurs for wave lengths of about 1800 km.

Hackett[6] applied the same model for barotropic trench waves as that developed by Mysak et al.,[16] in order to explain the low-frequency motion mentioned earlier. Instead of representing the western slope by a discontinuity as Martinsen[15] did, he introduced a continuous profile approximating fairly well the average echo-sounding depths along latitude 60.75°N. This profile was expressed in terms of a *tanh(x)* function and is replotted in Figure 3. The coast (Norwegian) was taken to be straight and the topography uniform in the longshore direction. As in Martinsen's work, Hackett did not introduce any coast limitations toward the west.

The governing equations in the model are the vorticity-balance equations for small-amplitude motions. By assuming the perturbations on the basic state to be barotropic and nondivergent, the mass transport can be expressed by means of a stream function, which is

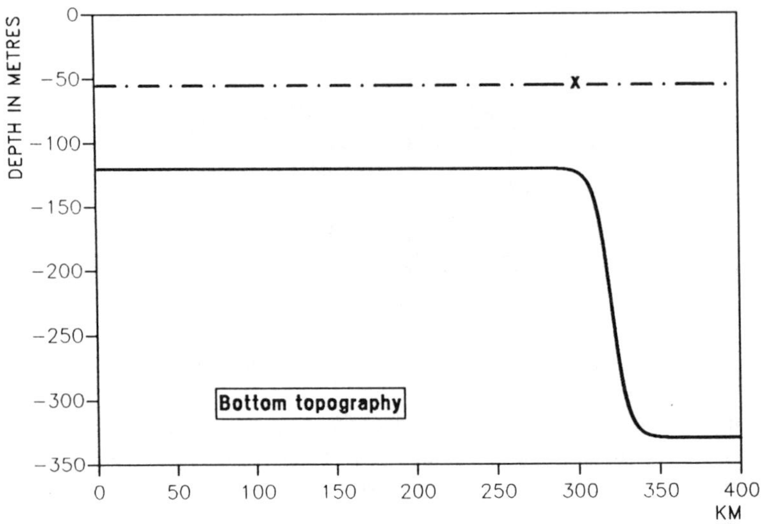


FIGURE 3. Model bottom profile used in this paper and in Hackett.[6] X denotes the position and depth (interface) in which time-series analyses are performed.

separated into a periodic alongshore part and an amplitude factor varying only with the offshore coordinate. The amplitude factor is then described by a differential equation in the offshore coordinate that, together with the described boundary conditions, forms a standard Sturm-Liouville problem for a given wave number. The individual modes can then be recovered from this problem. In Hackett's paper the dispersion relations for the first three modes are presented. The gravest mode is quite similar to that found by Martinsen[15] for the wave trapped at the western slope and denoted trench waves. There is a maximum frequency in the dispersion relation for this mode corresponding to a wave period of about 36 h. This is the zero-group-speed wave and has a wavelength of about 360 km. The gravest mode is essentially nondispersive over a large portion of the frequency range on the low wave number branch. It was also found by Hackett that there is a large difference in frequency between the gravest and the higher modes. A coastal trapped wave was not recovered in Hackett's analysis.

## C. RELATED STUDIES

The first reported observations of continental shelf waves are in sea-level records from the east Australian coast by Hamon.[17] In this area the bottom topographical conditions are of such a nature that an interpretation in terms of simple Kelvin wave theory could hardly be expected. Robinson[18] developed a simple theory of coastally trapped, nondispersive, and subinertial vorticity waves. Linearized shallow-water equations are considered and by introducing a prescribed shelf slope he showed that the topography acts as a wave guide for the propagation of low-frequency energy along the coast. Robinson termed this phenomenon "continental shelf waves". Today there are large numbers of reported observations of shelf waves from all over the world, including the Antarctic.[19] Several generation mechanisms have been proposed; the most likely seems to be that they are produced by the alongshore component of the wind stress as discussed by Gill and Schumann[20] and Allen.[21] Off the west coast of Scotland, on the other hand, shelf waves at or near zero group velocity periods appear to be tidally induced.[22,23] A recent paper by Grimshaw[24] presents a theory for a resonant formation of barotropic coastally trapped waves for the case when the resonance occurs on the low wave number side of the spectrum. His theory is developed to take account of the forcing due to interaction of a longshore current with a longshore topographic feature.

Barotropic trench waves as treated by the above-mentioned authors, and by several others (e.g., Mysak and Willmott[25]), have the same essential physics as barotopic continental shelf waves, but are confined to submarine trenches. For a very complete and systematic review of the numerous studies of continental shelf waves up to 1980, please see Mysak.[26]

The presence of stratification in the Norwegian Trench, the longshore mean coastal current as well as the Atlantic inflow, may significantly change the properties of the trench waves. The first studies of the effects of stratification on shelf waves used two-layer models with various different bottom topographies and interface depths. A discussion and reference to these papers can be found in Mysak.[26] It is evident from these papers that the effects of stratification are very much dependent upon the position of the interface between the upper and lower layers (pycnocline). If the stratification lies over the shelf (shallow stratification) the phase speed of the shelf wave increases by only a few percent. If it is below the shelf break (deep-sea stratification), the phase speed of the gravest mode can be increased twofold for typical density differences.

Huthnance[27] has developed a theory for shelf waves in a continuously stratified ocean covering any monotonic, gently sloping depth profile. He found that for weak stratification (defined by a nondimensional internal Rossby radius), the response is essentially barotropic. Further, he found that for a given wave number, increasing the stratification gives increased bottom trapping and higher frequencies for all modes. The effect on frequency can be considerable: increasing the nondimensional stratification parameter gives a twofold increase in frequency. Essentially the same results were found by Brink[28] in a study of effects of stratification on long, gravest mode trench waves in the Aleutian Trench.

In a recent paper, Huthnance et al.[29] give a comprehensive discussion of coastal trapped waves. The paper presents a wide aspect of shelf-wave physics, including effects of stratification, topography and geometry, mean flow, nonlinearity, and alongshore changes. The generation and dissipation mechanisms are also considered.

The following model studies are motivated by the work of Martinsen[15] and Hackett[6] and address the effect of density on trapped waves in the western slope of the Norwegian Trench. A two-layer model is developed,[30] in which the upper layer is continuously stratified and the lower layer is homogeneous. The model is driven by atmospheric forces and can deal with any distribution of the vertical basic-density profile. An eddy-viscosity formulation has been adopted: the eddy viscosity is made proportional to the reciprocal of the static stability. The model is coded to take account of a realistic bottom topography, but for the present study it was decided to use the same profile as Hackett:[6] an approximation to a section between Norway and Scotland with uniformity in the longshore direction. It differs from the model of Hackett[6] and Martinsen,[15] in that boundary conditions for the Scottish coast are included. The model is based on the linear three-dimensional hydrodynamical equations. By expanding the dependent variables in terms of continuous functions over the stratified part of the depth, a continuous representation of the current field and vertical displacement is obtained. In the deeper constant-density part of the water column, only a two-dimensional depth mean structure is recovered from the model.

## III. MODEL FORMULATION

Consider a two-layer fluid system in which the upper layer is continuously stratified with respect to density, and the lower layer is homogeneous. The basic equations used to describe this system are the linear equations of motion, with hydrostatic approximation in the vertical, and the equations for conservation of density and volume. These equations may be written in vector form as:

$$\rho \frac{\partial \vec{v}}{\partial t} + \rho f \vec{k} \times \vec{v} = -\nabla p + \frac{\partial}{\partial z}\left(\mu \frac{\partial \vec{v}}{\partial z}\right) \tag{1a}$$

$$\frac{\partial p}{\partial z} = -\rho g \tag{1b}$$

$$\frac{\partial \rho}{\partial t} + \vec{v} \cdot \nabla \rho + w \frac{\partial \rho}{\partial z} = 0 \tag{2}$$

$$\nabla \cdot \vec{v} + \frac{\partial w}{\partial z} = 0 \tag{3}$$

Equation 2 is identical to zero in the lower layer. In these equations $t$ denotes the time, $\nabla$ the horizontal operator $\vec{i}\frac{\partial}{\partial x} + \vec{j}\frac{\partial}{\partial y}$ and $x,y,z$ are Cartesian coordinates, forming a right-handed set, in which $x,y$ are in the horizontal plane of the interface between the two layers. Negative values of $z$ are depths below the interface and positive values are above, in the stratified fluid. $\vec{i},\vec{j},\vec{k}$ are unit vectors along the axes, $p$ is the hydrostatic pressure, $\vec{v}$ is the horizontal velocity vector, $\mu$ is the eddy viscosity, $f$ is the Coriolis parameter, and $g$ is the acceleration of the Earth's gravity.

It is assumed that $f$ and $g$ are constants, but $\vec{v}$ is a function of $x,y,z,$ and $t$.

The upper stratified layer and the lower homogeneous layer are now treated separately: in the upper layer are sought a three-dimentional description of the flow field, while in the lower layer a two-dimensional depth mean solution will be considered.

## A. THE STRATIFIED UPPER LAYER

The density at an arbitrary position and time is described by the density at rest, denoted $\rho_0(z)$ plus a small correction denoted $\rho_1(x,y,z,t)$. Applying the same consideration for the pressure field, the two quantities are expressed by:

$$\rho = \rho_0(z) + \rho_1(x,y,z,t)$$

and

$$p = p_0(z) + p_1(x,y,z,t)$$

Substituting $\rho$ and $p$ into Equation 1a and performing a full linearization, the *momentum equation* takes the form

$$\rho_0 \frac{\partial \vec{v}}{\partial t} + \rho_0 f \vec{k} \times \vec{v} = -\nabla p_1 + \frac{\partial}{\partial z}\left(\mu \frac{\partial \vec{v}}{\partial z}\right) \tag{4}$$

Introducing the vertical displacement (e.g., of an isopycnal) $\zeta(x,y,z,t)$, the relation between the vertical velocity component $w$ and $\zeta$ is, in linear form:

$$w = \frac{\partial \zeta}{\partial t}$$

Together with the hydrostatic approximation deduced from Equation 1b

$$\frac{\partial p_1}{\partial z} + \rho_1 g = 0$$

and

$$\frac{\partial \rho_1}{\partial t} + w \frac{d\rho_0}{dz} = 0$$

obtained by linearization of Equation 2, the following *pressure equation* is derived

$$\frac{\partial p_1}{\partial z} = g \frac{d\rho_0}{dz} \zeta \tag{5}$$

On substituting for $w$ into Equation 3 the *continuity equation* takes the form

$$\nabla \cdot \vec{v} + \frac{\partial}{\partial z}\left(\frac{\partial \zeta}{\partial t}\right) = 0 \tag{6}$$

## B. THE CONSTANT-DENSITY LOWER LAYER

In the following we will denote the vertical displacement of the interface between the layers as $\zeta_0 = \zeta(x,y,0,t)$ and take the density to be continuous across the interface. The equations describing the depth mean flow field in the lower layer are then obtained by integrating the equations of motion and continuity from the sea bed $z = -h(x,y)$ to $z = \zeta_0$. Hence

$$\frac{\partial \overline{\vec{v}}}{\partial t} + f\vec{k} \times \overline{\vec{v}} = -\frac{1}{\rho_2} \nabla p_1(x,y,0,t) + \frac{\vec{F}_0}{\rho_2 h} - \frac{\vec{F}_b}{\rho_2 h} \tag{7}$$

and for the continuity equation

$$\frac{\partial \zeta_0}{\partial t} + \nabla \cdot (h\overline{\vec{v}}) = 0 \tag{8}$$

In Equations 7 and 8, $\overline{\vec{v}}$ denotes the depth mean current

$$\frac{1}{h}\int_{-h}^{\zeta_0} \vec{v}dz$$

and $\vec{F}_o$ is the interfacial stress $\left(\mu\frac{\partial \vec{v}}{\partial z}\right)_z = 0$ acting from the upper layer on the lower layer; $\vec{F}_b$ is the frictional stress $\left(\mu\frac{\partial \vec{v}}{\partial z}\right)_z = -h$ on the sea bed.

## C. SOLUTION OF THE COUPLED EQUATIONS

We have now formulated the set of equations governing the considered wave problem. This set consists of seven equations with seven unknown, independent variables. The determination of the dependent variables as given in Equations 4 to 8 for $\vec{v}$, $p_1,\zeta$ and $\overline{\vec{v}},\zeta_0$, together with a set of prescribed boundary conditions, requires a numerical solution procedure. The boundary conditions are specified at the sea surface, and the interface between the upper and lower layers, as well as at the sea bed and at lateral boundaries. At the sea surface and the interface, two conditions are given, namely the dynamic boundary conditions for pressure and stress, which at the sea surface are related to the atmospheric forcing. At the sea bed the stress can be related to, for example, a quadratic law but in the present study $\vec{F}_b$ will be set to zero. On lateral land boundaries the conditions are obviously that the normal component of current is zero for all $t \geqslant 0$.

The coupled equations are solved on a finite difference staggered grid (Arakawa C type) in the horizontal space domain. In time, a forward second-order time-stepping procedure is employed. Regarding the depth variations of the upper layer, all the dependent variables are expanded in terms of continuous differentiable functions, $\varphi_n(z)$, expressed by (for details see Furnes and Mork[30])

$$\vec{v} = \sum_{n=1}^{\infty} \Phi_n \hat{v}_n \varphi_n' \tag{9}$$

$$p_1 = \rho_0(z) \sum_{n=1}^{\infty} \Phi_n \hat{p}_n \varphi_n' \tag{10}$$

$$\zeta = \zeta_0 + \sum_{n=1}^{\infty} \Phi_n \hat{\zeta}_n \varphi_n \tag{11}$$

Here $'$ denotes differentiation with respect to $z$ and $\Phi_n$ is a constant to be defined later. The expansion coefficients $\vec{v}_n$, $\hat{p}_n$, and $\zeta_n$ vary with respect to the horizontal coordinates and time only. Taking the eddy viscosity $\mu$ to be inversely proportional to the static stability [30] and expressed by

$$\mu = -\alpha(x,y,t) \frac{\rho_0^2}{\frac{d\rho_0}{dz}} \tag{12}$$

in which $\alpha$ describes how the eddy viscosity should vary in the horizontal and in time, and $\rho_0$ determines its depth profile. It is then easy to show that the functions $\varphi_n$ satisfy

$$\frac{d}{dz}(\rho_0 \varphi_n') = \lambda_n \rho_0' \varphi_n \tag{13}$$

Equation 13, together with a set of homogeneous boundary conditions for $\varphi_n$, leads to an eigenvalue problem with infinitely many solutions, where $\lambda_n$ denotes the ascending eigenvalues and $\varphi_n$ the corresponding eigen functions. According to Furnes and Mork,[30] the appropriate boundary conditions for $\varphi_n$ are:

$$\varphi_n(0) = 0 \tag{14}$$

and

$$\varphi_n'(H) - \lambda_n \varphi_n(H) = 0 \tag{15}$$

for all *n*. Here *H* refers to the undisturbed sea surface and 0 to the interface. For scaling purposes the derivative of the eigenfunctions at the interface are set to unity by

$$\varphi_n'(0) = 1 \tag{15a}$$

For any given basic density distribution it is now possible to calculate the set of eigenfunctions and eigenvalues. Having done that, the next step is to calculate $\Phi_n$, appearing in the expansions Equations 9 to 11, which is given by

$$\Phi_n^{-1} = \int_0^H \rho_0 \cdot (\varphi_n')^2 dz \tag{16}$$

The $z$ independent expansion coefficients $\vec{v}_n$, $\hat{p}_n$, and $\zeta_n$ as well as $\zeta_0$ and the depth mean current over the lower layer, are then computed numerically from the following set of equations[30]

$$\frac{\partial \hat{\zeta}_n}{\partial t} = \nabla \cdot \vec{v}_n \tag{17}$$

$$\hat{p}_n = \frac{g}{\lambda_n} \hat{\zeta}_n + \frac{g\rho_0(0)\zeta_0}{\lambda_n} \tag{18}$$

$$\frac{\partial \hat{v}_n}{\partial t} + \alpha\lambda_n \hat{v}_n + f\vec{k} \times \hat{v}_n = -\nabla \hat{p}_n + \vec{R}_n \tag{19}$$

$$\frac{\partial \zeta_0}{\partial t} = -\nabla \cdot (h\vec{\bar{v}}) \tag{20}$$

$$\frac{\partial \vec{\bar{v}}}{\partial t} + f\vec{k} \times \vec{\bar{v}} = -\frac{1}{\rho_2} \nabla p_1(x,y,0,t) + \frac{\vec{F}_0}{\rho_2 h} \tag{21}$$

where

$$p_1(x,y,0,t) = \rho_0(0) \sum_{n=1}^{\infty} \Phi_n \hat{p}_n$$

$$\vec{R}_n = \vec{F}_H \varphi_n'(H)$$

$$\vec{v}_0 = \sum_{n=1}^{\infty} \Phi_n \hat{v}_n$$

$$\vec{F}_0 = \alpha\rho_0(0)\vec{v}_0$$

Here $\vec{F}_H$ is the wind stress acting on the sea surface. The expression for $\vec{F}_0$ is in accordance with Furnes and Mork,[30] the natural stress law at the interface. As seen, the stress depends upon the turbulence parameter $\alpha$, and the velocity at $z = 0$.

The vertical velocity component can now be calculated from the equation of continuity (Equation 3) by integration over $z(0 \leqslant z \leqslant H)$ and application of prescribed boundary conditions. Hence, to the lowest order

$$w = \frac{\partial \zeta_0}{\partial t} - \sum_{n=1}^{\infty} \nabla \cdot (\Phi_n \hat{v}_n \varphi_n)$$

We have now presented the complete set of equations which have been applied in the present study of bottom trapped waves over the western slope of the Norwegian Trench.

## IV. APPLICATION TO THE WESTERN SLOPE

The set of equations presented above is now used to study waves trapped over the western slope of the Norwegian Trench. Although any depth distribution can be handled in the model we will consider a topography that only varies in one direction, here taken to be

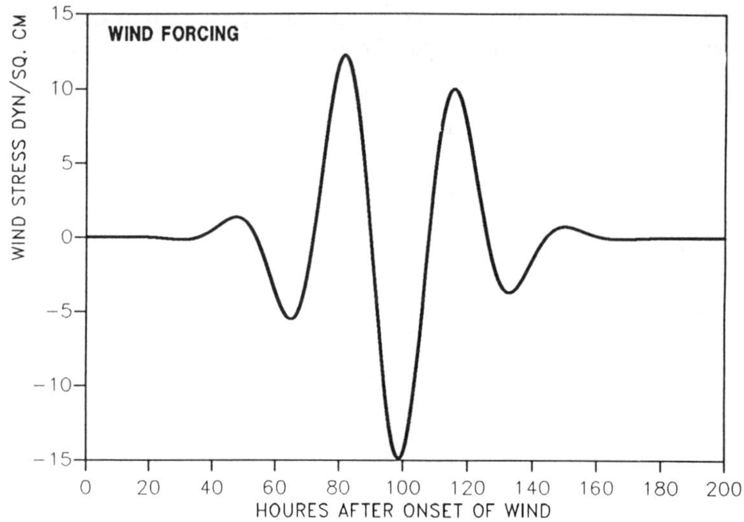


FIGURE 4. Time series of alongshore wind forcing used in present study.

the $x$-direction. The $y$-axis is then aligned along an isobath. The bottom profile is taken to approximate a section between Norway and Scotland along latitude 60.75°$N$ and is the same as that employed by Hackett,[6] see Figure 3. In this region the trench runs essentially north-south so that $y$ is pointing northward and $x$ eastward. In the trench the maximum depth is 330 m and the minimum on the shelf is 120 m. Since we are looking for waves trapped over the slope, we assume periodicity in the $y$ direction. By this assumption it is possible to transform the equations into an $x - z - t$ dependency only, in which the wave number or the wavelength along the slope is a given free parameter in the problem. For a single given wavelength, the model is started from a state of rest by imposing a wind stress given by

$$\vec{F}_H = \vec{j} F sin \frac{2\pi t}{T}$$

in which T is chosen to be 36 h, and the amplitude $F$ is Gaussian distributed in time but constant in space. Excitation of inertial oscillations is of less interest in this study. The applied wind forcing is plotted in Figure 4. After 192 h the wind forcing is set to zero. Calculations are performed with two density fields; first with a relatively weak density distribution and then with a strong one, depicted in Figure 5 and Figure 6, respectively. For simplicity, $\rho_0$ and $\alpha$ were chosen to give a linear profile of eddy viscosity ($N = \frac{\mu}{\rho}$), such that $N(H) = 500\ cm^2/s$ and $N(0) = 100\ cm^2/s$. The interface between the upper and lower layer is situated 55 m below the sea surface and hence represents a ''shallow sea stratification''. The length of the section is 400 km, chosen to represent the distance between Norway and Scotland, and the space and time steps are $\Delta x = 2$ km and $\Delta t = 20$ s, respectively. The Coriolis parameter $f$ is taken to represent the rotation at the actual latitude, giving $f = 1.27 \cdot 10^{-4}$ rad/s.

The models were run for 1200 h (50 d) for each selected wavelength. Time series was then recovered from several positions and depths but analysis for the determination of periods was only performed at the interface near the shelf break, point x in figure 3. After 192 h the model was in a state of free oscillation. Having ignored this part of the data, time-series

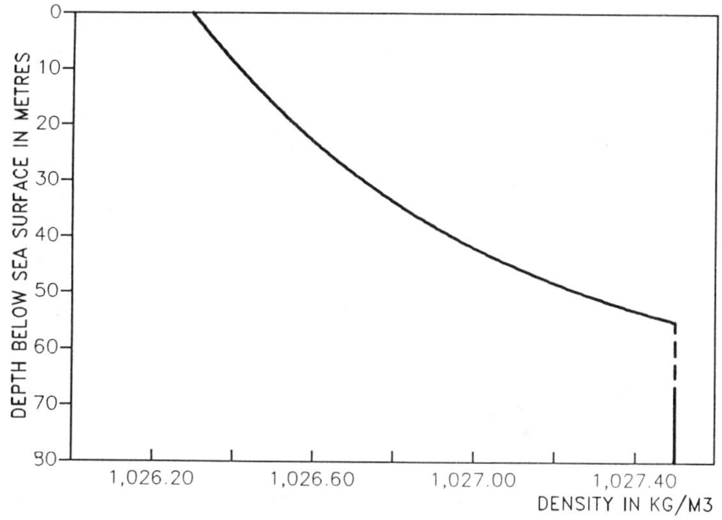


FIGURE 5. Basic density profile of the upper layer; weak stratification.

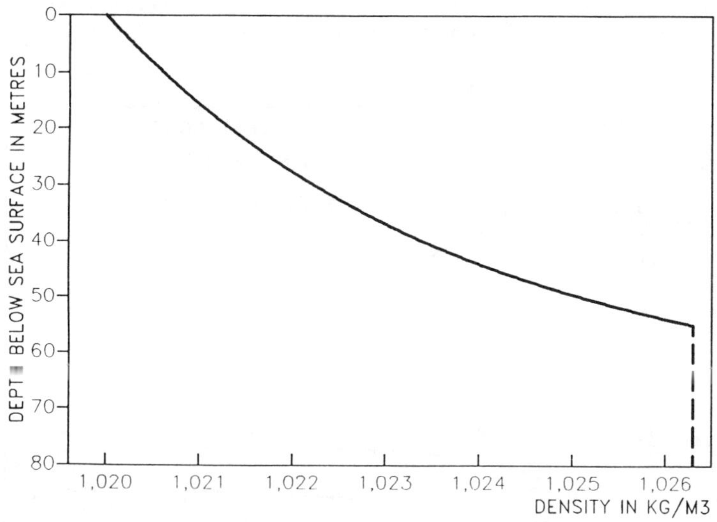


FIGURE 6. Basic density profile of the upper layer; enhanced stratification.

analysis was performed to determine all the significant periods excited in the system. These periods could be determined accurately by fitting exponentially damped cosine functions to the data and minimizing the error through a nonlinear least squares iterative procedure.

## V. DISCUSSION OF RESULTS

Starting with the weak stratification (Figure 5), the model was run with 18 different wavelengths spanning the interval 25 to 20000 km. for each wavelength the frequencies were determined and the results plotted in a frequence-wave number plane to describe the despersion relation for the problem. The frequency is scaled against the inertial frequency and the wave number against the width of the trench, taken as $L = 80$ km. The results are plotted in Figures 7 and 8.

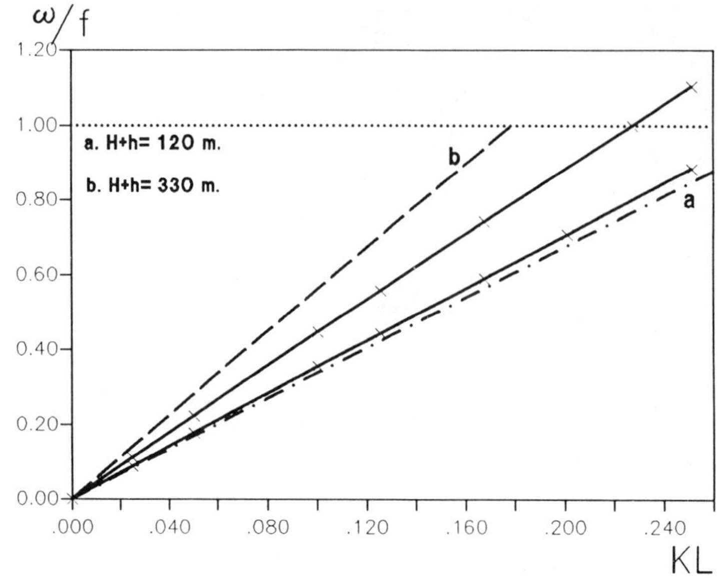


FIGURE 7. Dispersion relation for long coastal trapped waves. X: computed from the model; a and b: calculated from $\omega = k\sqrt{g(h + H)}$.

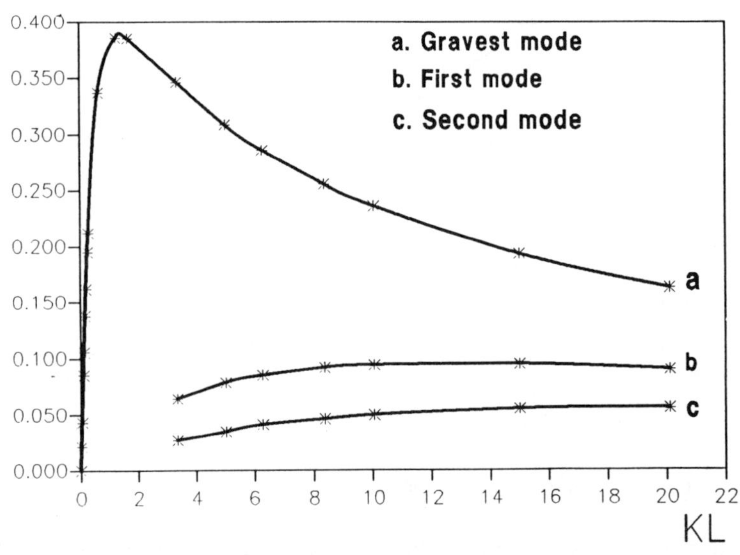


FIGURE 8. Dispersion relation for the trapped wave over the western slope.

It is evident from the time-series analysis that for each given wavelength there is a well-defined set of frequencies. At the long wavelength branch (2000 to 20,000 km) there are in particular two frequencies that exhibit an accurate linear distribution with the wave number, as displayed in Figure 7. These waves are recognized as barotropic modified Kelvin waves trapped at the two coasts, and propagate with the coast to the right. For comparison purposes, the "ideal" Kelvin wave is also plotted in the same figure, deduced from the phase speed $\sqrt{gD}$, in which $D$ is 120 and 330 m, similar to the shelf and trench depth, respectively. The shelf width is approximately four times that of the trench, which taking account of the barotropic Rossby radius of deformation, reflects the degree of modification of the frequency seen in Figure 7.

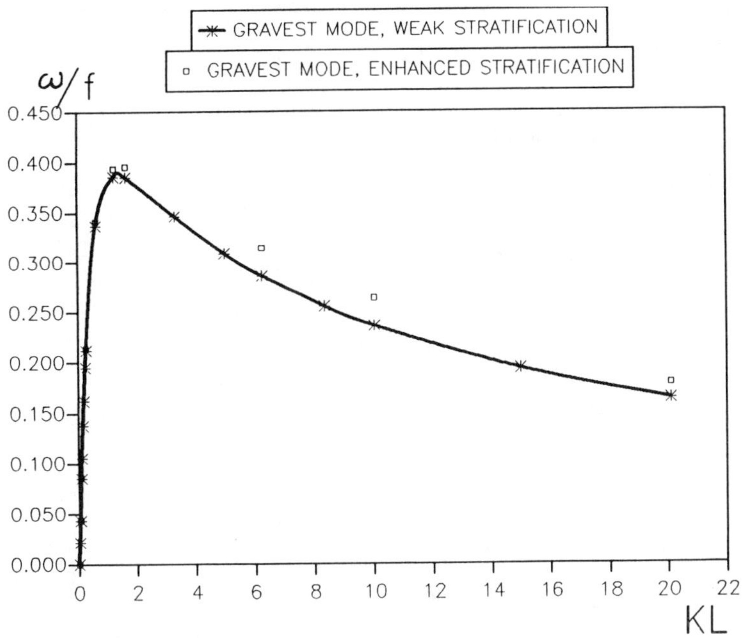


FIGURE 9. Comparison of dispersion relation for the gravest mode shelf wave calculated from weak and enhanced stratification.

A third frequency was also recovered from the analysis. This frequency has a near-linear behavior for very long wavelengths but it is lower than the two discussed above. The difference between them becomes more and more pronounced as the wavelength is reduced. This wave is the gravest mode bottom trapped wave over the western slope of the Norwegian Trench. Together with the next two modes, its dispersion relation is plotted in Figure 8. By comparing Figures 7 and 8 a marked difference can be seen between the properties of the coastal trapped waves and the wave over the western slope. The frequency of the coastal trapped modified Kelvin waves continues to grow linearly towards about the inertial frequency, corresponding to a wavelength of about 2000 km. For shorter wavelengths they gradually diminish and change completely in character from a well-determined trapped wave structure with exponential offshore decay, to nontrapped waves with offshore nodes. The wave above the western slope, however, increases its frequency with wave number up to a maximum value corresponding to about 36 h at a wave-length of 360 km. From this point on, the frequency decreases slowly with wave number. The dispersive nature of this wave leads to the conclusion that the phase and the group velocities are in the same direction at long wavelengths and opposite directions at short wavelengths. The dividing wavelength is at about 360 km which corresponds to the zero group velocity. The results achieved here correspond quite well with comparable results obtained by Hackett[6] and Martinsen.[15]

In addition to the wave frequencies discussed above, the analysis also recovered some long periodic oscillations with frequencies below those of the three modes given in Figure 8. It is outside the scope of this paper to go into more details on these phenomena.

In the next experiments, the strong stratification (Figure 6) was introduced and the value of $\alpha$ adjusted to keep the same magnitude of the eddy viscosity as before. Otherwise the same parameters as above were employed in order to look at the effect of increased density alone, on the wave over the western slope. The result is plotted in Figure 9 to Figure 11 for the first three modes. For comparison purposes, the dispersion relation for the previous experiment is also included. It is evident from these plots that the influence on the frequency

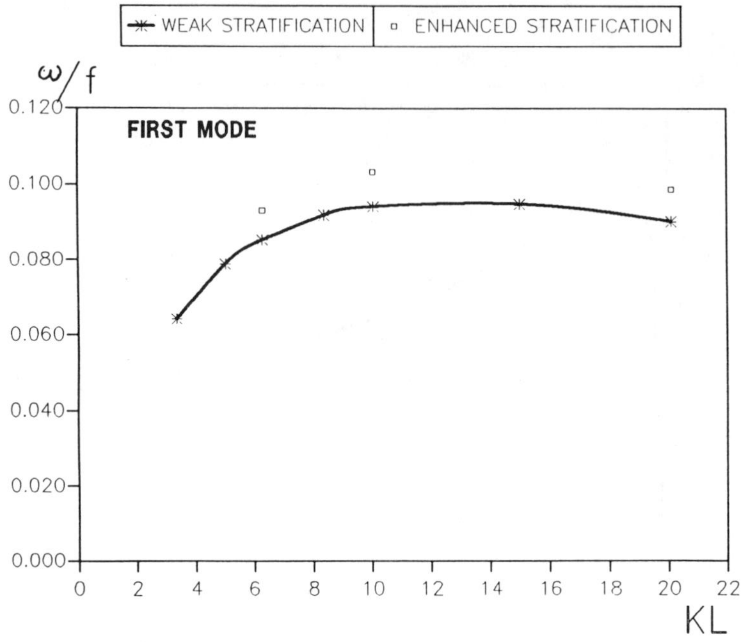


FIGURE 10. Comparison of dispersion relation for the first mode shelf wave calculated from weak and enhanced stratification.

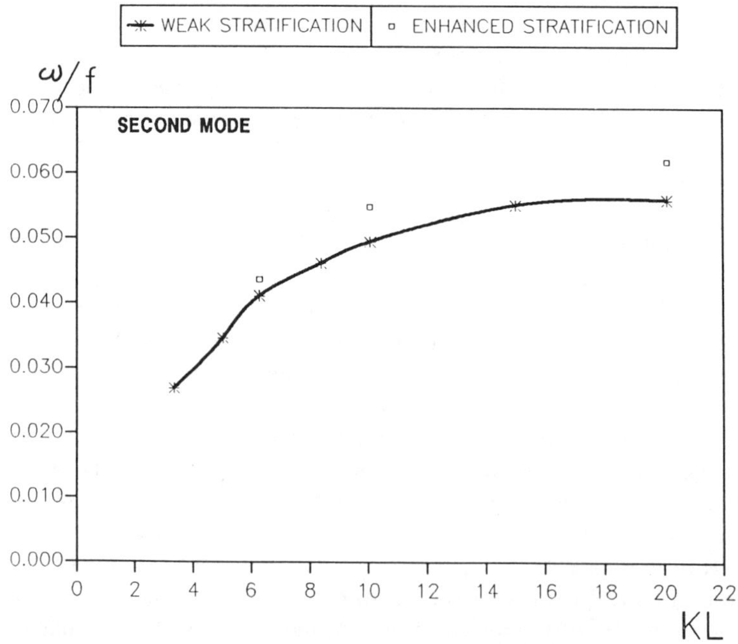


FIGURE 11. Comparison of dispersion relation for the second mode shelf wave calculated from weak and enhanced stratification.

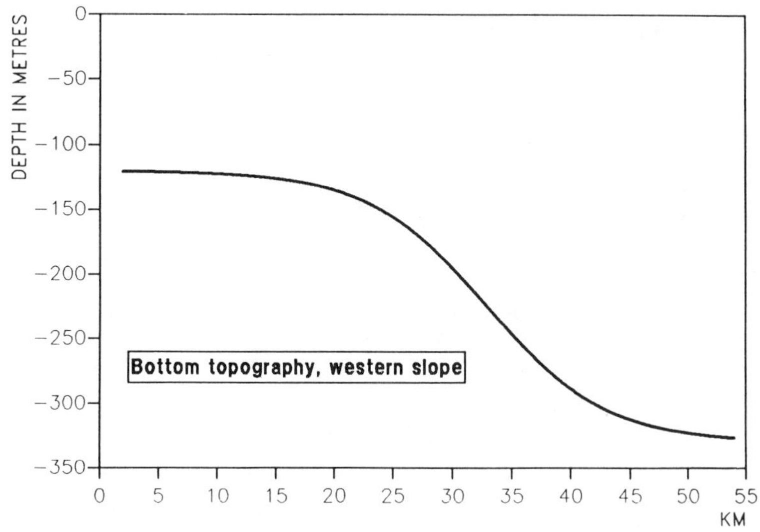


FIGURE 12. Extracts of Figure 3; bottom topography over the western slope.

of increasing the density gradients is not dramatic. For a shallow stratification as studied here, the wave period of the gravest mode is decreased by about 10% for short wavelength ($\approx$100 km and less). For longer wavelengths, however, the influence of stratification becomes less and less important. Of special interest is the zero-group-speed (ZGS) wave which has a period near that observed in the western slope.[6,9] The increase in frequency at this wavelength (Figure 9) corresponds to a reduction of the period by about 2%, which is a difference of less than 1 h.

The vertical displacement and velocity components generated at a prescribed wavelength of 360 km, as calculated at the surface and interface, are plotted in Figures 13 to 15. Although higher modes are generated in the model, the ZGS wave mode is by far the dominant mode, for the prescribed windstress. The bottom topography in the region considered on the western slope is plotted in Figure 12 with the same horizontal scale. It is apparent from Figures 12 and 13 that the vertical displacement is greatest near the strongest gradient of the slope, and decays on both sides. Further, it is seen that the displacement is in phase at the two levels but that the amplitudes are very different. Over the strongest bottom slope, the displacement of the interface is nearly 60 times that of the surface but decreases towards the trench and the shallower plateau. Due to the double exponential behavior of this wave, it is sometimes called a "double Kelvin wave". The along-slope velocity component $v$ (north) and the transverse component $u$ (east), are plotted in Figures 14 and 15 for the sea surface and interface. Profiles are presented every ninth hour over a 36 h period. We note immediately that the major defining characteristic of ordinary Kelvin waves ($u = 0$) is not found in this wave. The value of $u$ varies along the section with strongest amplitude on the plateau. Also the $v$-component exhibits the strongest amplitude on the shallower part of the section. An interesting result appears when we compare current components at the surface with those at the interface. It is apparent that the velocity increases with depth. This is particularly prominent in the $v$-component and leads to an enhancement of the lateral shear with depth. Finally, it is seen, by comparing the time variations of $u$ and $v$, that the rotational properties of the wave are consistent with Hackett[6] and Eide[9]: clockwise over the shallow side of the topography, counterclockwise over the deep side. The dividing position is near, but on the deep side of, the maximum vertical displacement.

It was briefly mentioned above that the wave trapped along the two coastal boundaries

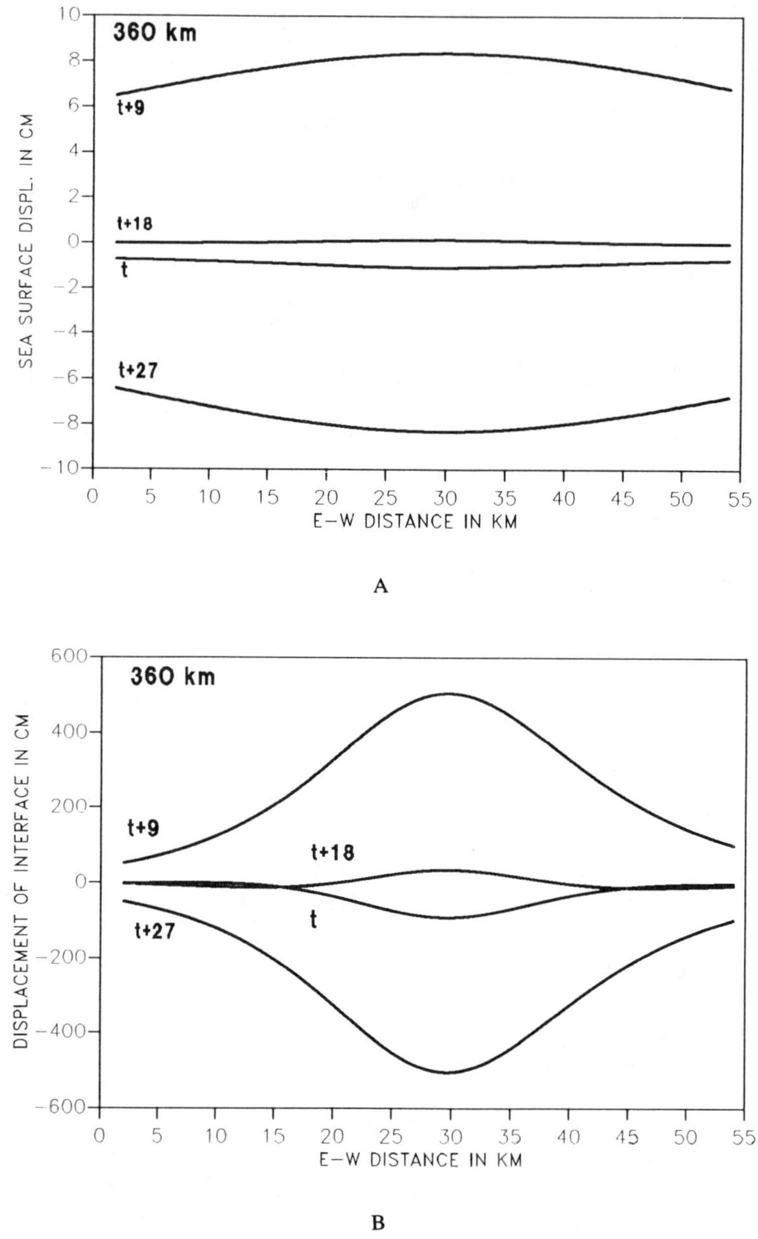


FIGURE 13. Displacement of (A) sea surface and (B) interface over the topography in Figure 12, at 9-h interval, 360-km wavelength.

gradually changes character and disappears as the wave length is reduced down to about 2000 km; for shorter wavelengths the only free oscillations recovered from the time-series analysis are the shelf wave modes trapped over the western slope. In order to demonstrate the structure of the new "untrapped" solutions, the frequency of the wind forcing is changed to $\omega = 0.9\,f$ but otherwise the conditions are unchanged (using the enhanced stratification). The time series now reveal a "cascade" of frequencies around and below the inertial frequency, and also, of course, those already found for the shelf wave (at this wavelength, the lowest mode has a period of $\approx$70 h). Figure 16 shows the position of the interface in

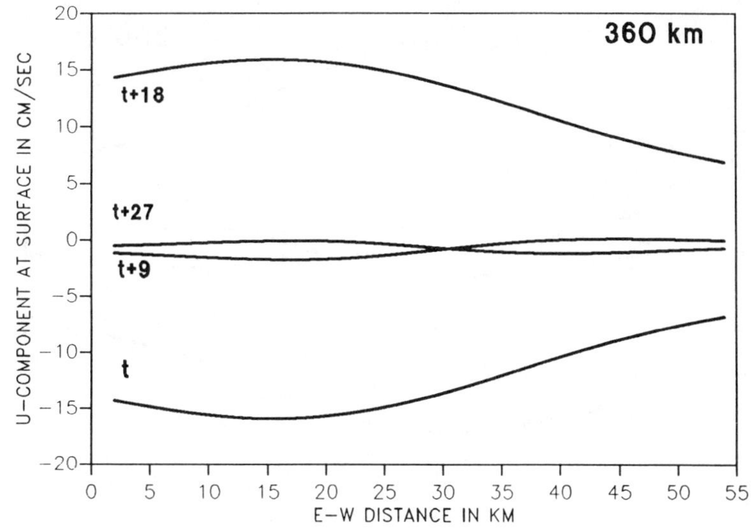


A

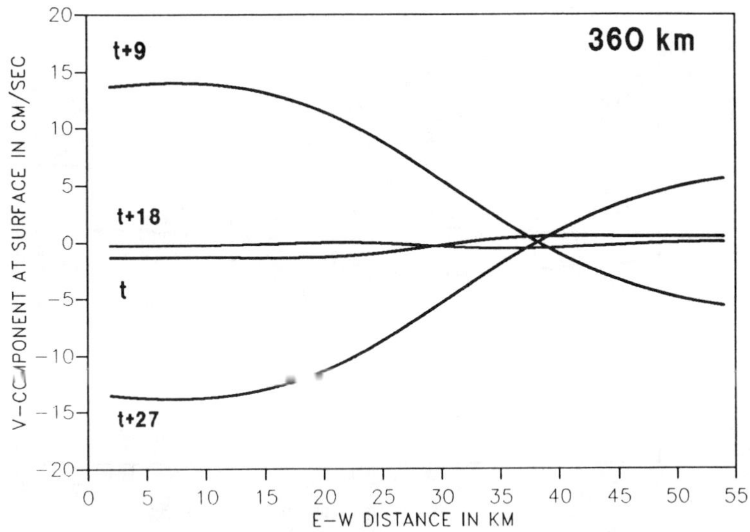


B

FIGURE 14. The (A) U- and (B) V-velocity components at surface over the topography in Figure 12, at 9-h interval, 360-km wavelength.

a state of free oscillation, for every fourth hour over a 20-h period. The horizontal axis covers the distance between the coasts using the same scale as in Figure 3. It is seen from Figure 16 that the baroclinic response appears to exhibit the form of near inertial Poincaré oscillations which spread into the section from the boundaries. The unsymmetrical distribution is produced through interaction with the bottom topography. A contrast to this along-section undulation is seen in Figure 17 where the interface is plotted from a computation with a 4000-km wavelength. The period of the wind stress was chosen to be the same as in the first experiment, namely 36 h, but all other parameters including the stratification were as in the latest calculation.

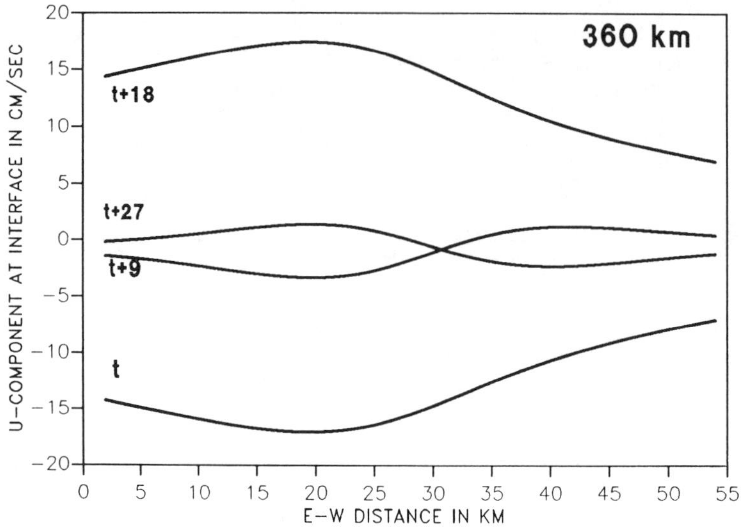


A

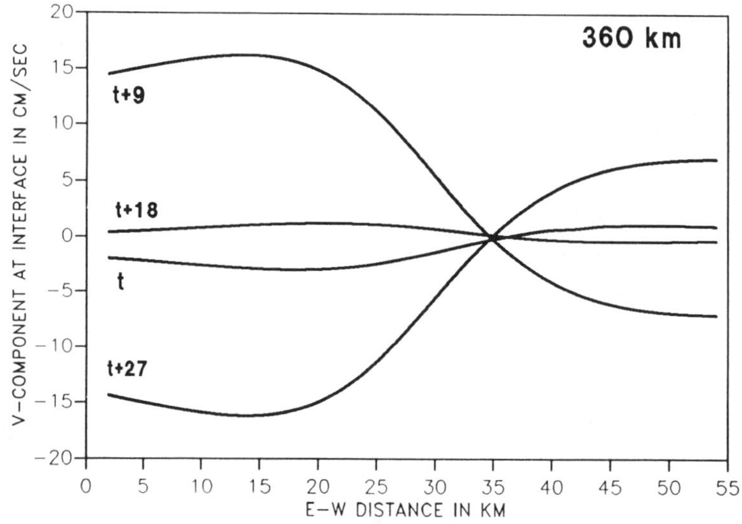


B

FIGURE 15. The (A) U- and (B) V-velocity components at interface over the topography in Figure 12, at 9-h inverval, 360-km wavelength.

Considering Figure 17 it is evident that trapped waves are set up at the two coasts and over the western slope. It is apparent from the excursion of the interface that the applied wind forcing favors the generation of a lowest mode shelf wave, and a (modified) Kelvin wave trapped along the western boundary. Along the eastern boundary, however, the amplitude of the trapped modified Kelvin wave is much lesser. This is as expected for a linear system as studied here. The period of a free Kelvin wave at a 4000-km wavelength is about 32 h for a depth of 120 m, and 20 h for 330 m. Hence, by running the same model setup with a wind period of 20 h, the wave trapped along the eastern coast is by far the most dominating, as seen in Figure 18.

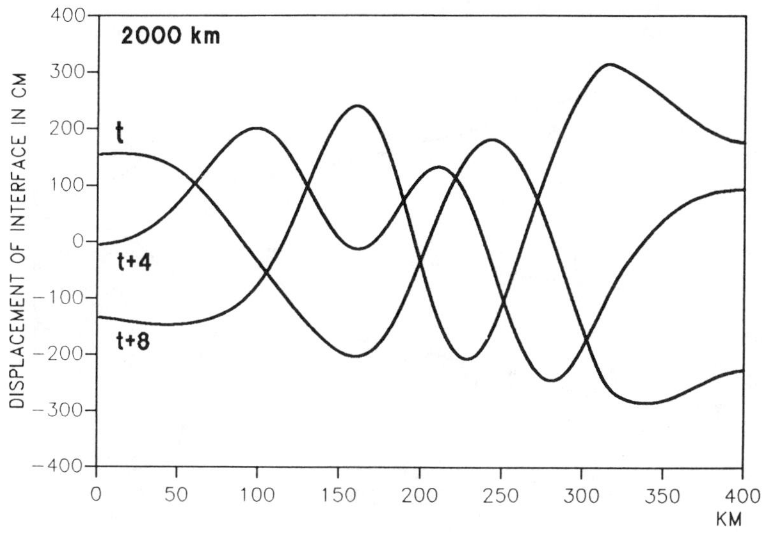


A

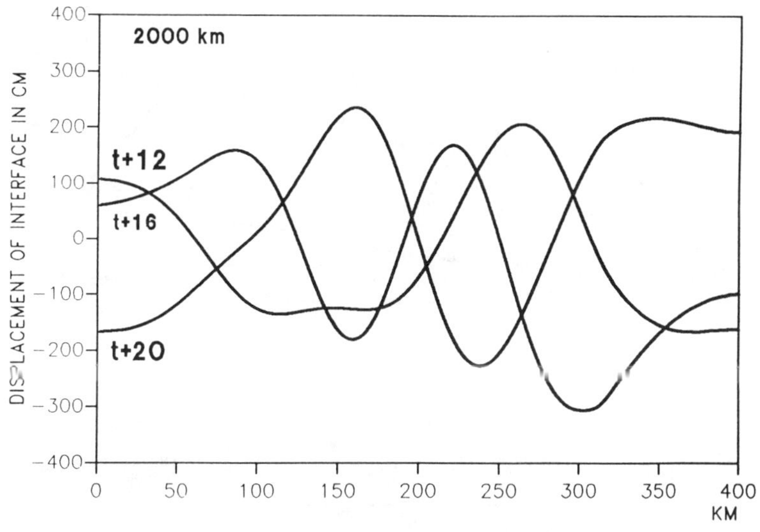


B

FIGURE 16. Vertical displacement of the interface at 4-h intervals (A and B) over the topography in Figure 3; 2000-km wavelength and wind stress frequency of 0.9*f*.

Finally, in Figure 19 we show an example of the amplitude distribution over the western slope, using a wind forcing that favors the excitation of a higher mode. For this particular experiment a wavelength of 80 km was arbitrarily chosen, and the wind-field period was 275 h. The three-time exposure of the interface with 32-h intervals indicates that the second mode dominates. The period of this mode has been determined earlier to be 315 h for the enhanced stratification as employed here. It is apparent from the figure that the main contribution to the vertical displacement takes place above and in the vicinity of the steepest bottom topography gradients.

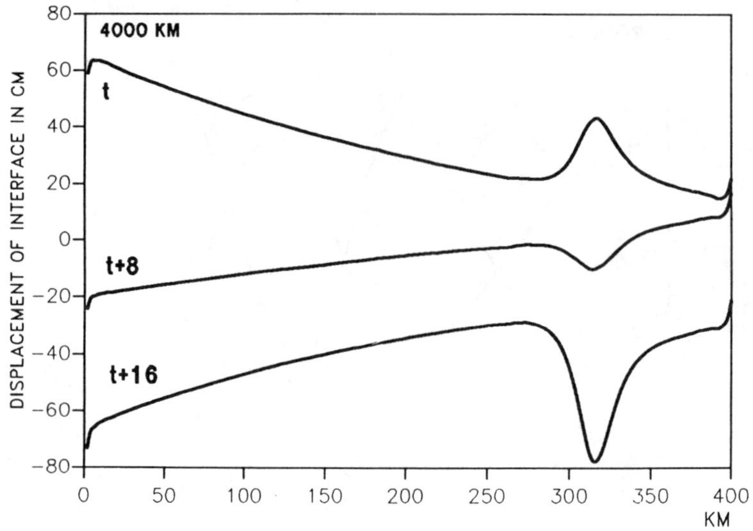


FIGURE 17. Vertical displacement of the interface at 8-h interval over the topography in Figure 3; 4000-km wavelength and wind-stress period of 36 h.

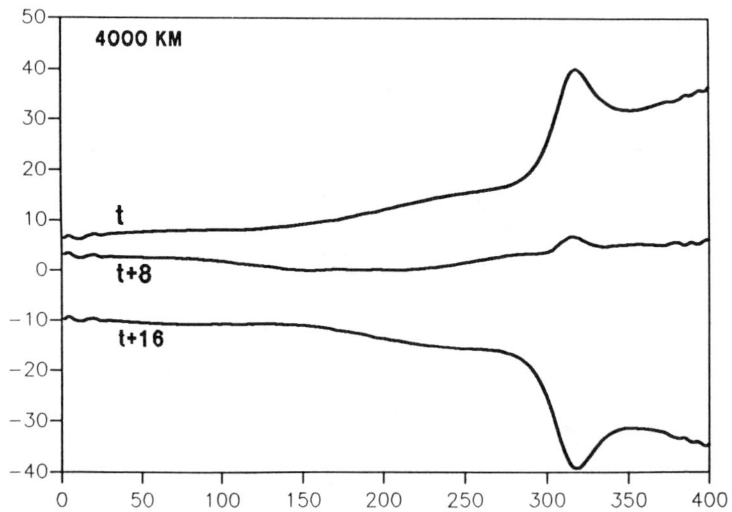


FIGURE 18. Vertical displacement of the interface at 8-h interval over the topography in Figure 3; 4000-km wavelength and wind-stress period of 20 h.

# VI. SUMMARY REMARKS

In this paper a review has been given of some of the major studies addressing low-frequency motions connected to the western slope of the Norwegian Trench. Literature references to some other relevant studies are also included. While earlier studies have dealt with processes in a homogeneous sea, this paper focuses on effects of stratification on waves trapped over the western slope. A two-layer model is formulated in which the upper layer is continuously stratified and the lower layer homogeneous. By relating the eddy viscosity of the upper layer to the static stability, it is possible to give a three-dimensional description of the upper layer in terms of continuous vertical structure functions. These functions are

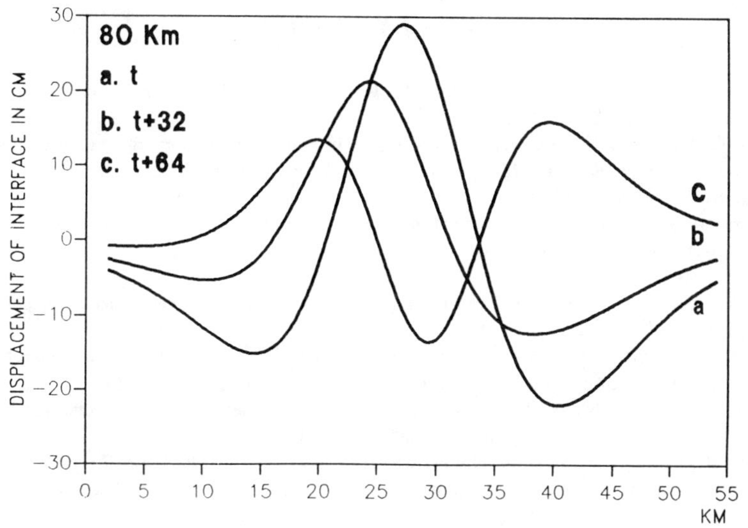


FIGURE 19. Vertical displacement of the interface at 32-h interval over the topography in Figure 12; 80-km wavelength and wind-stress period of 275 h.

eigenfunctions determined by the density field, together with a set of homogeneous boundary conditions.

Several current-meter and tidal-gauge observations have clearly shown the existence of low-frequency, subdiurnal wave activity, with an energy intensification towards the western slope. In particular there is one period that is found to be rather persistent. This is a 36-h period, which, during specific events, can have an amplitude above that of the dominating tidal component at the given position. The model studies indicate that this is the zero group speed, gravest mode wave (double Kelvin wave) trapped above the steepest part of the western slope. Analysis of observed data and model considerations clearly demonstrates an anticyclonic rotation at the deep side (trench) of this position and a cyclonic rotation on the shallow (shelf) side. Further, the wave propagates its phase with shallow water to the right, and the wavelength corresponding to this frequency was calculated to be 360 km. The influence of stratification on the period seems to be small, about 2%. This conclusion is based on a shallow stratification, e.g., an early spring situation. For a deep stratification the effects are expected to be much more important.

In the present model studies the motions were set up by an alongshore wind field with a prescribed period and duration. In the future work with this model it is of interest to look at the generation mechanisms behind these waves, with respect to different forms of the wind field. It is also of interest to include realistic bottom topography (in both the *x*- and *y*-directions). This is already allowed for in the coding of the model. The work that has been carried out recently by Slørdal[31] on nesting of this model into a barotropic model, may lead to interesting further experiments on low-frequency waves in the Norwegian Trench.

## ACKNOWLEDGMENTS

Thanks are due to:

- IBM — Bergen Scientific Centre (BSC) for providing first-class computer facilities.
- Jan Helge Salvesen for teaching me how to use them.
- Norsk Hydro for leave to work at BSC.
- Øistein Bøe for giving me lessons in $\mathrm{T_EX}$ and hence making it possible to acknowledge myself for typing the manuscript.
- Alastair Jenkins for linguistic consultancy.

# REFERENCES


1. **Furnes, G. K., Hackett, B., and Sætre, R.,** Retroflection of Atlantic water in the Norwegian Trench, *Deep-Sea Res.,* 33(2), 247, 1986.
2. **Furnes, G. K. and Sælen, O. H.,** Currents and hydrography in the Norwegian Coastal Current off Utsira during JONSDAP-76, Report 2/77, The Norwegian Coastal Current Project, 1977.
3. **Alcock, G. A.,** private communication.
4. **Koltermann, K. P.,** Low frequency variations of currents and water-levels in the northern North Sea during JONSDAP-76, in *The Norwegian Coastal Current,* Sætre, R. and Mork, M., Eds., University of Bergen, Norway, 1980, 72.
5. **Mork, M.,** Circulation phenomena and frontal dynamics of the Norwegian coastal current, *Philos. Trans. R. Soc. London Ser. A,* 302, 635, 1981.
6. **Hackett, B.,** Low frequency motions in the Norwegian Trench, Report 2/83, Norwegian Marine Pollution Research and Monitoring Program, Norwegian Ministry of Environment, 1983.
7. **Johannessen, O. M., Mork, M., et al.,** Remote Sensing Experiment in the Norwegian Coastal Waters, Report 3/79, The Norwegian Coastal Current Project, 1979.
8. **Hackett, B.,** Currents and hydrography off western Norway during NORSEX-79, Report 3/82, Norwegian Marine Pollution and Monitoring Program, Norwegian Ministry of Environment, 1982.
9. **Eide, L. I.,** private communication.
10. **Gordon, R. B., Flather, R. A., and Wolf, J.,** Modeling the current response of continental shelf waters to winter storms: comparison with data, OTC Paper No. 5417, presented 19th Annu. Offshore Technol. Conf., Houston, April 1987.
11. **Johannessen, J. A., Svendsen, E., Sandven, S., Johannessen, O. M., and Lygre, K.,** Three dimensional oceanic and current structure of mesoscale eddies in the Norwegian Coastal Current. *J. Phys. Oceanogr.,* 19, 1989.
12. **Kelvin, Lord W.,** On gravitational oscillations of rotating water, *Proc. R. Soc. Edinburg,* 10, 92, 1879.
13. **Longuet-Higgins, M. S.,** On the trapping of waves along a discontinuity of depth in a rotating ocean, *J. Fluid Mech.,* 31, 417, 1968.
14. **Longuet-Higgins, M. S.,** Double Kelvin waves with continuous depth profiles, *J. Fluid Mech.,* 34, 49, 1968.
15. **Martinsen, E. A.,** The barotropic current due to storm surges along the coast of Norway, in *The Norwegian Coastal Current,* Sætre, R. and Mork, M., Eds., University of Bergen, Norway, 1981, 500.
16. **Mysak, L. A., LeBlond, P. H., and Emery, W. J.,** Trench waves, *J. Phys. Oceanogr.,* 9, 1001, 1979.
17. **Hamon, B. V.,** The spectrums of mean sea level at Sydney, Coff's Harbour, and Lord Howe Island, *J. Geophys. Res.,* 67, 5147, 1962; Correction, *J. Geophys. Res.,* 68, 4635, 1963.
18. **Robinson, A. R.,** Continental shelf waves and the response of the sea level to weather systems, *J. Geophys. Res.,* 69, 367, 1964.
19. **Middeleton, J. H., Foster, T. D., and Foldvik, A.,** Low-Frequency currents and continental shelf waves in the southern Weddel Sea, *J. Phys. Oceanogr.,* 12, 618, 1982.
20. **Gill, A. E. and Schumann, E. H.,** The generation of long shelf waves by wind, *J. Phys. Oceanogr.,* 4, 83, 1974.
21. **Allen, J. S.,** Models of wind-driven currents on the continental shelf, *Annu. Rev. Fluid Mech.,* 12, 389, 1980.
22. **Cartwright, D. E.,** Extraordinary tidal currents near St. Kilda, *Nature,* 223, 928, 1969.
23. **Cartwright, D. E., Huthnance, J. M., Spencer, R. S., and Vassie, J. M.,** On the St. Kilda shelf tidal regime, *Deep-Sea Res.,* 27, 61, 1980.
24. **Grimshaw, R.,** Resonant forcing of barotropic coastally trapped waves, *J. Phys. Oceanogr.,* 17, 53, 1987.
25. **Mysak, L. A. and Willmott, A. J.,** Forced trench waves, *J. Phys. Oceanogr.,* 11, 1481, 1981.
26. **Mysak, L. A.,** Recent advances in shelf wave dynamics, *Rev. Geophys. Space Phys.,* 18(1), 211, 1980.
27. **Huthnance, J. M.,** On coastal trapped waves: analysis and numerical calculation by inverse iteration, *J. Phys. Oceanogr.,* 8, 74, 1978.
28. **Brink, K. H.,** Some effects of stratification on long trench waves, *J. Phys. Oceanogr.,* 13, 496, 1983.
29. **Huthnance, J. M., Mysak, L. A., and Wang, D.-P.,** Coastal trapped waves. Baroclinic processes on continental shelves, *Coastal and Estuarine Science,* Vol. 3, American Geophysical Union, Washington, D.C., 1986.
30. **Furnes, G. K. and Mork, M.,** Formulation of a continuously stratified sea model with three-dimensional representation of the upper layer, *Coastal Eng.,* in press.
31. **Slørdal, L. H.,** Undersøkelse av en Baroklin Numerisk Havmodell, Thesis, University of Oslo, Norway, 1988.

Chapter 4

# AN INVESTIGATION OF THE STORM SURGE OF FEBRUARY 1, 1983 USING NUMERICAL MODELS

**Roger Proctor and Judith Wolf**

## TABLE OF CONTENTS

## I. ABSTRACT

A meteorological depression passing eastwards across the north of Britain on February 1 and 2, 1983 caused sea level heights on the British coast of the northern North Sea to approach those attained during the storm surge of February 1, 1953. The prediction of these levels during the 1983 event was only partially successful. Numerical model studies show the critical nature of the correct meteorological input with respect to the sea model response. Energy input to the sea by the storm event is seen to take place mainly in the northern part of the North Sea but dissipation occurs mainly in the southern North Sea. The energy budget is examined in some detail.

## II. INTRODUCTION

On February 1, 1983 a meteorological depression passed eastwards across the north of the British Isles causing water levels along the northwest coast of Scotland and along the Scottish and English coasts of the northern North Sea to reach similar levels to those recorded during the great storm of February 1, 1953.[21] However, the depression of 1983 kept on an easterly track towards Scandinavia instead of, as in 1953, taking a southeasterly track into the North Sea. Major surge conditions in the southern North Sea, attained in 1953, were consequently avoided in 1983 although they caused the largest surge that winter and were considered serious enough to result in the first operational closure of the recently commissioned Thames barrier.[18]

However, unlike the situation in 1953, predictions of the impending high water levels were available and proved accurate, thus allowing the Storm Tide Warning Service[23] to issue sufficient warning to coastal sites in the North Sea, and also in the Irish Sea, for defensive action to be taken. The predictions were provided, in part, by a storm surge forecasting scheme developed at Proudman Oceanographic Laboratory, Bidston[5] as part of the research studies prompted by the surge event in 1953. The prediction scheme, based on numerical models of the sea and of the atmosphere, has operated in real time at the British Meteorological Office during each winter season since 1978 and has provided, throughout this time, useful forecasts of surge heights up to 30 h ahead. On occasions, unfortunately, poor forecasts are produced and in the case of the surge on February 1, 1983, although high water levels were well predicted, in the southern North Sea the magnitude of a negative surge preceding the positive surge was overestimated and the positive surge itself was prolonged.

The purpose of this chapter is twofold: (1) to briefly review the development of numerical models for storm surge forecasting in the U.K. and (2) to investigate the dynamics and energy budget of the 1983 storm event. From (2), an insight into the significant time periods during the storm and the important geographical regions on the North West European Continental Shelf, in terms of energy input and dissipation, may be found. These are compared with similar results given by Flather[8] for the storm surge of February 1, 1953.

## III. NUMERICAL MODELS FOR STORM SURGE FORECASTING IN THE U.K.

Since 1953 a great deal of progress has been made in the understanding and prediction of storm surges around the British coast. Initially, the main effort was into the analysis of observed surges and the production of empirical formulae for surge prediction which followed on from earlier work.[1] The advent of digital computers allowed the solution of the equations of motion, previously only treated analytically for simplified cases, to be solved for real sea areas. Some of the earliest applications of numerical methods were by Hansen.[13] The use of computers has become widespread since the 1960s and the application of finite difference methods to the storm surge problem has been made by many workers. Heaps[15,17] has given comprehensive reviews of the work done on storm surge prediction and was himself the instigator of many lines of research. The initial solution of the linearized two-dimensional (2-D) equations[14] was followed by the use of three-dimensional (3-D) equations allowing the prediction of extreme currents.[16] Heaps' methods have been followed and extended by other workers (e.g., Davies[2], and Flather[4]). In particular, Flather continued the development of the Heaps[14] model to the current operational 2-D model for storm surge prediction described below.

Storm surge prediction is still a topic of further research. Work has begun on incorporating the effects of surface wind waves (see Wolf et al.[24]), as suggested by Heaps.[17] Other problems which he highlighted, such as the specification of meteorological forcing and the form of vertical eddy viscosity for 3-D models, are still prime research areas.

A 2-D numerical sea model for storm surge prediction has been run at the British Meteorological Office at Bracknell throughout the winter months of September to April since 1978. This model forecasts the changing sea level on the Northwest European Continental Shelf in response to meteorological forcing (in the form of sea-surface atmospheric pressure gradients and wind stresses) derived from air pressures and surface winds predicted by the numerical atmospheric model of the Meteorological Office. Two forecasts per day are produced, with initial data times 0000 GMT and 1200 GMT, which provide surge predictions up to 36 h ahead. However, due to computer scheduling, the first 6 h are of no practical value. Modifications to the system have been introduced from time to time aimed at improving the surge forecasts and eliminating possible sources of error introduced, on occasions, by the meteorological data derived from the atmospheric model. The development of the system from its beginnings up to 1982 are described by Flather[6] and Flather and Proctor.[9]

In 1982 the Meteorological Office acquired a new computer, a CDC CYBER® 205E vector processor. This led to the introduction of a 15-level atmospheric model[3] and major changes in the surge forecasting procedure. The new system, introduced during the 1982 to 1983 surge season, is described by Proctor and Flather.[20]

Three significant modifications were made to the system:

1. The spatial coverage of the sea model was extended to include some continental shelf slope regions and the adjacent Atlantic Ocean. This is to include likely deep water areas of surge generation which will help in the correct description of external surge propagation into shallower areas, e.g., the North Sea and the English Channel. The finite difference mesh of the sea model is shown in Figure 1, along with the air pressure points (x) of the fine mesh, 15-level atmospheric model.
2. Six constituents $M_2$, $S_2$, $N_2$, $K_2$ $O_1$, and $K_1$ instead to two ($M_2$ and $S_2$) were used to describe the open boundary tidal forcing. In shallow-water areas, because of tide-surge interaction, the better tidal reproduction will lead to improve surge predictions. The open boundary reproduction will lead to improved surge predictions. The open boundary tidal input was taken from a model of the northeast Atlantic Ocean.[7]
3. The model equations were modified to allow the inclusion of horizontal viscosity terms (see Equations 1 and 2). This was partly for completeness and partly to avoid local instabilities (grid-scale oscillations) being generated at the shelf edge in the extended model grid.

Comparisons of the model tide, before and after inclusion of six constituents and horizontal viscosity, were made with the predicted tide at six ports. The chosen ports were Milford Haven, Avonmouth, and Liverpool on the west coast of the U.K. and Aberdeen, Immingham, and Sheerness on the east coast.

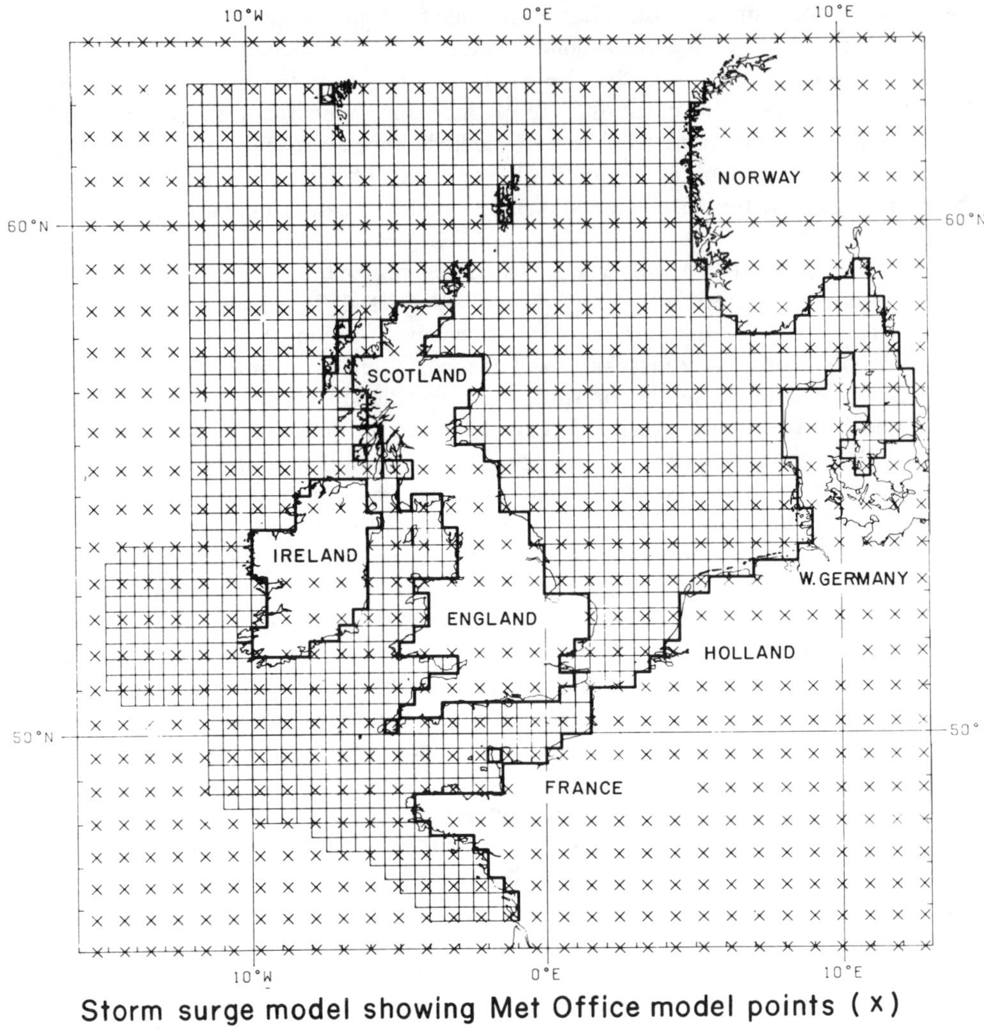


FIGURE 1. Sea model grid with pressure points (x) of the 15-level atmospheric model.

The inclusion of six rather than two constituents brought the model high water (HW) and low water (LW) times generally to within 10 min of the predicted tide, correcting times by as much as half an hour. The HW and LW heights were improved by about 30 cm on average, with HW at Avonmouth and Liverpool raised by 50 cm and LW at Avonmouth, Liverpool, and Immingham lowered by 50 cm. Overall, the increase in the number of constituents brought the model to within 10 cm of predicted HW and LW height, except at Liverpool and Avonmouth where there are still differences of up to 50 cm.

The basic equations and boundary conditions used in the numerical sea model are as follows:

$$\frac{\partial u}{\partial t} + \frac{u}{R\cos\phi}\frac{\partial u}{\partial \chi} + \frac{v}{R\cos\phi}\frac{\partial}{\partial \phi}(u\cos\phi) - 2v\omega\sin\phi = \frac{-g}{R\cos\phi}\frac{\partial(\zeta - \bar{\zeta})}{\partial \chi} - \frac{1}{\rho R\cos\phi}\frac{\partial p_a}{\partial \chi} + \frac{1}{\rho D}(F_S - F_B) + A_H\nabla^2 u \quad (1)$$

$$\frac{\partial v}{\partial t} + \frac{u}{R\cos\phi}\frac{\partial v}{\partial \chi} + \frac{v}{R}\frac{\partial v}{\partial \phi} + \frac{u^2\tan\phi}{R} + 2u\omega\sin\phi =$$

$$\frac{g}{R}\frac{\partial(\zeta - \bar{\zeta})}{\partial \phi} - \frac{1}{\rho R}\frac{\partial p_a}{\partial \phi} + \frac{1}{\rho D}(G_S - G_B) + A_H\nabla^2 v \tag{2}$$

$$\frac{\partial \zeta}{\partial t} + \frac{1}{R\cos\phi}\left\{\frac{\partial}{\partial \chi}(Du) + \frac{\partial}{\partial \phi}(Dv\cos\phi)\right\} = 0 \tag{3}$$

where the notation is

| | |
|---|---|
| $\chi,\phi$ | east longitude and latitude, respectively |
| $t$ | time |
| $\zeta$ | elevation of the sea surface |
| $\bar{\zeta}$ | equilibrium tide |
| $u,v$ | components of depth mean current, $\underset{\sim}{q}$ |
| $D$ | total water depth ($= h + \zeta$) |
| $h$ | undisturbed water depth |
| $R$ | radius of the Earth |
| $\omega$ | angular speed of rotation of the Earth |
| $g$ | acceleration due to gravity |
| $\rho$ | density of sea water, assumed constant |
| $p_a$ | surface atmospheric pressure |
| $F_S,G_S$ | components of wind stress $\underset{\sim}{\tau}_s$ on the sea surface |
| $F_B,G_B$ | components of bottom stress |
| $A_H$ | coefficient of horizontal viscosity |
| $\nabla^2 \equiv$ | $\frac{1}{(R\cos\phi)^2}\frac{\partial^2}{\partial\chi^2} + \frac{1}{R^2\cos\phi\,\partial\phi}\left(\cos\phi\frac{\partial}{\partial\phi}\right)$ approximately[7] |

A quadratic law is used for bottom friction

$$\underset{\sim}{\tau}_B = k\rho\underset{\sim}{q}|\underset{\sim}{q}| \qquad \text{with} \qquad k = 0.0025 \tag{4}$$

and also for the surface wind stress

$$\underset{\sim}{\tau}_S = C_D\rho_a\underset{\sim}{w}|\underset{\sim}{w}| \tag{5}$$

where $\rho_a$ is the density of air (assumed constant), $\underset{\sim}{w}$ is the surface wind vector and the drag coefficient

$$C_D = 10^{-3}(0.63 + 0.066|\underset{\sim}{w}|) \tag{6}$$

with $|\underset{\sim}{w}|$, the wind speed in $ms^{-1}$.[22]

The horizontal viscosity was chosen to be $A_H = 5D$, where D is the total water depth in meters. This gave little extra damping in the shallow water on the continental shelf, while damping out oscillations in the deep water at the shelf edge.

Equations 1 to 3 are solved starting from a prescribed initial state of elevation and motion and subject to boundary conditions on land and open sea boundaries. At land boundaries the component of depth mean current, $q_n$, along the outward-directed normal to the boundary is set to zero, and on an open sea boundary a "radiation" condition is used,

$$q_n = q_n^M + q_n^T + ch^{-1}(\zeta - \zeta^M - \zeta^T) \tag{7}$$

where $c = (gh)^{1/2}$ and $q_n^M$, $\zeta^M$ and $q_n^T$ and $\zeta^T$ describe input of meteorological (M) and tidal (T) origin, respectively. The open-boundary tidal input, $q_n^T$ and $\zeta^T$, expressed in standard harmonic form, was derived from earlier numerical model computations.[10] The surge input, $q_n^M$ and $\zeta^M$, is assumed to be

$$q_n^M = 0, \qquad \zeta^M = (\bar{p}_a - p_a)/\rho g \tag{8}$$

where $\bar{p}_a = 1012$ mbar.

The equations were solved for tide alone, omitting the meteorological terms, then in the full form for tide plus surge. The residual elevations and currents were obtained by subtracting the results for tide alone from those for tide plus surge.

## IV. THE STORM SURGE

### A. METEOROLOGICAL DEVELOPMENT

A depression, formed over the mid-Atlantic, moved gradually ENE, deepening slowly to 970 mbar, to reach a position NW of Ireland by 1200 GMT on January 31. Over the following 18 h the depression deepened rapidly to 950 mbar as it passed, still in a ENE direction, across Scotland and into the North Sea. Gale-force winds were experienced over most of northern Scotland with winds of 40 to 50 knots recorded at 0600 GMT on February 1 at Cape Wrath on the north Scottish coast. Between 0600 and 1800 GMT as the depression passed across the North Sea, storm-force winds acted over a large part of the western North Sea. By 0000 GMT January 2, the depression had passed across Denmark and filled to 972 mbar. The movement of the depression is illustrated in the synoptic charts in Figure 2.

### B. OBSERVED SURGE RESPONSE

Observed storm surge elevations at Scottish and English ports in the North Sea and English Channel are shown in Figure 3 (+ + +). The surge itself was not exceptionally large, but, coupled with the high spring tides, extreme levels were recorded along the northeast coast of Britain. At Buckie, a small fishing port on the Scottish coast, the highest ever levels were observed, and at Immingham in the Humber Estuary, the second highest ever. Recognized danger levels were exceeded at five locations on the coast, at North Shields, Immingham, Lowestoft, Felixstowe, and Southend. Surge residuals at high water are given in Table 1, together with those observed on February 1, 1953 for comparison. The fact that extreme levels were not reached in the southern North Sea was due to the continued ENE movement of the depression towards Scandinavia, taking the strong surge-generating winds away from the shallower areas of the sea. The surge itself appears to have been an external surge which, after being generated in the northern part of the North Sea, traveled southwards down the east coast of Scotland and England.

### C. THE SEA MODEL SURGE FORECASTS

Predictions of surge heights at the time of the highest observed levels, made by the real-time forecasting system, are also given in Table 1. The predictions are taken from what is considered the most useful part of the forecast, hours 6 to 18 of a forecast beginning at 0z (0000 GMT) or hours 18 to 06 (on the following day) of a forecast beginning at 12z (1200 GMT). At ports in the northern North Sea the tendency is to underpredict the surge while the opposite is true in the southern North Sea. The predictions were, however, accurate enough to provide sufficient warning of impending dangerously high levels and to allow the necessary defensive action to be taken. Undoubtedly, these warnings provided a contribution

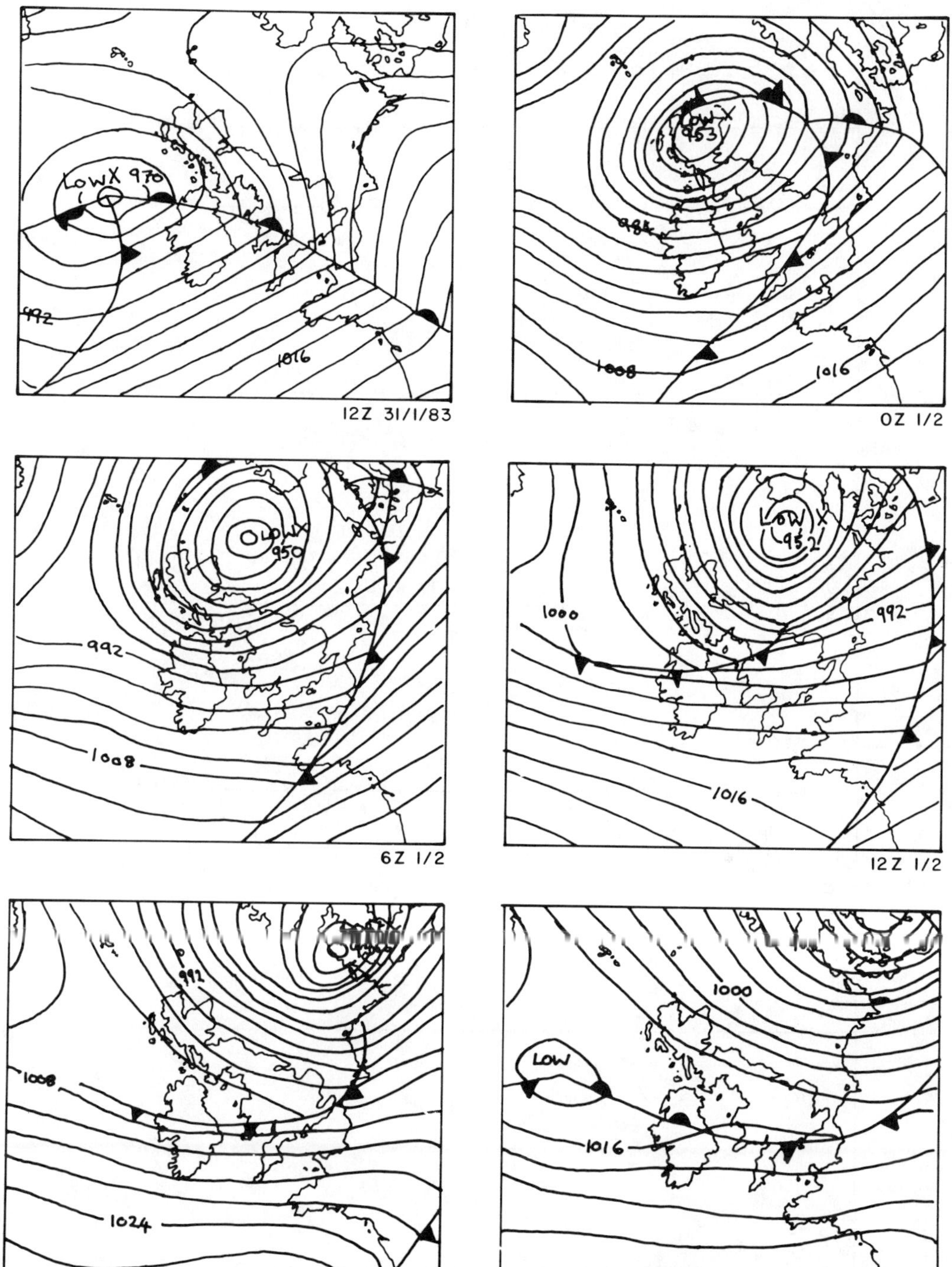


FIGURE 2. Meteorological development from 1200 GMT January 31 to 0000 GMT February 1, 1983.

to averting a potentially disastrous situation. At both Hull and London surge barriers were closed as precautionary measures.

Examining the surge forecasts (solid line, Figure 3) reveals that, while the trends of the actual surge were reproduced by the model forecasts, the positive surge was prolonged and the preceding negative surge was overestimated. Discrepancies between the last value of

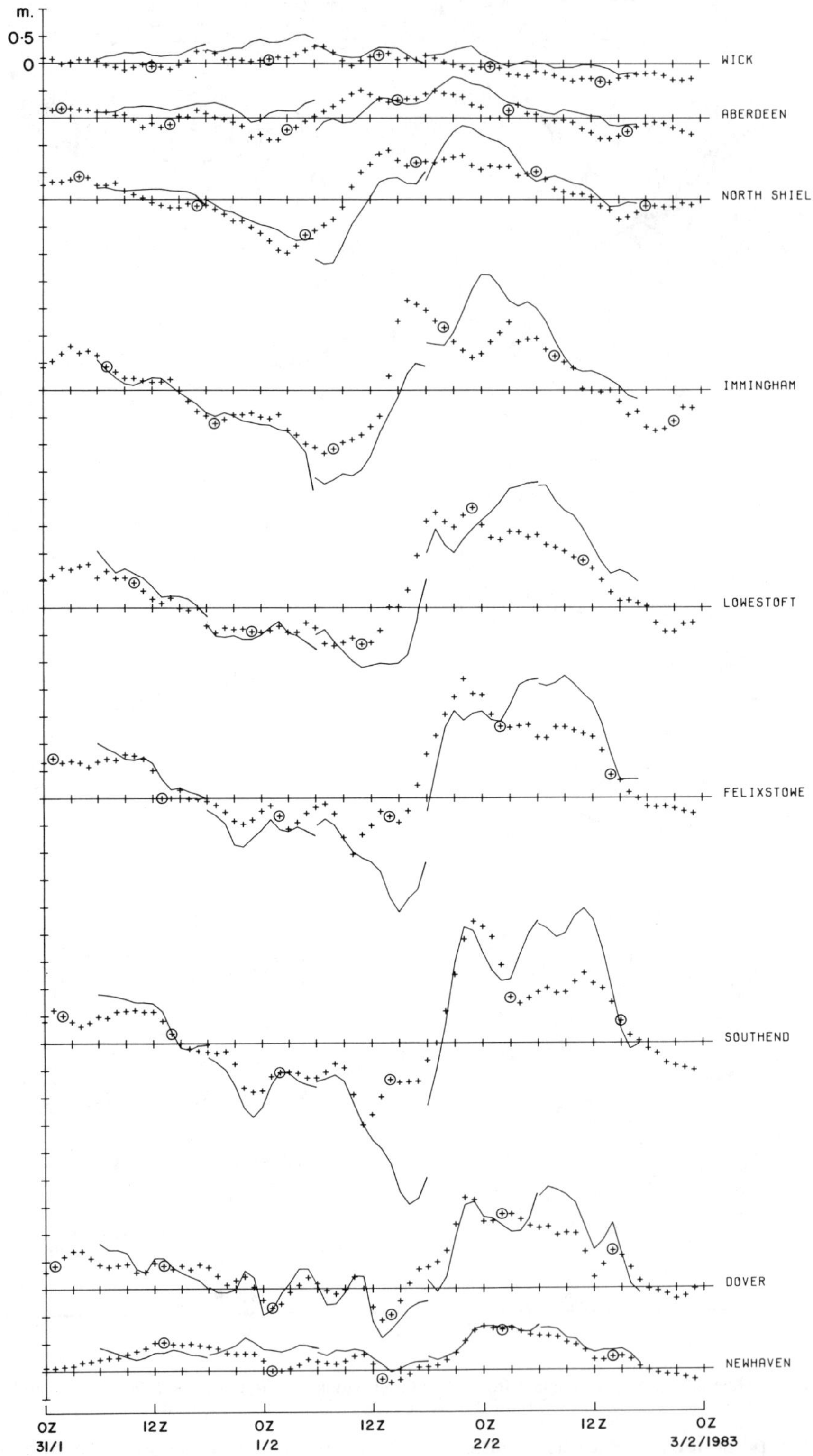


FIGURE 3. (+ + +) Observed surge and (—) computed forecast surge at U.K. ports in the North Sea and English Channel. 0 denotes high water residual.

**TABLE 1**
**Predictions at Time of Maximum Water Level**

| Location | Residual (m) at time of highest hourly level 1953 | 1983 | Forecast residual at this time, 1983 | Hindcast residual from observed pressures, 1983 |
|---|---|---|---|---|
| Wick | | 0.19 | 0.29 | 0.34 |
| Buckie | | 1.40 | | |
| Aberdeen | 0.60 | 0.35 | 0.25 | 0.64 |
| Tyne | 1.42 | 0.70 | 0.34 | 0.77 |
| Immingham | 1.37 | 1.15 | 0.82 | 0.86 |
| Lowestoft | | 1.84 | 1.45 | 1.56 |
| Felixstowe | | 1.31 | 1.40 | 1.35 |
| Southend | 2.29 | 0.84 | 1.17[a] | 1.25[a] |
| Sheerness | | 0.96 | 1.17[a] | 1.25[a] |
| Dover | 1.89 | 1.36 | 1.15 | 1.16 |
| Newhaven | 0.80 | 0.75 | 0.77 | 0.66 |
| Vlissingen | 3.24 | 1.36 | 2.22 | 1.44 |
| Hoek v. Holland | | 1.42 | 2.17 | 1.53 |
| Denhelder | 2.57 | 1.91 | 1.50 | 1.60 |
| Delfzijl | 1.96 | 2.38 | 2.34 | 2.23 |

[a] Same model point.

one forecast and the first value of the next, i.e., at times 6z and 18z each day, give some indication of the corrections applied to the atmospheric model forecasts. In particular, it is evident that large corrections were made between forecasts beginning 0z and 12z on February 1.

# V. MODEL EXPERIMENTS

## A. USE OF ANALYZED METEOROLOGICAL DATA

In attempting to improve the performance of the model to predict the surge, several aspects may be considered. In the real-time forecasts, the system shown in Figure 4 was used to achieve continuity between forecasts. Prior to the implementation of the CYBER® computer, a hindcast period lasting 12 h was run in advance of each forecast. The hindcast meteorological data was obtained from the previous forecast but modified at three hourly intervals by available observations. The use of the hindcast helped to eliminate errors from being passed forward in time from one forecast to the next. Unfortunately, hindcast data of this type was unavailable on the new computer at the time of this surge event. However, at a later date, the Meteorological office was able to supply objective analyses of surface wind components and atmospheric pressures at 12 hourly intervals. By comparing these data with the forecast data at the corresponding times, the error, $\epsilon$, in the forecast at time $t_0$ + 12 h can be calculated (where $t_0$ is the start time of the forecast). A simple way of incorporating these corrections into the forecast data is to apply the correction to the forecast data at time $t_0$ + 12 h and to linearly decrease the correction backwards through time, resulting in $\epsilon = 0$ at time $t_0$.

Taking the first 12 h of each forecast and applying corrections to both the pressure and the wind fields, a revised data set was constructed for the recomputation of the 5-d surge period. The resulting surge elevations are given in Figure 5. It can be seen, particularly at Southend, that the agreement between observed and computed surge is much better than achieved with forecast data alone. The negative surge on February 1 is also reproduced more accurately. The overestimate of the positive surge on February 2 is reduced but the values are still prolonged for too long a time.

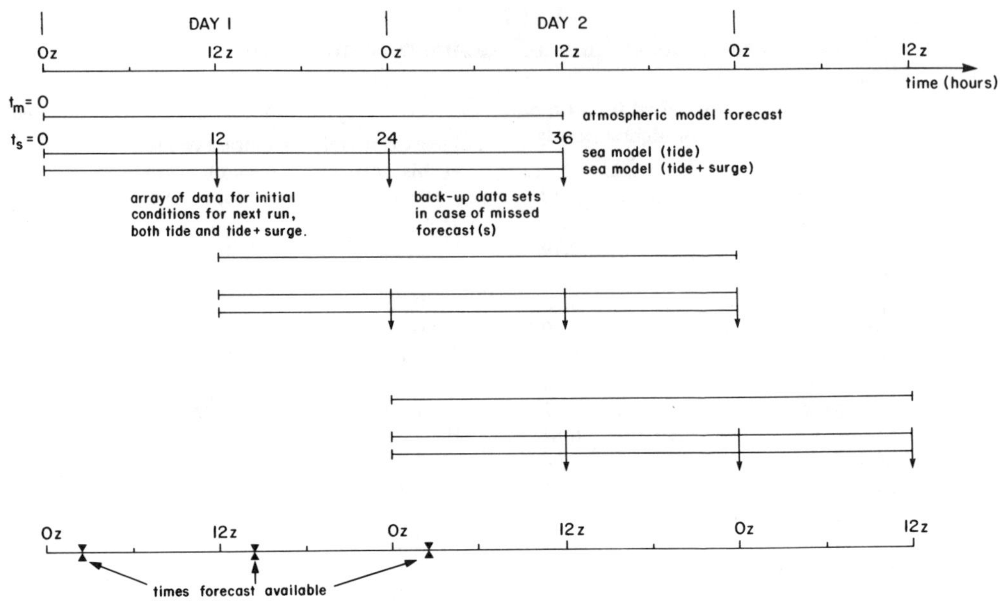


FIGURE 4. Routine operation of the storm surge prediction scheme.

## B. USE OF OBSERVATIONAL DATA

Instead of using meteorological information in the form of numerical model forecasts or objective analyses, an alternative is to base the metorological input on observed data. In surge hindcast studies this is usually done by deriving a set of surface pressure fields from a set of time series of observed pressures evenly distributed over the domain of interest.[8] From these pressure fields, pressure gradients can be obtained and surface wind stresses can then be calculated by empirical means and both used as forcing in a model. Flather,[8] in computing the spatial pressure distributions for the surge of February 1, 1953, used a second order piecewise polynomial function which fitted the observed data exactly, matching first and second order derivatives at the data points. This method requires the careful selection of locations to ensure an even distribution of observations in the area considered, since it can develop spurious oscillations away from the data points.

For the surge of February 1, 1983 an alternative for fitting the pressure surfaces was considered. At three hourly intervals for the period 0z January 29 to 0z February 3, all available observations of surface pressure within the area bounded by 45°N to 65°N and 20°W to 13°E (both land and sea) were obtained from the British Meteorological Office. The majority of these observations were at land stations but it was hoped that there would be a sufficient number of ship observations to provide adequate coverage in all the open sea areas. This did not prove to be so for a number of reasons:

1. The number of reported ship observations was surprisingly few; several data times had no observations at all while the maximum reported at any time was 20.
2. The spatial distribution of the observations was confined to the major shipping lanes, i.e., in the English Channel and the west coast of France. Virtually no observations were available north of 52°N.
3. The observations proved unreliable both in reported position and pressure value.

The land stations proved to be, in the most part, very reliable. After preliminary data screening, approximately 240 observations were available at each hourly interval. An abortive

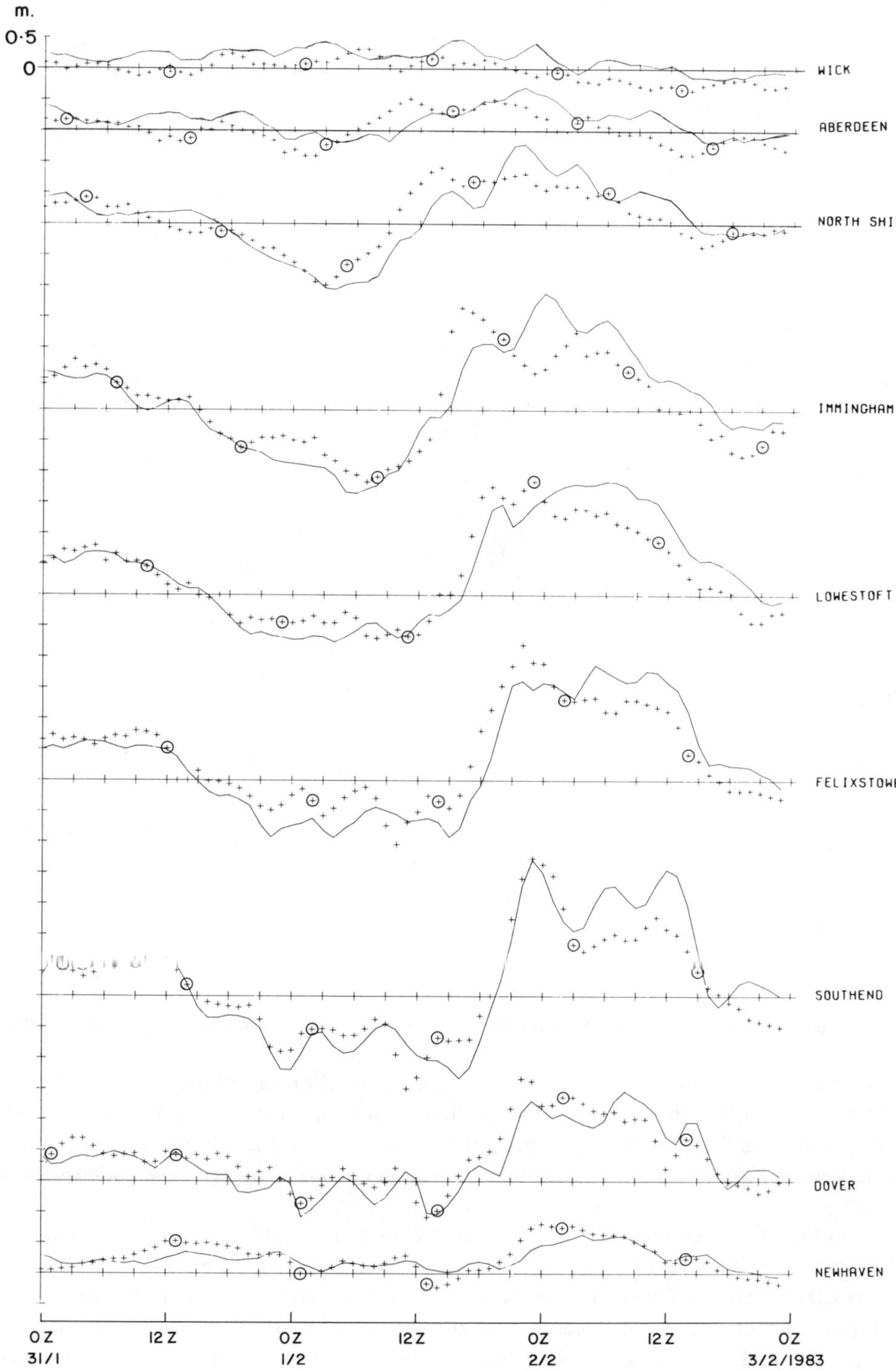


FIGURE 5. (+ + +) Observed surge and (—) computed corrected forecast surge at U.K. ports in the North Sea and English Channel.

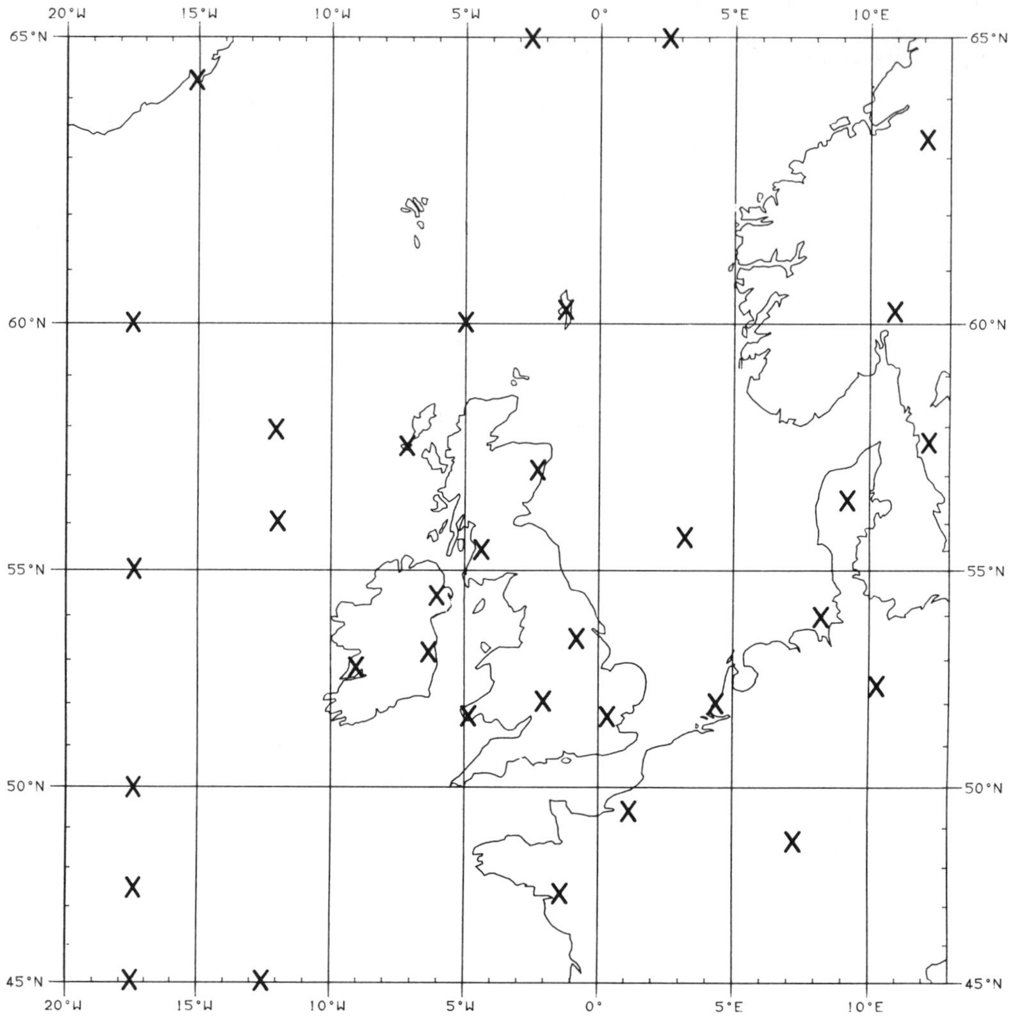


FIGURE 6. Network of locations of observations used in estimation of pressure distributions (polynomial function).

attempt was made to produce pressure maps using all these observations, by fitting bicubic splines with a least-squares method. The method was found to produce smoother isobars than the polynomial fitting method but this had the undesirable effect of reducing the intensity of the depression. The maximum winds were thus reduced, which in turn gave underestimates of the surge.

Taking time series of pressures at 34 of the locations plus additional values to mark the depression centers, the pressure distributions were recomputed with the polynomial function used by Flather.[8] The 34 chosen locations were shown in Figure 6. A more realistic reconstruction of the meteorological development at the time of its most rapid change (between 12z January 31 and 6z February 1) was obtained than that using the bicubic spline function. The storm surge computation was repeated using the improved pressure distributions and the model response is shown in Figure 7. A marked improvement in the development of the negative surge and in the decay of the positive surge can be seen when compared with both the forecast and hindcast simulations, shown in Figures 3 and 5, respectively.

Insight into the storm surge development can be gained from an examination of the spatial distributions of surge elevation and current shown in Figure 8. As the depression approaches the British Isles at 12z January 31 (see Figure 2A), southwesterly winds drive

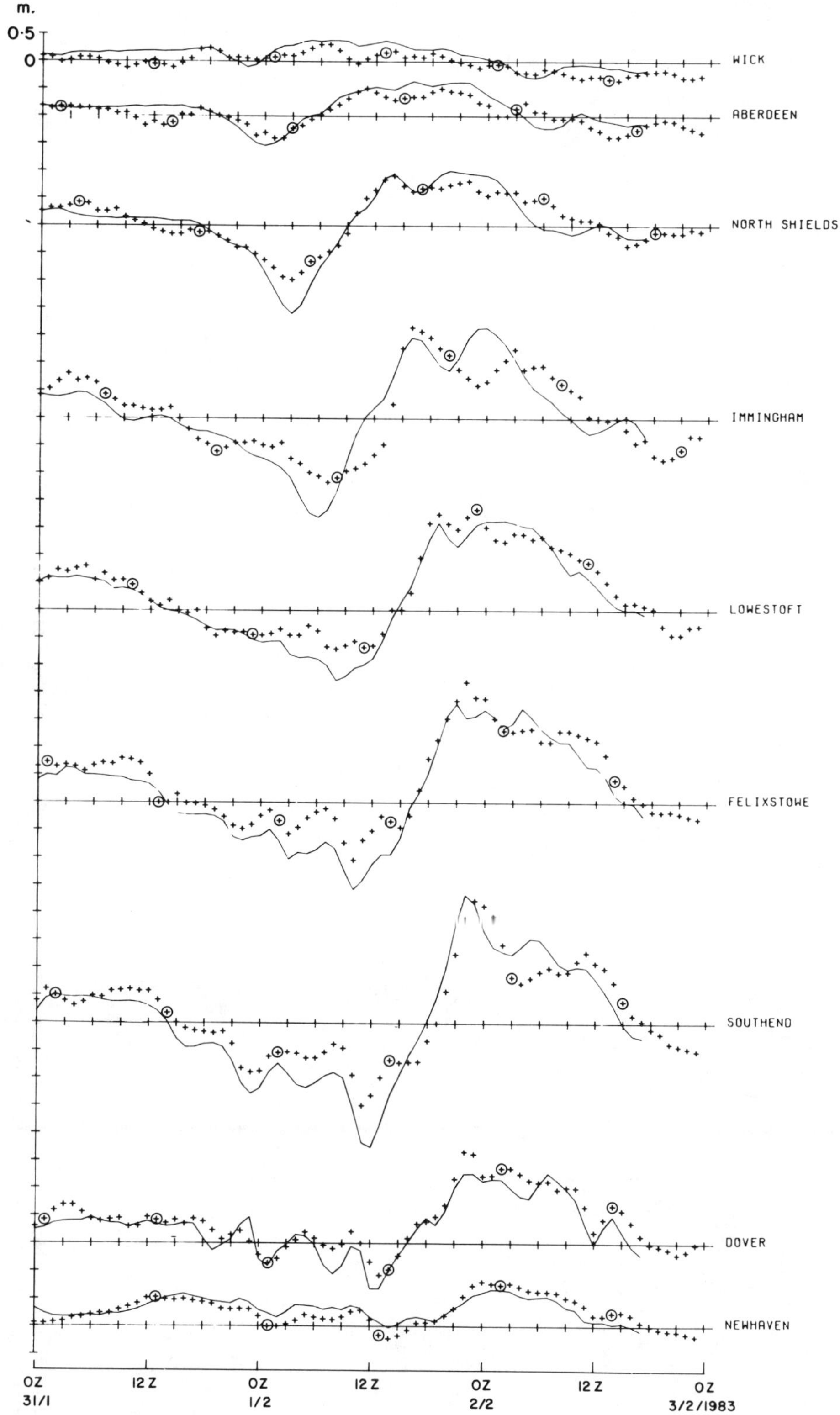


FIGURE 7. (+ + +) Observed and (—) computed surge using meteorological forcing derived from observed pressures.

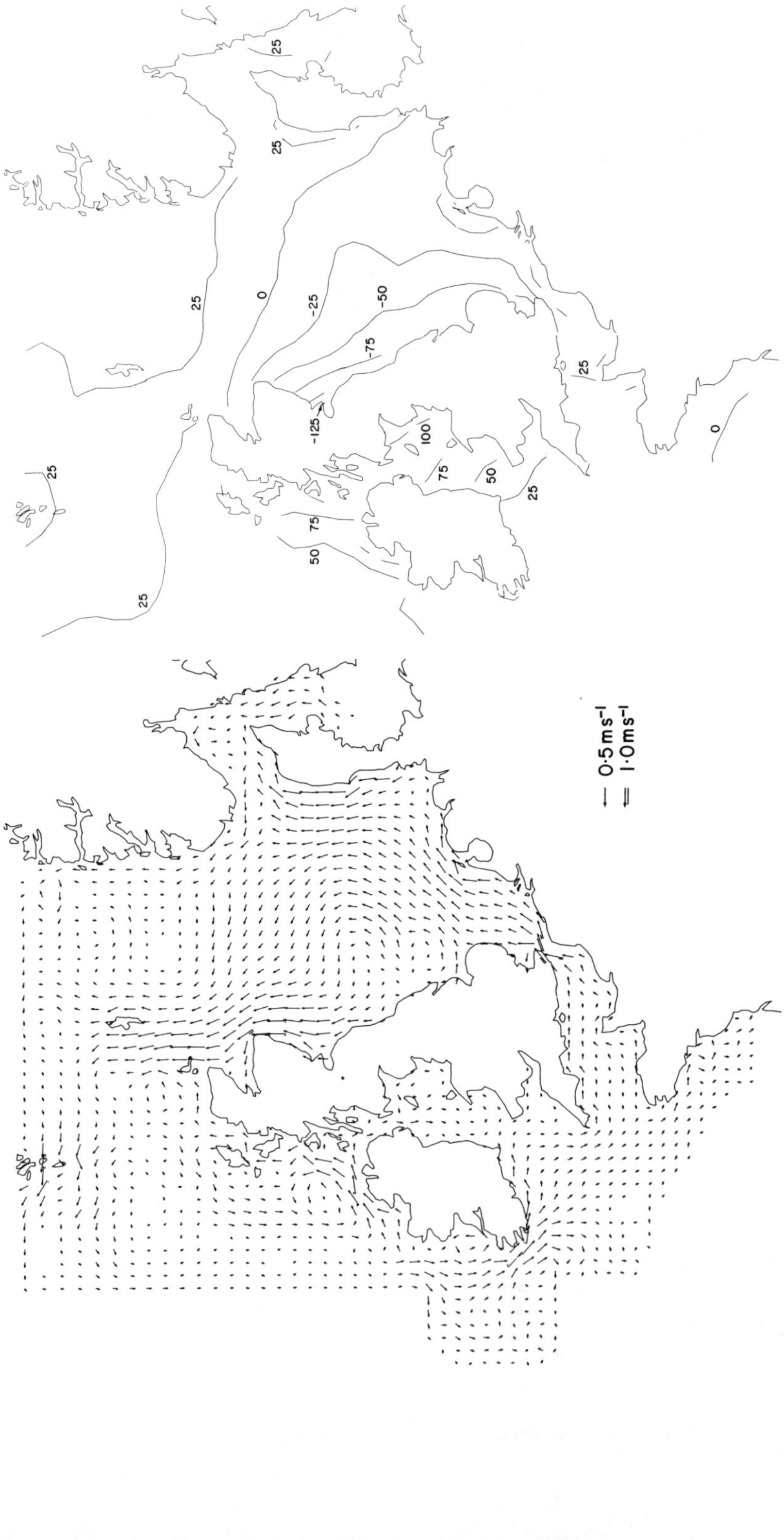


FIGURE 8. Distributions of surge elevation and current from model calculations at (A) 0000 GMT February 1, 1983, (B) 0600 GMT February 1, 1983, (C) 1800 GMT February 1, 1983, (D) 0000 GMT February 2, 1983. Elevations in cm.

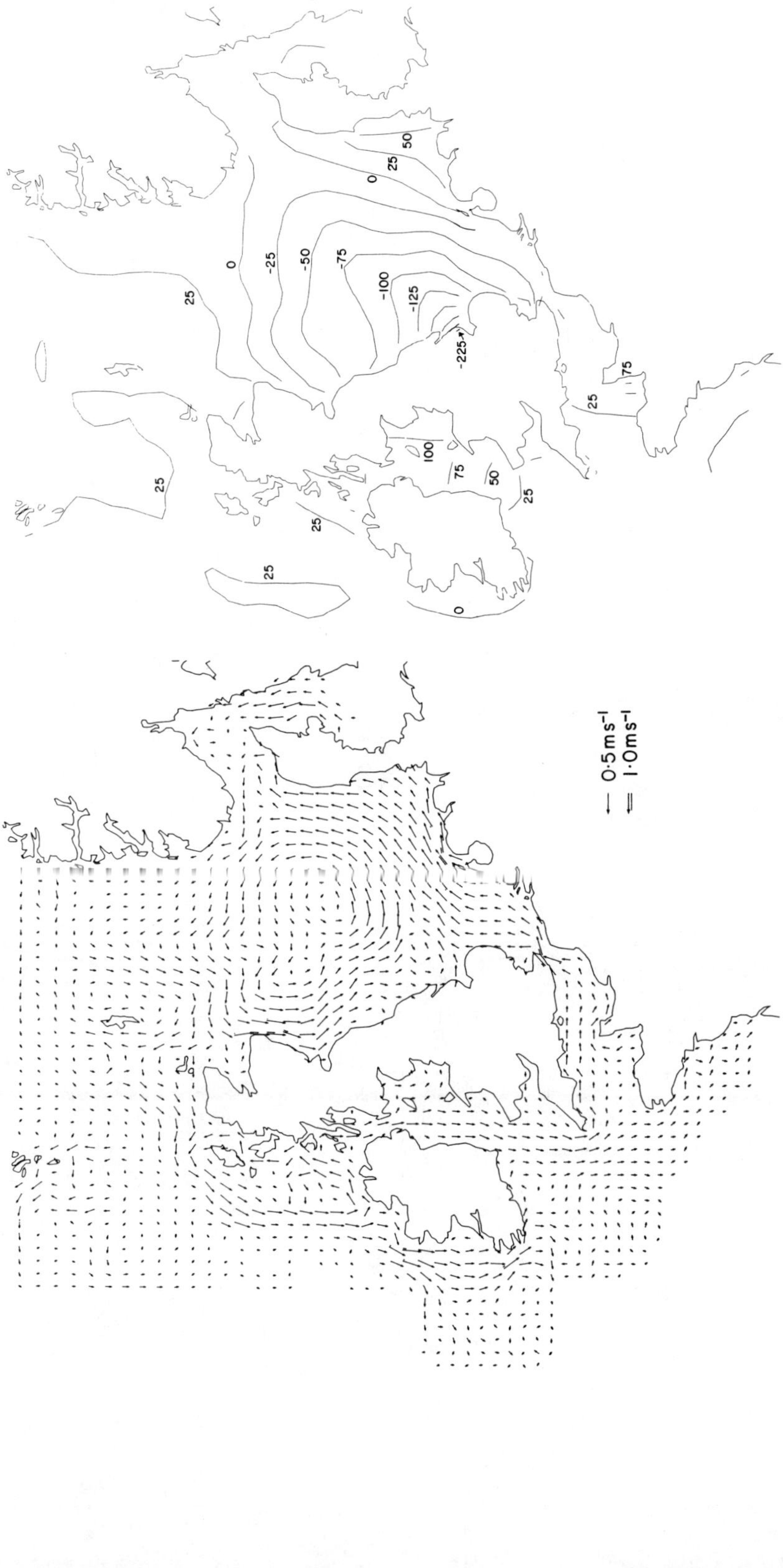


FIGURE 8B.

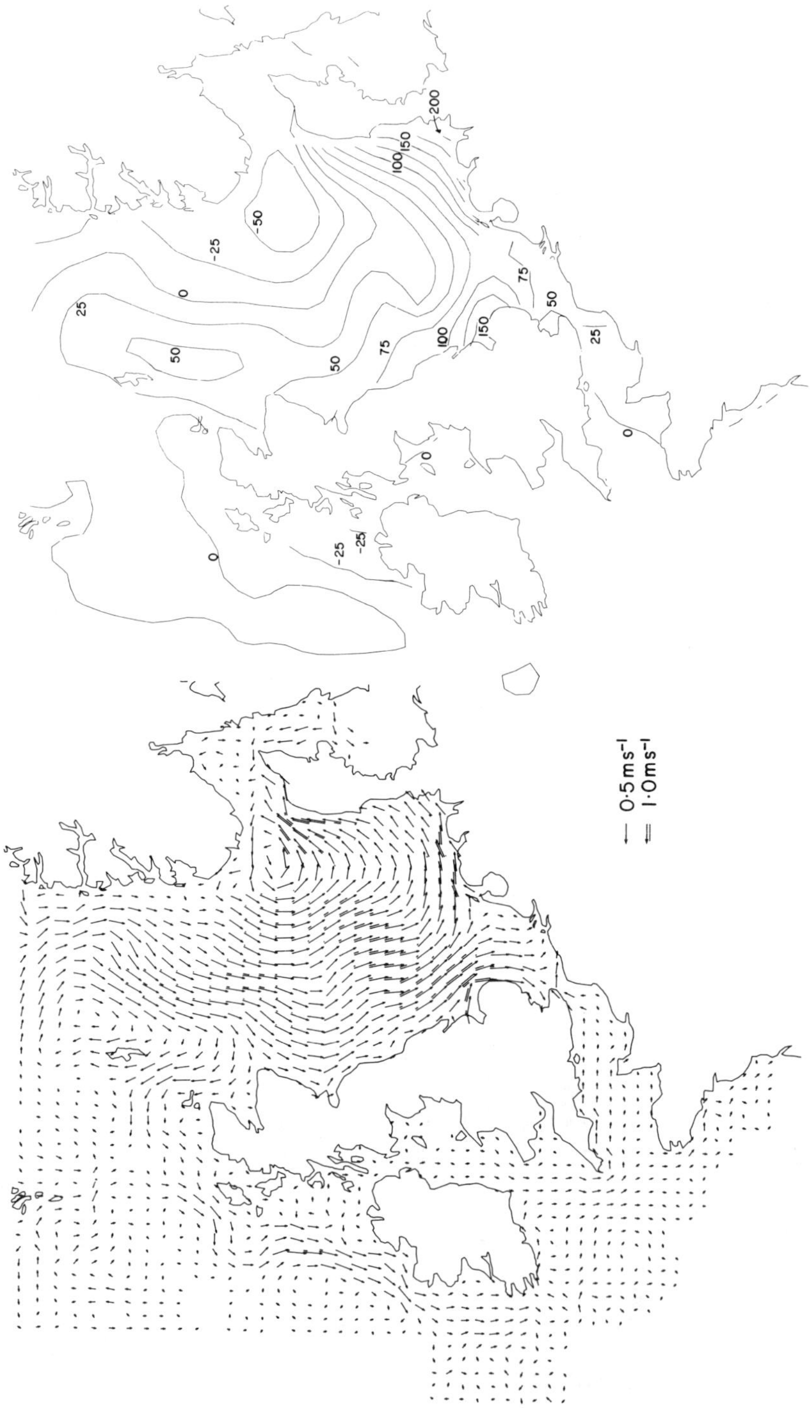


FIGURE 8C.

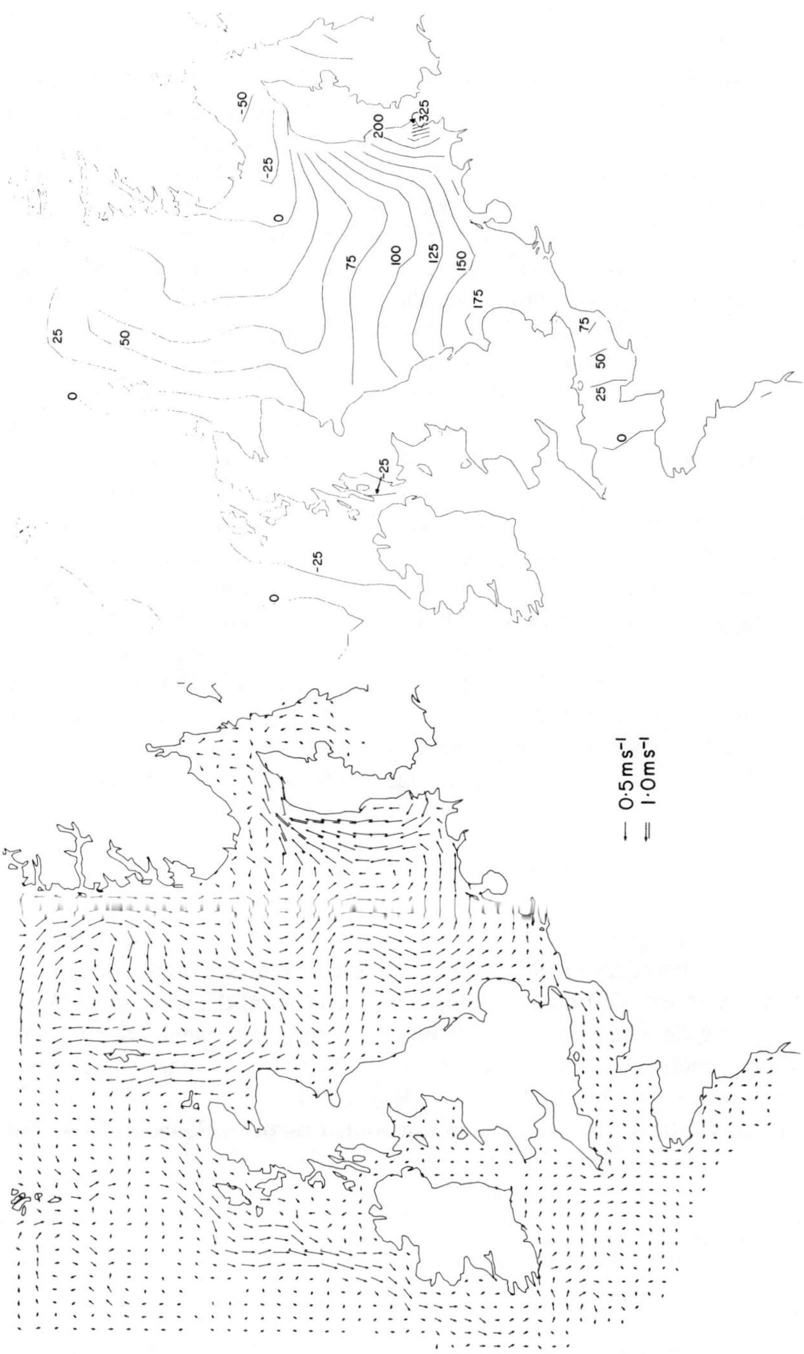
-50
200
325
-25
0
75
100
125
150
175
25
50
0
75
50
25
0
-25
-25
0
0·5 ms$^{-1}$
1·0 ms$^{-1}$

FIGURE 8D.

water onto the shelf south of Ireland, northwards into the Irish Sea and into the English Channel. At 0z February 1, with the depression located over northern Scotland, Figure 2B, northwesterly winds lead to a positive surge on the west coast of Britain, Figure 8A, and to a negative surge on the east coast, particularly in southern Scotland. At the same time there was a strong flow through the Dover Straits into the southern North Sea and a flow northwards out of the North Sea between Scotland and the Shetlands driven by the southerly winds leading the depression. By the time the depression had deepened to its lowest value (950 mbar) at 6z, it had moved eastwards to the center of the northern North Sea, Figure 2C. The slight southeasterly movement in its track took the strongest west/northwesterly winds southwards over Britain, causing the negative surge to have its lowest values between Immingham and East Anglia, Figure 8B. In the Irish Sea the positive surge reached its maximum with 1.30 m being recorded at Heysham, Lancashire. While the westerly winds drove water eastwards through the Dover Straits, northerly winds, to the west of the depression center, forced water into the North Sea along the Scottish coast. By 12z February 1, westerly winds were still driving water through the Dover Straits and offshore in the Thames Estuary, while, at the same time northerly winds were causing an inflow of water into the northern North Sea resulting in the buildup of a positive surge on the east coast of Scotland. Over the next 12 h northerly winds over the North Sea drove water southwards down the east coast of England (Figure 8C, 18z February 1) and caused a setup over the whole of the southern North Sea (Figure 8D, 0z February 2). By 12 z February 2, major surge conditions had subsided in the North Sea except along the Danish coast where the northwesterly winds were maintaining setup with an associated northerly coastal flow into the Skaggerak. Some water escaped from the North Sea through the Dover Straits but most appears to have rotated anticlockwise around the southern North Sea into the Skaggerak and Norwegian Trench. Additionally, the northerly winds caused a southwards directed current to flow along the shelf edge to the west of Scotland from 6z February 1 to 12z February 2. This flow has been seen in the residual currents from IOS mooring A1[25] which were recorded as part of the CONSLEX experiment on the shelf edge northwest of the U.K. in 1982 to 83.[12]

## VI. ENERGY BUDGET

Previous analyses of the energy budget of storm surges in the North Sea have concentrated on the surge of January 31 to February 1, 1953. Using numerical models, Prandle[19] computed the energy budget in the southern North Sea, and Flather[8] considered the work done by the meteorological forcing over the whole of the North West European Continental Shelf. Here, we examine in detail the energy budget for January 31 to February 1, 1983 surge period over the extended shelf region covered by the model (Figure 1).

The energy equation is obtained by multiplying Equation 1 by $\rho Du$, Equation 2 by $\rho Dv$, and Equation 3 by $\rho g\zeta$ and summing them.[11] In vector form this gives

$$\underbrace{\frac{\partial}{\partial t}\left(\frac{\rho D}{2}|\underset{\sim}{q}|^2\right)}_{(a)} + \underbrace{\frac{\partial}{\partial t}\left(\frac{\rho g\zeta^2}{2}\right)}_{(b)}$$

$$= \underbrace{-\rho\underset{\sim}{\nabla}\cdot\left(\frac{D\underset{\sim}{q}|\underset{\sim}{q}|^2}{2}\right)}_{(c)} \underbrace{-\ \rho g\underset{\sim}{\nabla}\cdot(D\zeta\underset{\sim}{q})}_{(d)} \underbrace{-\ \underset{\sim}{\tau}_B\cdot\underset{\sim}{q}}_{(e)} + \underbrace{\rho DA_H\underset{\sim}{q}\cdot\nabla^2\underset{\sim}{q}}_{(f)}$$

$$+ \underbrace{\underset{\sim}{\tau}_s\cdot\underset{\sim}{q}}_{(g)} \underbrace{-\ D\underset{\sim}{q}\cdot\underset{\sim}{\nabla}p_a}_{(h)} + \underbrace{\rho gD\underset{\sim}{q}\cdot\underset{\sim}{\nabla}\zeta}_{(i)} \qquad (9)$$

where $\nabla \equiv \left(\frac{1}{R\cos\phi}\frac{\partial}{\partial\chi} \quad \frac{1}{R}\frac{\partial}{\partial\phi}\right)$

Per unit area, the terms (a) to (i) in Equation 9 may be described as:

(a) The rate of change of kinetic energy $\partial(KE)/\partial t$
(b) The rate of change of potential energy $\partial(PE)/\partial t$
(c) The divergence of the energy transport due to advection
(d) The energy flux divergence
(e) The energy dissipated by bottom friction
(f) The energy dissipated by the horizontal viscosity
(g) The work done by the surface wind-stress
(h) The work done by the atmospheric pressure gradient
(i) The work done by the tide-generating force (t.g.f.)

Approximate values for each of these terms may be obtained by applying Equation 9 to the finite difference equations used in the model. An exact balance between the left-hand side (LHS) and right-hand side (RHS) of Equation 9 is not obtained in the finite difference form used here because some of the terms in Equations 1 and 2 do not cancel exactly. The terms which are affected are the advection and flux of energy, (c) and (d), and an extra term which appears because the Coriolis terms do not cancel out. An exact energy budget was carried out on the first 25 h of the tide-plus-surge run, using elevations and currents at every time step and including the extra terms for advection, flux, and Coriolis to reproduce the energy balance in the finite difference momentum equations. The energy terms are calculated at each elevation point of the model grid and integrated over the whole area to give a mean rate of change of energy density at each time step. The sum of the error terms generally has a magnitude of 1 to 2% of the LHS of Equation 9 at each time step, the largest term being the Coriolis error. The energy budget was repeated for the same period using hourly elevation and current data, which reduced the accuracy of the calculation of each term, particularly the time derivatives. Each term was compared with the exact calculation. From this comparison, the energy dissipation terms (bottom friction and horizontal viscosity) were seen to be very accurately estimated from hourly data (<0.1% mean error). The meteorological forcing terms and work done by t.g.f. were also accurate (<1% error). The flux term was slightly less accurate (<2% error) with the advection and Coriolis error being about 5% in error. However, more significant are the total errors in the RHS and LHS. The dissipation tends to balance the energy input terms, so the net energy input (RHS) is more sensitive to the cumulative effect of the errors in the individual terms. At each time step the RHS of Equation 9 is more accurately estimated than the LHS, but taking a mean over 25 h gives an error of over 50% in the estimation of the RHS, while the LHS error is only about 5%.

Separate energy budgets for the tide and for the tide plus surge were calculated, using hourly data. The former was then subtracted from the latter to obtain a balance for the residual energy due to the surge hour by hour. These residual energy time series are plotted in Figure 9 a to i for the period 16z January 31 to 17z February 2, 1983. The error due to the discretization of the Coriolis terms is plotted as term j in Figure 9. The above-mentioned estimation errors account for the discrepancy between the sum of the RHS terms (which will be called the energy input) and the LHS terms (the rate of increase of energy). Figure 9k shows the energy input plus the Coriolis error term and 9l is the rate of increase of energy. The contributions of advected energy (c) and the tide-generating forces (i) are seen to be negligible. The amount of energy dissipated by horizontal viscosity (f) is about 1% of that dissipated by bottom friction (e). Several of the terms, mainly the energy flux divergence (d) and the rate of change of kinetic (a) and potential (b) energy contain a strong

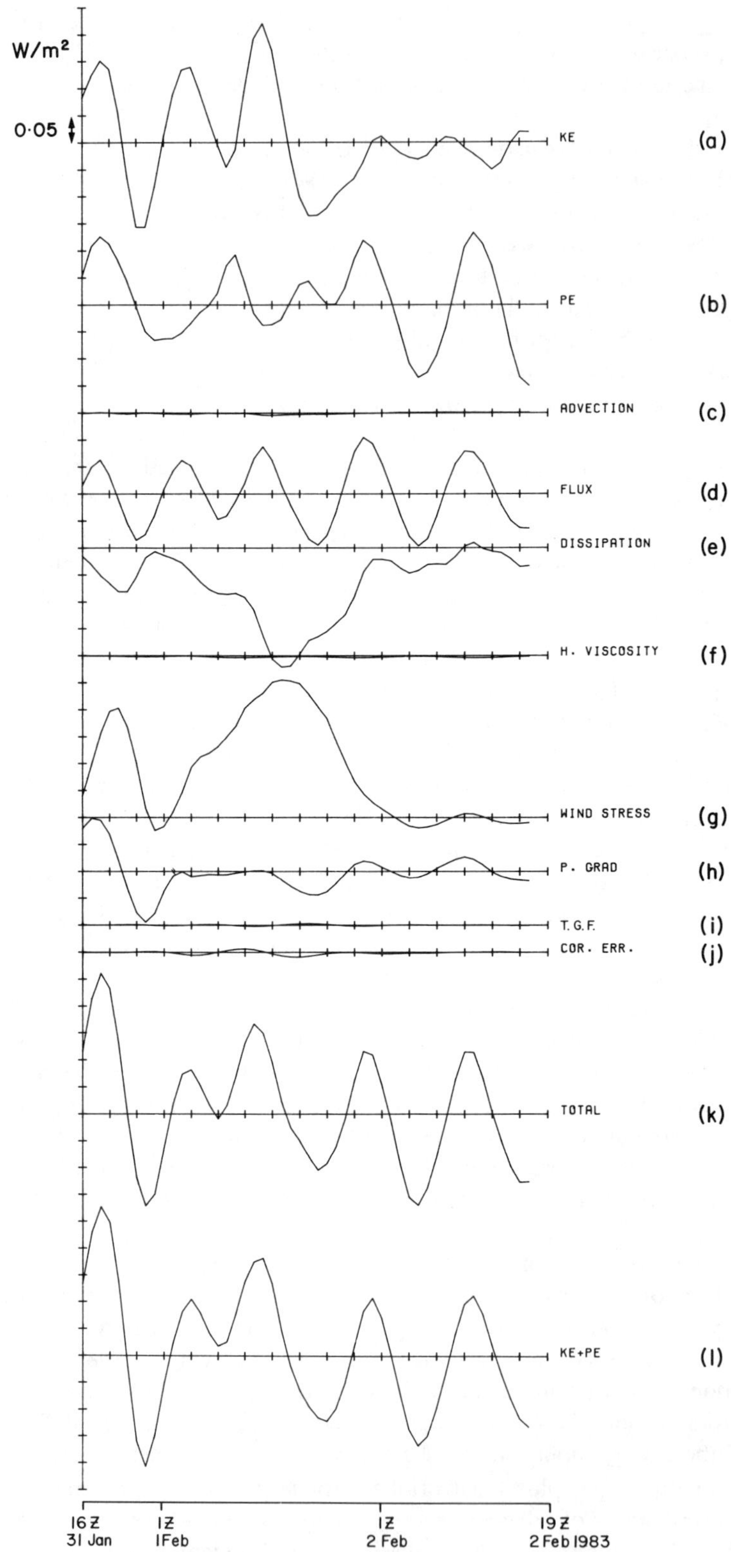


FIGURE 9. Time series of energy components averaged over the model grid.

tidal component, due to modification of the tidal energy balance by the surge. The work done by the wind stress (g) was far greater than that done by the pressure gradients (h) and reached a maximum of 0.273 W/m$^2$ at 13z on February 1. The two time series (g) and (h) can be compared with those calculated by Flather[8] for the January 1953 surge. Similar features are seen, the wind dominating the pressure gradient and reaching its maximum rate of work when the depression was located off the Danish coast. The maximum rate of work in the 1983 case (0.27 W/m$^2$), however, was only half that calculated for the 1953 surge (0.6 W/m$^2$), showing the winds to have been markedly weaker than in 1953. The duration of energy input was also less, about 24 h instead of 30, so that the total energy input in the 1953 case was considerably greater. The rate of dissipation of energy by bottom friction (e) is approximately in phase with the rate of work done by the wind, but slightly smaller. The maximum dissipation lags the maximum wind effect by about 1 h. This may be contrasted with the case in the southern North Sea in the 1953 surge where Prandle[19] deduced a phase lag of several hours between energy input and dissipation.

The spatial distributions of the energy terms were also investigated. Time averages of each term a to i in Figure 9 were calculated for each model gridpoint. The averages were performed over the 50-h period, 16z January 31 to 17z February 2, in order to eliminate most variations of tidal period. The most significant terms are plotted in Figure 10. Here Figures 10A and 10B are the average rate of increase of kinetic and potential energy, respectively. Over a sufficiently long interval the net increase of kinetic plus potential energy should be zero, provided that the model system conserves energy. In Figures 10A and B there are small but significant changes in residual energy. In particular a loss of kinetic energy occurs in areas of high tidal current, e.g., Bristol Channel, Southern North Sea, and a large increase in potential energy is seen at the French coast near Cherbourg.

The spatial distributions of work done by pressure gradients ([H] in 9), wind stress (g), and energy dissipated by bottom friction (e) are shown in Figures 10C to E, respectively. Distributions of (g) and (h) were also computed by Flather[8] for the 1953 surge and it is of interest to compare them with the present ones. The work done by the pressure gradients is quite small in both cases. The distributions of (g) are quite different. In the 1953 surge most of the energy input occurred on the western side of the North Sea with maxima (1) just north of Scotland, (2) off Wick, (3) off Aberdeen, and (4) on the Dogger Bank, reaching values slightly more than 1.[illegible] W/m$^2$. In Figure 10D a maximum occurs off Aberdeen but the other main areas are located in the middle of the northern North Sea and off the north coast of Holland. The maximum values are much smaller, up to 0.5 W/m$^2$. The differences reflect the different tracks of the depressions in the two cases; the 1953 depression moved south into the North Sea, whereas the 1983 depression moved continually eastwards, reaching its maximum intensity just west of Denmark (see Figure 11). The strongest winds in 1953 were very close to the east coast of the U.K., whereas in 1983 they were over the central northern North Sea. The distribution of energy dissipation, Figure 10E, shows that there are three main areas of dissipation (1) the Dover Straits, (2) off the Danish Coast, and (3) on the Dogger Bank, which are all shallow water areas. The dissipation in the Dover Straits may be compared with that calculated by Prandle[19] for the 1953 surge. He obtained maximum values in excess of 7 W/m$^2$, whereas a maximum of just over 2 W/m$^2$ was found here, again illustrating that much less energy was input into the southern North Sea in the 1983 surge.

A net energy budget was obtained by taking a 50-h time average of the terms in Equation 9, averaged over the whole model area. All the terms were scaled by the mean dissipation rate for tide plus surge (0.327 W/m$^2$, given by the sum of bottom friction and horizontal viscosity). Table 2 compares the energy budget for (1) the tide alone, (2) the tide plus the surge, and (3) the residual surge (2 − 1), ignoring terms which contribute less than 1%. The maximum discrepancy in these approximate budgets is 1%, due to neglect of small terms and the inaccuracy of the estimates of energy terms from the finite difference scheme, which seems acceptable. The results may be summarized as follows:

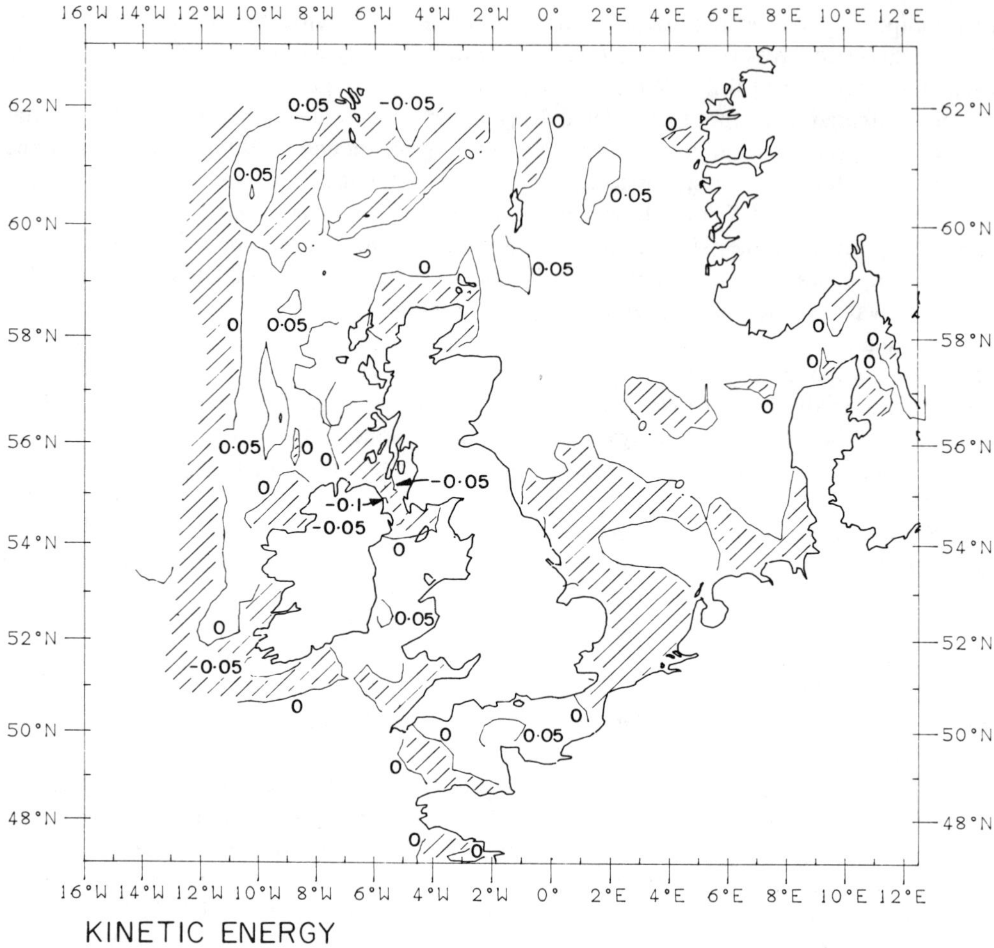


A

FIGURE 10. Time-averaged spatial distributions of mean rate of change of (A) kinetic and (B) potential energy, rates of working of (C) pressure gradients, (D) wind stress, and (E) bottom friction. (Hatched areas <0.)

1. For the tide alone, the influx of energy across the open boundaries plus the work done by the tide-generating forces is approximately balanced by the dissipation plus a net energy loss of 4%.
2. For the tide plus surge, the influx of tidal energy across the boundaries plus the work done by the tide-generating forces, plus the energy input by the storm is approximately balanced by the dissipation plus a net energy loss of 1%.
3. For the residual surge, the energy input by the storm is balanced by the increase in dissipation due to the surge plus a net increase in energy of about 3%.

The net loss of energy by the tide over the 50 h is presumably due to the tides beginning to reduce after springs. The surge provides a net increase of energy which has not all been dissipated after 50 h. The dissipation rate is increased by about 30% due to the presence of the surge.

More detail is provided by dividing the model area into sub-areas as shown in Figure 12. The energy balance has different characteristics in different areas. Table 3 shows the mean energy budget over 50 h for each sub-area for (1) tide alone and (2) residual energy

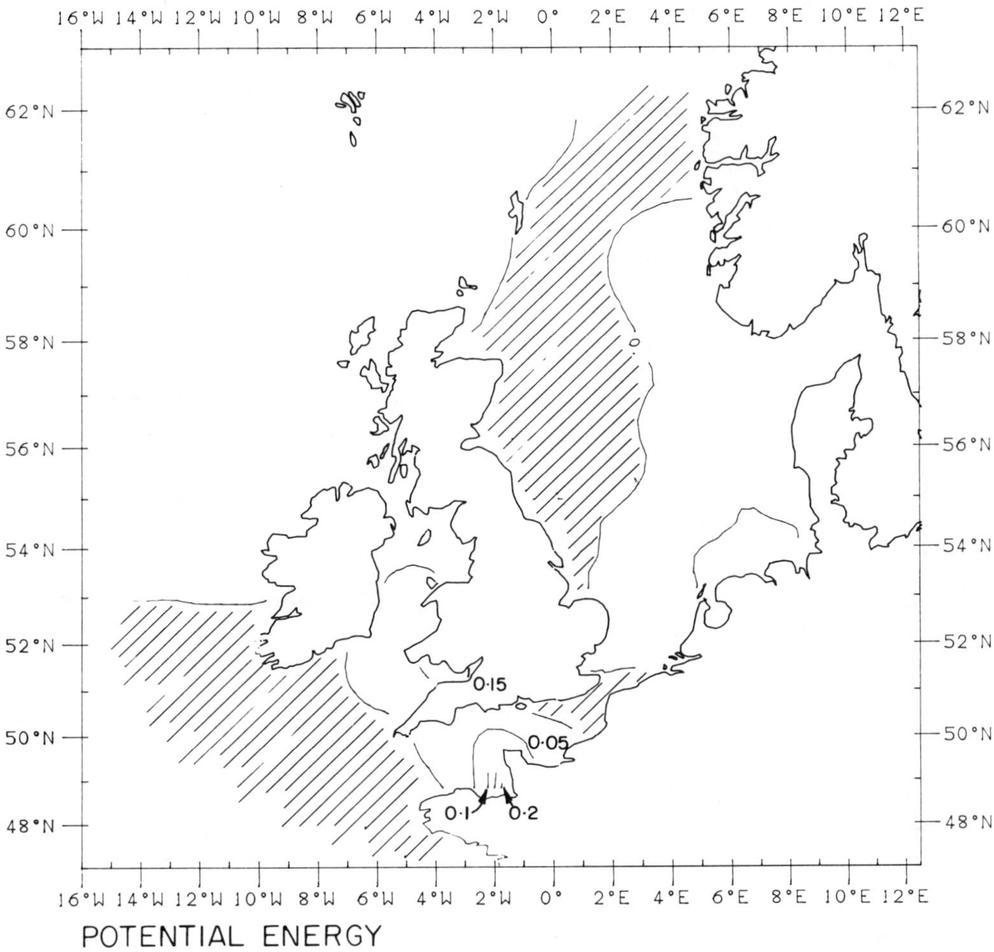


FIGURE 10B.

(change in energy balance due to the occurrence of the storm). These two added together give the total energy budget for the tide-plus-surge case. In each area the input and dissipation terms cancel out except for a net increase or decrease of energy, plus the numerical discretization error terms. Various points may be noted; first for the tide alone:

1. The main balance is between the influx of energy and dissipation, the latter being greater in all areas through this 50-h period, leading to a net reduction in energy.
2. The tide-generating forces act mainly on the outer shelf, being very small even there, although there is some energy input into the Irish Sea.
3. The maximum flux and dissipation of energy occurs in the Bristol Channel, a total of about 2 W/m$^2$, with significant values occurring in the English Channel, Irish Sea, and southern North Sea.

For the excess energy due to the occurrence of the surge event:

1. Energy is introduced by the wind-stress. The atmospheric pressure gradient increases the energy input on the Faeroe-Shetland Shelf but reduces it on the shelf northwest of the U.K. The pressure effect is much smaller than the wind effect except on the outer shelf.

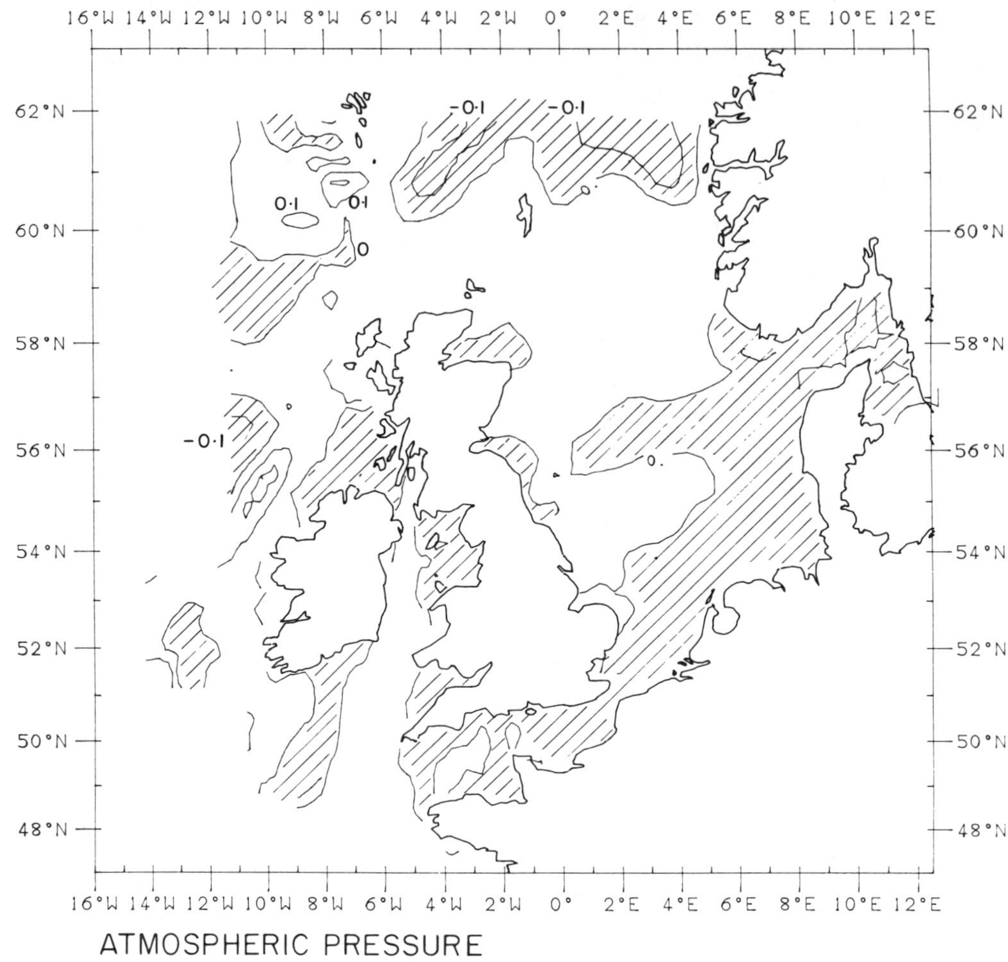


FIGURE 10C.

2. The main work done by the wind is over the northern part of the North Sea, almost all of which is removed locally by increased dissipation, with about 10 to 15% being transported out by the flux term, some to the Southern North Sea but mainly to the Skaggerak and Kattegat.
3. Dissipation is increased by the surge everywhere except in the Bristol Channel where there is a large net increase in potential energy.

In general the changes in kinetic and potential energy are of similar magnitude but the changes in kinetic energy are significantly larger on the western Ireland Shelf and in the Irish Sea for both tide and residual. The potential energy is significantly larger (by up to a factor of 10) in the Bristol Channel. In the southern North Sea the tidal energy budget shows a factor of 10 greater increase in potential energy than kinetic energy, but this is reversed for the residual energy budget when most of the increase is in the kinetic energy (this is a factor of 12 larger, in agreement with Prandle[19]).

## VII. CONCLUSIONS

A brief review of the use of numerical models for storm surge forecasting in U.K.

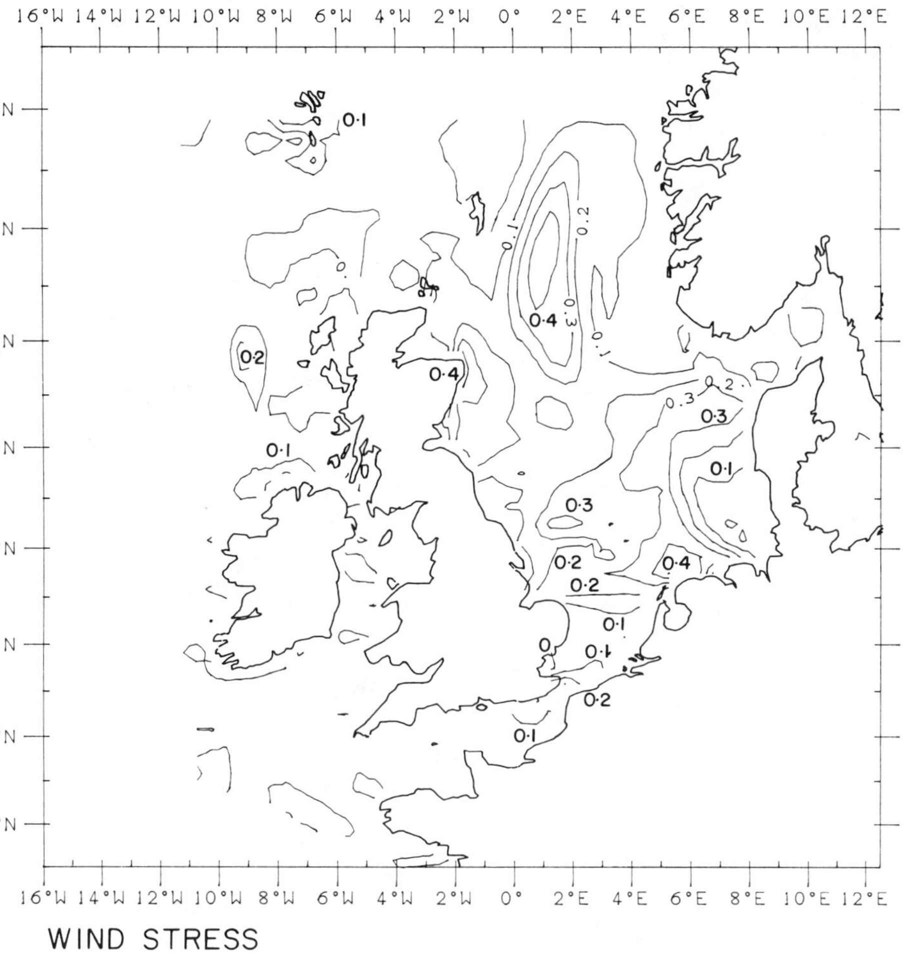


FIGURE 10D.

waters, culminating in a description of the current operational forecast scheme, has been given.

Recent revisions in the storm surge forecasting scheme based on numerical models, which is run routinely at the British Meteorological Office, have been described. These include extending the sea-model grid out over the continental shelf edge, using six tidal constituents instead of only two, and allowing horizontal viscosity to be specified in the equations of motion. Some improvements in tidal prediction have been obtained.

The revised model was used to examine sea level forecasts for the storm surge of the 1st and 2nd of February, 1983. The original forecast was poor on either side of high water and this was shown to be mainly the result of erroneous meteorological predictions. The atmospheric model forecasts underestimated the depth of the depression and large corrections were made between the successive atmospheric model forecasts used in the real-time surge prediction. Further hindcast model runs were performed using observed atmospheric pressure distributions to improve the surge reproduction. Ways of fitting pressure surfaces through the data were investigated and the features responsible for the initial forecast errors were identified.

Examination of the events of February 1 and 2, 1983 showed that a negative surge occurred first as a result of westerly winds moving southwards down mainland Britain and

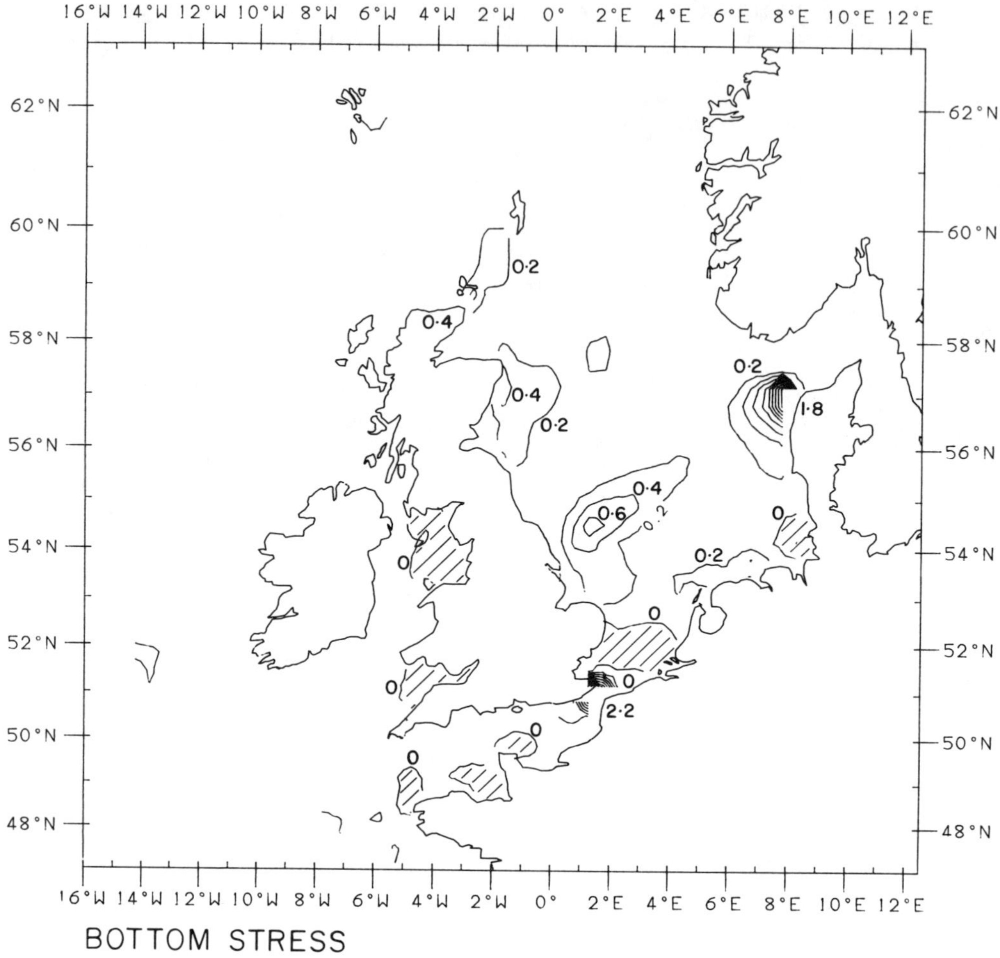


FIGURE 10E.

causing offshore flow at the coast. A positive surge then developed in the northern North Sea and initially propagated south along the east coast of Britain. Subsequently, the depression moved eastwards towards Denmark resulting in surge setup over the entire southern North Sea with simultaneous maxima on all coasts.

The calculation of an energy balance within the model was instructive in providing an overview of the surge event. The work done by the wind acting on the sea surface provided the largest input of energy into the sea, much larger than that due to the action of the atmospheric pressure gradients. Integrated over the whole area, the dissipation of energy by bottom friction increased by 30% over that dissipated by the tide alone, and was seen to correlate quite well with the work done by the wind stress. The spatial distribution of work done by the wind stress showed that the major inputs of energy occurred in a few places only but these differed from those which were important in the surge of February 1, 1953. This was because the different track taken by the 1983 depression meant that the strongest winds occurred much further offshore in the northern North Sea and most energy dissipation took place locally. In the southern North Sea the energy magnitude was considerably less than in the 1953 surge. An energy budget over 50 h showed that most of the energy input by the storm was dissipated but there was a net increase in energy of about 3% over that time. It is suggested that the energy balance is a good way of quantifying the difference

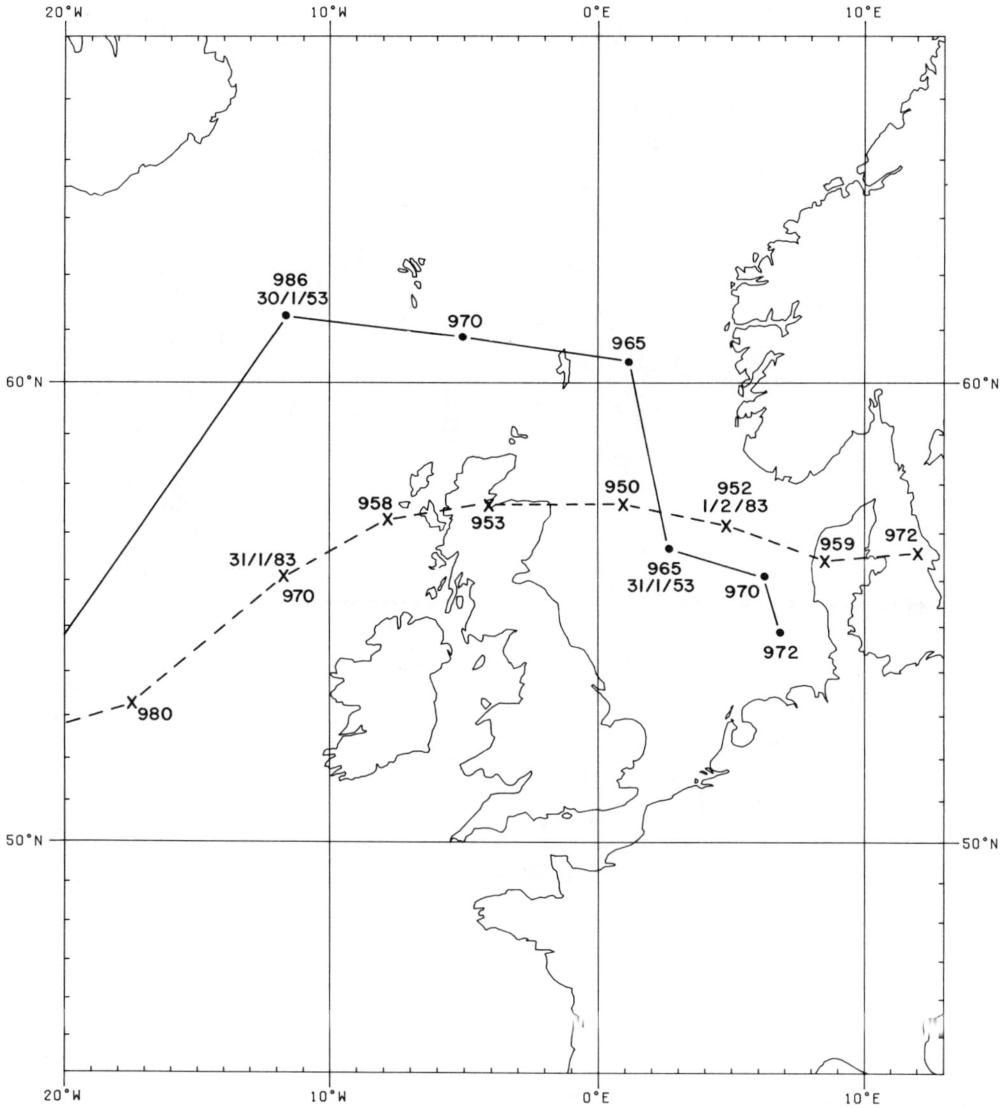


FIGURE 11. Depression tracks for the February 1983 and February 1953 storm surges.

## TABLE 2
## Time-Averaged Energy Budget (Scaled by Total Dissipation Rate 0.327 W/m²)

| | Tide alone | Tide + surge | Residual |
|---|---|---|---|
| Flux divergence | 0.71 | 0.71 | |
| Meteorological energy input | | 0.26 | 0.26 |
| Work done by t.g.f. | 0.01 | 0.01 | 0.00 |
| Dissipation rate | −0.77 | −1.00 | −0.23 |
| Rate of increase in total energy | −0.04 | −0.01 | 0.03 |
| Discrepancy (numerical) | −0.01 | −0.01 | 0.00 |

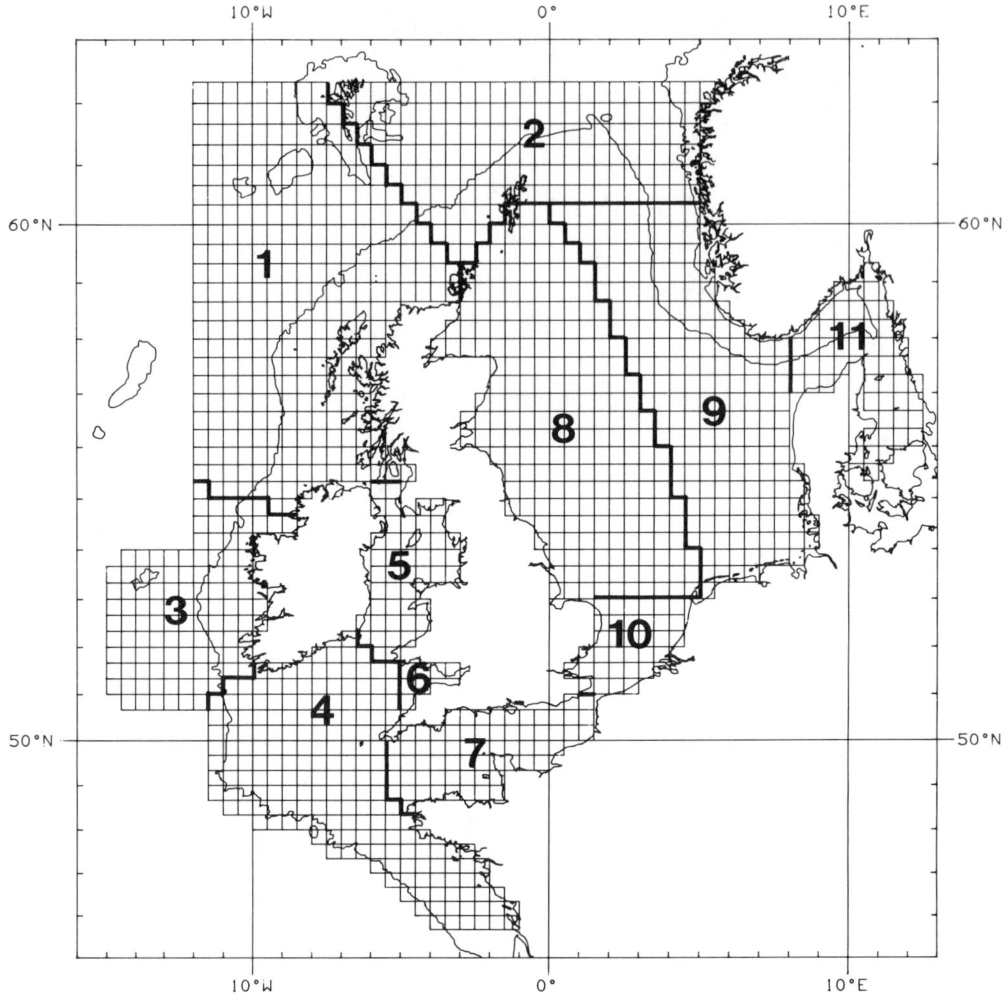


FIGURE 12. Sub-areas of model grid used in energy budget calculation.

between two surges and of locating the areas which are important for surge generation and decay. Energy budgets for sub-areas of the model grid enabled the study of the relative importance of various terms in the equation of motion in different areas, for the tide alone, and for the additional effect of a storm event.

## ACKNOWLEDGMENTS

We would like to thank Dr. R. A. Flather for valuable comments during the course of this study. We gratefully acknowledge the influence and guidance given to us by the late Dr. N. S. Heaps. Thanks are due to Mrs. J. Huxley for typing the manuscript and Mr. R. Smith for preparing the diagrams. The work was funded by the Ministry for Agriculture, Fisheries and Food.

**TABLE 3**
**Mean Energy Budget for Sub-Areas (energy density, W/m²) (50-h average)**

| Area | Tide alone | | | | | Residual (= increase in energy due to surge) | | | | | |
|---|---|---|---|---|---|---|---|---|---|---|---|
| | Flux + advection | Work done by t.g.f. | Dissipation | Increase in KE + PE | Discretization error | Flux + advection | Work done by wind | Work done by p. grad | Dissipation | Increase in KE + PE | Discretization error |
| Hebrides Shelf | 0.03 | 0.01 | −0.07 | −0.03 | 0.00 | 0.04 | 0.03 | −0.02 | −0.03 | −0.03 | −0.01 |
| Faeroe-Shetland Shelf | 0.04 | 0.01 | −0.03 | 0.00 | 0.02 | −0.05 | 0.07 | 0.03 | −0.04 | 0.01 | 0.00 |
| W. Ireland Shelf | 0.00 | 0.00 | −0.01 | 0.00 | −0.01 | 0.01 | 0.04 | −0.02 | −0.03 | 0.00 | 0.00 |
| SW Approaches | 0.16 | −0.01 | −0.16 | 0.00 | −0.01 | 0.00 | 0.03 | −0.01 | −0.02 | 0.00 | 0.00 |
| Irish Sea | 0.77 | 0.01 | −0.83 | −0.04 | −0.01 | 0.02 | 0.02 | 0.00 | −0.03 | 0.01 | 0.00 |
| Bristol Channel | 1.98 | 0.00 | −2.13 | −0.11 | −0.04 | −0.05 | 0.03 | 0.00 | −0.08 | 0.06 | 0.00 |
| English Channel | 1.54 | 0.00 | −1.62 | −0.06 | −0.02 | 0.03 | 0.05 | 0.01 | −0.05 | 0.04 | 0.00 |
| Western North Sea | 0.14 | 0.00 | −0.15 | −0.01 | 0.00 | −0.02 | 0.20 | −0.01 | −0.17 | 0.00 | 0.00 |
| Eastern North Sea | 0.03 | 0.00 | −0.03 | 0.00 | 0.00 | −0.03 | 0.19 | 0.00 | −0.14 | 0.01 | 0.01 |
| Southern North Sea | 0.70 | 0.00 | −0.73 | −0.02 | −0.01 | 0.01 | 0.08 | 0.01 | −0.11 | −0.01 | 0.00 |
| Skaggerak/ Kattegat | 0.00 | 0.00 | 0.00 | 0.00 | 0.00 | 0.07 | 0.02 | 0.00 | −0.05 | 0.03 | 0.01 |
| Total area | 0.23 | 0.00 | −0.25 | −0.01 | −0.01 | 0.00 | 0.09 | −0.01 | −0.07 | 0.01 | 0.00 |

*Note*: Maximum value of each term is underlined.

# REFERENCES


1. **Corkan, R. H.,** Storm surges in the North Sea, 2 Vol., H. O. Misc. 15072, U.S. Hydrographic Office, 1948.
2. **Davies, A. M.,** Three-dimensional modelling of surges, in *Floods due to High Winds and Tides,* Peregrine D. H., Ed., Academic Press, London, 1981, 45.
3. **Dickinson, A. and Temperton, C.,** Operational numerical weather prediction model, Met.0.11 Technical Note No. 183, Meteorological Office Library, unpublished, 1984.
4. **Flather, R. A.,** A tidal model of the north-west European continental shelf, *Mem. Soc. R. Sci. Liege 6 Ser.,* 10, 141, 1976.
5. **Flather, R. A.,** Recent results from a storm surge prediction scheme for the North Sea, *Marine Forecasting, — Predictability and Modelling in Ocean Hydrodynamics,* Proc 10th Int. Liège Colloq. Ocean Hydrodynamics, Nihoul, J. C. J., Ed., Elsevier, Amsterdam, 1979, 835.
6. **Flather, R. A.,** Practical surge prediction using numerical models, in *Floods due to High Winds and Tides,* Peregrine, D. H., Ed., Academic Press, London, 1981, 21.
7. **Flather, R. A.,** Results from a model of the north east Atlantic Ocean relating to the Norwegian Coastal Current, in *The Norwegian Coastal Current,* Saetre, R. and Mork, M., Eds., University of Bergen, Norway, 1981, 427.
8. **Flather, R. A.,** A numerical model investigation of the storm surge of 31 January and 1 February 1953 in the North Sea, *Q. J. R. Meteorol., Soc.,* 110, 591, 1984.
9. **Flather, R. A. and Proctor, R.,** Prediction of North Sea storm surges using numerical models: recent deveopments in the UK, in *North Sea Dynamics,* Sundermann, J. and Lenz, W., Eds., Springer-Verlag, Berlin, 1983, 229.
10. **Flather, R. A., Proctor, R., and Wolf, J.,** A tidal model of the north-east Atlantic Ocean, in preparation.
11. **Garratt, C. J. R.,** Tides in gulfs, *Deep-Sea Res.,* 22, 23, 1975.
12. **Gould, W. J.,** CONSLEX: a progress report on current measurements on the continental slope north-west of the UK, *MIAS News Bull.,* 7, 12, 1984.
13. **Hansen, W.,** Theorie Zur Errechnung des Wasserstandes und der Stromungen in Rand meeren nebst Anwendungen, *Tessul,* 8(3), 287, 1956.
14. **Heaps, N. S.,** Storm surges on a continental shelf, *Philos. Trans. R. Soc. London, Ser. A,* 257, 351, 1965.
15. **Heaps, N. S.,** Storm surges, *Oceanogr, Mar. Biol. Annu. Rev.,* 5, 11, 1967.
16. **Heaps, N. S.,** On the numerical solution of the three-dimensional hydrodynamic equations for the tides and storm surges, *Mem. Soc. R. Sci. Liege, 6 Ser.,* 2, 143, 1972.
17. **Heaps, N. S.,** Storm surges, 1967—1982, *Geophys. J. R. Astronom. Soc.,* 74, 331, 1983.
18. **Horner, R. W.,** Flood prevention works with specific reference to the Thames Barrier, in *Floods due to High Winds and Tides,* Peregrine, D. H., Ed., Academic Press, London, 1981, 95.
19. **Prandle, D.,** Storm surges in the southern North Sea and River Thames, *Proc. R. Soc. London, Ser. A,* 344, 509, 1975.
20. **Proctor, R. and Flather, R. A.,** Routine storm surge forecasting using numerical models: procedures and computer programs for use on the CDC CYBER 205E at the British Meteorological Office, Rep. No. 167, Inst. Oceanographic Science, 1983.
21. **Rossiter, J. R.,** The North Sea storm surge of 31 January and 1 February 1953, *Philos. Trans. R. Soc. London,* 246, 371, 1954.
22. **Smith, S. D. and Banke, E. G.,** Variation in the sea surface drag coefficient with wind speed, *Q. J. R. Meteorol. Soc.,* 101, 665, 1975.
23. **Townsend, J.,** Storm surges and their forecasting, in *Floods due to High Winds and Tides,* Peregine, D. H., Ed., Academic Press, London, 1981, 1.
24. **Wolf, J., Hubbert, K. P., and Flather, R. A.,** A feasibility study for the development of a joint surge and wave model, Rep. No. 1, Proudman Oceanographic Laboratory, 1988.
25. **Howarth, M. J.,** private communication.

Chapter 5

# LAYERED MODELS OF COASTAL UPWELLING: A CASE STUDY OF THE SOUTH AUSTRALIAN REGION

**Cai Wenju, R. B. Schahinger, and G. W. Lennon**

## TABLE OF CONTENTS

# I. INTRODUCTION

In the marine dynamics of continental shelf waters the phenomenon of upwelling is of considerable significance. On the one hand, it represents the product of a complex interaction between the atmosphere and ocean on a rotating earth, and, on the other, it has great practical implications in both climate modification and in its ability to inject nutrients into the surface layers. The latter feature is particularly true of limited sections of the southern coastal zone of Australia. For the greater part of each year this vast zone experiences a persistent west to east current, trapped on the shelf in the characteristic southern hemisphere downwelling condition.[1] It is not surprising then, that the coastal waters of much of this region have often been termed a biological desert. This unfavorable condition seems to be relieved intermittently during the summer period, when the latitudinal shift of zonal barometric pressures allows the eastward-tracking high pressure systems to linger occasionally in the Great Australian Bight (Figure 1 inset). Given stability over a number of days, the associated southeasterly wind field in the eastern Bight region (detail shown in Figure 1) may halt the southeast-going current and reverse its flow to create conditions suitable for upwelling. That such upwelling features do occur has been established beyond doubt by observation of water properties and by satellite remote sensing of sea-surface temperatures.[2] It is also clear that a rock lobster fishery is supported by the system.[3] However, the evidence suggests that prolonged favorable conditions occur infrequently and that there are preferred localities for upwelling cells to develop. It is therefore important to understand the physics of the upwelling mechanism in greater detail.

Based upon repetitive Conductivity/Temperature/Depth surveys and the deployment of current meter and thermistor strings off the coast of southeast South Australia near 37.5°S, 139.5°E, Schahinger[2] has established a preliminary data set which is relevant to the problem (hereafter referred to as S87). He has also interpreted the data through simple conceptual upwelling models. The purpose of the present study is to examine the effectiveness of layered hydrodynamic numerical models in attempts to identify and reproduce the interactions of the physical mechanisms.

The application of modeling techniques to the baroclinic signals of shelf seas is already well established. The modern thrust was spearheaded in the early 1970s by O'Brien and Hurlburt[4] who developed a two-layered, nonlinear, and time-dependent model which attempted to reproduce conditions on a cross-shelf transect. Although successful in producing an equatorward baroclinic surface jet, the model failed to reproduce the normally observed poleward undercurrent. This deficiency was overcome by Hurlburt and Thompson,[5] who made the extension to upwelling on a $\beta$-plane. They utilized a similar model geometry to the former paper but added an alongshore pressure gradient. By this means a more realistic solution was obtained, including the reproduction of the poleward undercurrent.

One might then cite Suginohara,[6] who recognized that the earlier neglect of alongshore derivatives proved to be undesirable in the consequent exclusion of many shelf waves, for example both the barotropic and baroclinic Kelvin variety. Although Suginohara operated in conventional three dimensions, no internal or bottom stress terms were included. He employed a straight coastline but was able to study alongshore variability through a variable wind stress. In the following year, with Kishi,[7] both coastline geometry and bottom topography were introduced so that their effect upon localizing the upwelling signal in the process of the generation and propagation of internal Kelvin and shelf waves could be investigated.

By the 1980s, regional studies of the effects of bottom topography were more common, notably the Preller and O'Brien[8] study of the Peruvian shelf, and topography was accepted as a major influence in determining the preferred locality for upwelling. Then Crépon et al.,[9,10] using a two-layered shelf sea in full three-dimensional form, included both interfacial and bottom stress terms, while recognizing the influence of coastal features. In so doing,

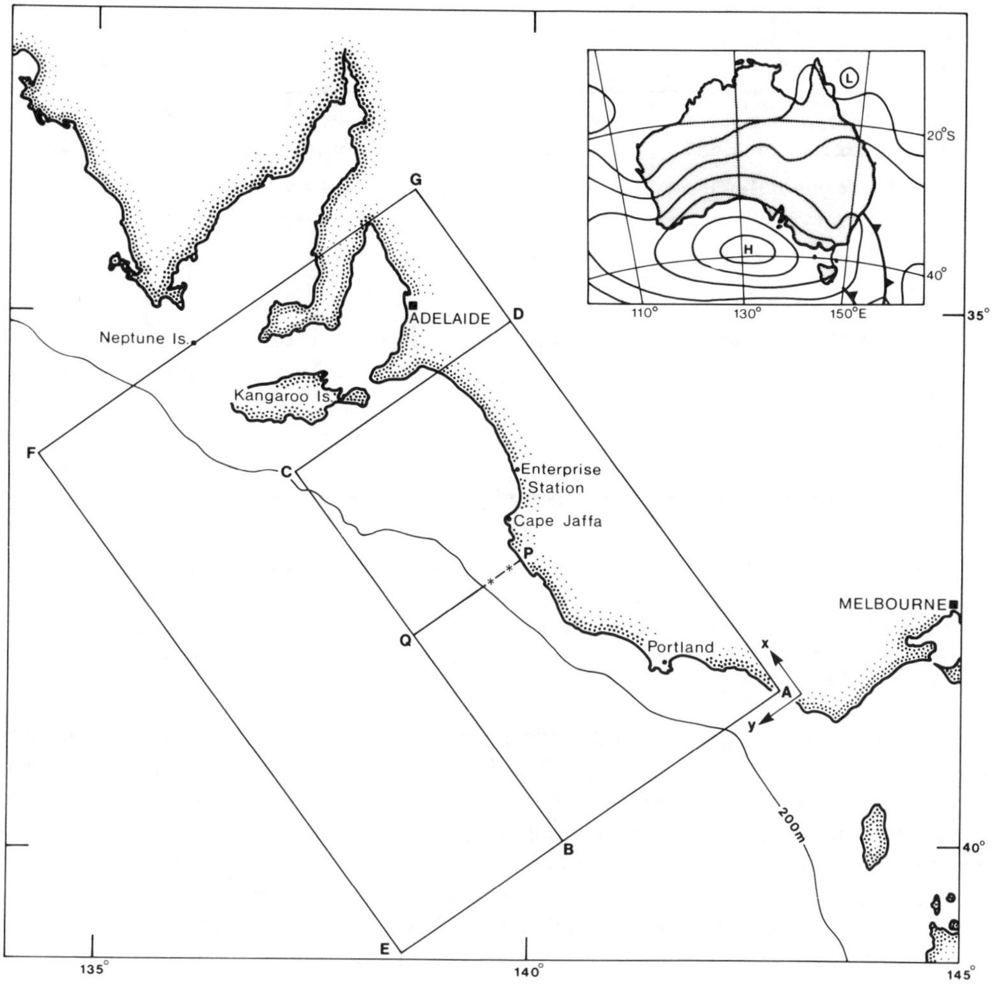


FIGURE 1. Inset: typical summer barometric pressure pattern over Australia, giving rise to southeasterly winds in the eastern Bight region. Detail: South Australian shelf region showing rectangular model areas. The three- and two-layer model areas are delineated by ABCD and AEFG, respectively. The asterisks along the *y*-directed line PQ indicate the location of the current meter moorings described in Schahinger.[2]

Crépon et al. set the scene for Hua and Thomasset[11] working in the Gulf of Lions. They effectively showed that preferred upwelling locations could arise as a function of coastal variability alone. The thermodynamics of the system were also included by Hua and Thomasset.

Then by 1983, O'Brien and Heburn[12] were prepared to comment upon the state of the art in coastal ocean modeling and to proclaim the virtues of a model which represented "the most complete study of ocean coastal circulation ever done. It is an excellent example of our actual capability to simulate ocean circulation and variables on a timescale of days to weeks". They incorporated thermodynamics and interfacial mixing into a three-dimensional model of upwelling off Peru, and also introduced a fluctuating alongshore undercurrent using a parameterization based on coastal-trapped wave dynamics.

The course of numerical marine science in the last quarter of a century has been influenced greatly by the late Norman S. Heaps and the productive group of modelers in the U.K. When Heaps turned his attention to upwelling, in fact an analytical treatment of shelf-break upwelling[13] he was able to deploy a wealth of experience gained in the address of a broad

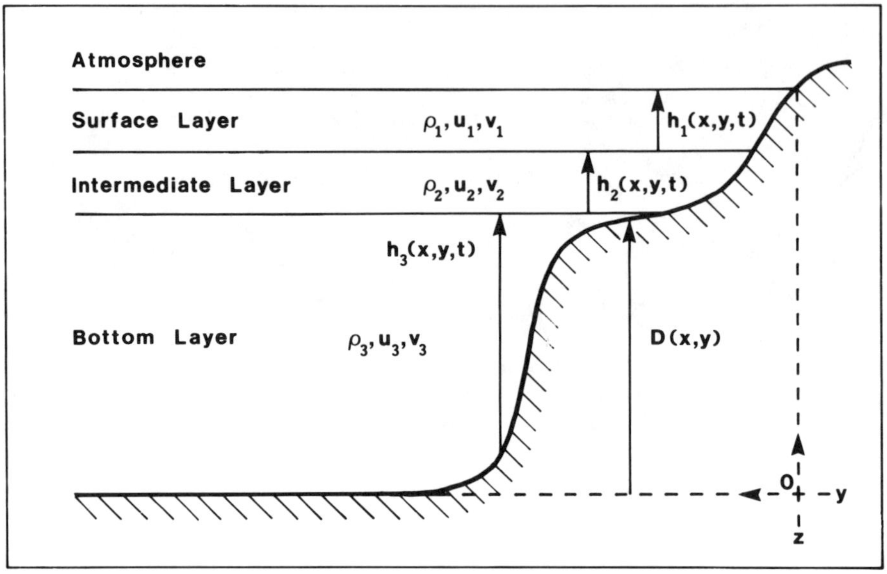


FIGURE 2. Typical geometry for a three-layer model with a free surface and two interfaces. The bottom topography is given by $D(x,y)$ and $h_1$, $h_2$, and $h_3$, are the thicknesses of each layer of density $\rho_1$, $\rho_2$, and $\rho_3$, respectively. The velocities $(u_i, v_i)$ are layer-averaged.

range of models of shelf seas in two and three dimensions, of ocean-atmosphere interactions and the complications of drying banks inter alia. It was then characteristic of the man that his development of a coastal-ocean model in 1983 proved to be an elegant classic.[14,15] He incorporated all of the hydrodynamic terms discussed above, though he neglected the thermodynamics of the system, yet he did introduce the vertical structure within each layer.

The present authors cannot match the sophistication of the Heaps model, which in any case requires computational facilities in excess of those available to them. Nevertheless, there is experience of modeling a baroclinic shelf. In Cai and Lennon[16] the case of the Taiwan Strait was considered. Despite the absence of detailed observational material it was possible to address the problem of the location of upwelling features in a seasonal sense and to contribute to the local controversy as to the regional winter-time behavior of the Kurishio Current. The exercise had numerous interesting features, viz., a major offshore island, varied topography, and a monsoonal regime which could be represented by a stable wind field of long duration. Internal and bottom stresses were present in a two-layered model with coastal and bottom topography. Although the working equations had the option for thermodynamics, this option was not used since no detailed data was to hand.

The interest in the southeast coastal zone of South Australia is a natural progression. Here some useful experimental data is available, the coastal configuration is complex, and again includes a major offshore island. The interest focuses upon the transience of upwelling features in a nonstable windfield and on setup times but, like the Taiwan Strait, the main question to address is why the system selects preferred locations. It is then intended to investigate the respective roles of wind direction and stability, botom topography, and interfacial stress.

## II. MODEL FORMULATION

The equations and geometry used in this study follow those of Cai and Lennon.[16] We consider a multilayer ocean with a right-handed Cartesian coordinate system, $z$ measuring distance upward, as depicted in Figure 2 (for the three-layer case). The primitive equations are as follows:

$$\frac{\partial \vec{v}_j}{\partial t} + (\vec{v}_j \cdot \nabla)\vec{v}_j + \hat{k} \times f\vec{v}_j = -g \sum_{i=1}^{j} \nabla h_i - g\nabla D$$

$$+ g \sum_{i=1}^{j-1} \epsilon_{ij}\nabla h_i + K_v \frac{\partial^2 \vec{v}_j}{\partial z^2} + A\nabla^2 \vec{v}_j \tag{1}$$

$$\frac{\partial h_j}{\partial t} + \nabla \cdot (h_j \vec{v}_j) = 0 \tag{2}$$

Here $j = 1,2,3 \ldots n$ refers to the layer number, counting down from the surface layer, t is time, and $\vec{v}_j = u_j\hat{\imath} + v_j\hat{\jmath}$ is the velocity vector, where $(u_j,v_j)$ are the $(x,y)$ components of current velocity; $f$ is the Coriolis parameter, $g$ the acceleration of gravity, $h_j$ the instantaneous local thickness of each layer, $D(x,y)$ the height of the bottom topography above the reference level $z = 0$, $A$ and $K_v$ the horizontal and vertical eddy viscosity coefficients, respectively, and $\epsilon_{ij} = (\rho_j - \rho_i)/\rho_j, i = 1, j - 1$ where $\rho_j$ is the water density of each layer.

The model is to be formulated in terms of volume transports, and hence equations 1 and 2 are integrated from the bottom to the top of each layer, i.e., from $D + \Sigma_{i=j+1}^{n} h_i$ to $D + \Sigma_{i=j}^{n} h_i$. Note that the horizontal eddy viscosity term $A\nabla^2\vec{v}_j$, is neglected due to its small magnitude, and that the components of stress at the horizontal surfaces are expressed in the form $\vec{\tau}_j = \rho K_v \frac{\partial \vec{v}_j}{\partial z}$. The depth integrated equations are:

$$\frac{\partial \vec{V}_j}{\partial t} + \left(\frac{\vec{V}_j}{h_j} \cdot \nabla\right)\vec{V}_j + \hat{k} \times f\vec{V}_j = -gh_j \sum_{i=1}^{j} \nabla h_i - gh_j\nabla D$$

$$+ gh_j \sum_{i=1}^{j-1} \epsilon_{ij} \nabla h_i + \frac{\vec{\tau}_j^T}{\rho} - \frac{\vec{\tau}_j^B}{\rho} \tag{3}$$

$$\frac{\partial h_j}{\partial t} + \nabla \cdot \vec{V}_j = 0 \tag{4}$$

Here $\vec{V}_j = U_j\hat{\imath} + V_j\hat{\jmath}$ is the transport vector, where $(U_j,V_j)$ are the $(x,y)$ components of transport; $\rho$ is a reference density. The stresses at the top and the bottom of each layer, $\vec{\tau}_j^T$ and $\vec{\tau}_j^B$ are formulated so as to be continuous across each interface. That is:

$$\vec{\tau}_j^T = \rho \cdot C_I \cdot \overline{q}_j^T(\vec{v}_{j-1} - \vec{v}_j), \qquad j \neq 1$$

$$\vec{\tau}_j^B = \rho \cdot C_I \cdot \overline{q}_j^B(\vec{v}_j - \vec{v}_{j+1}), \qquad j \neq n$$

$$\vec{\tau}_j^B = \rho \cdot C_B \cdot q_j\vec{v}_j, \qquad j = n$$

In the above formulation, $\overline{q}_j^T = (q_{j-1} + q_j)/2$ and $\overline{q}_j^B = (q_j + q_{j+1})/2$, where $q_j = (u_j^2 + v_j^2)^{1/2}$ $C_I$ and $C_B$ are the drag coefficients for interfacial and bottom stress, respectively. When $j = 1$, $\vec{\tau}_1^T = \vec{\tau}^S$, i.e., the wind stress vector at the sea surface. The surface stress is considered to be the driving force for upwelling and is assumed to have the form $\vec{\tau}^S = \rho_a \cdot C_D \cdot | \vec{U}_a | \vec{U}_a$, where $\rho_a$ is the air density *(1.2 kg m$^{-3}$)*, $C_D$ the drag coefficient ($1.3 \times 10^{-3}$), and $\vec{U}_a$ is the wind vector.

The present model has major differences from the one formulated by Heaps.[14,15] In his

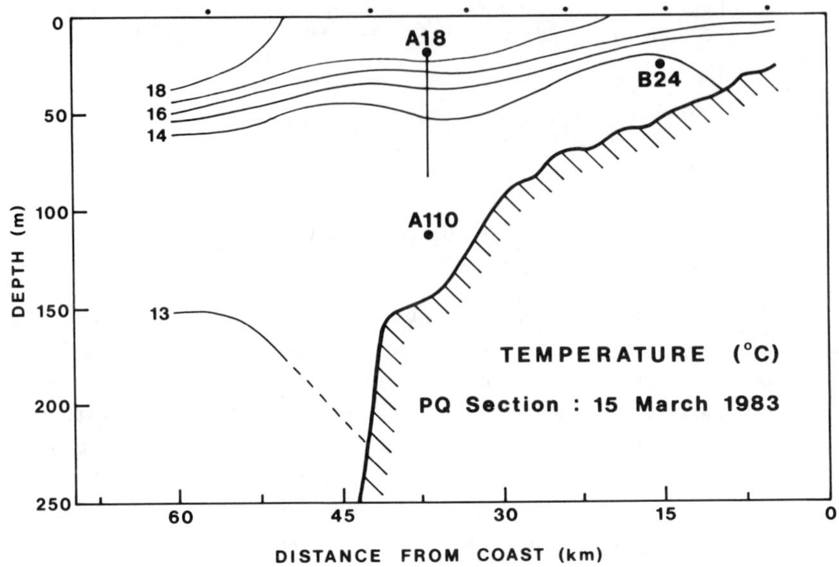


FIGURE 3. Temperature section along the line PQ shown in Figure 1. The large solid dots show the location of Aanderaa RCM4 current meters, denoted by A18, A110, and B24. The solid vertical line indicates the extent of a SeaData thermistor string, and the dots above the upper horizontal axis represent CTD stations (after S87).

model the equations of motion are integrated vertically using a Kernal function, thus allowing the velocity in each layer to vary with depth. Since the velocities in each layer are three dimensional, the continuity of both current and stress is satisfied. In our model, as in that of O'Brien and Hurlburt,[4] the interfacial stresses require formulation in terms of the depth-averaged current in adjacent layers.

## A. BOUNDARY AND INITIAL CONDITIONS

It is assumed that the surface area of the shelf sea model is rectangular, and that one side is aligned with the coast. Hence, there will be a coastal boundary and three open boundaries. For the coastal boundary the normal component of the volume transport is set to zero. A radiation condition is applied to the two cross-shelf open boundaries to allow disturbances from the interior of the model to propagate outwards. For the off-shelf open boundaries $h_j$ are taken to be undisturbed, and $(U_j, V_j)$ are obtained through the continuity equations 2 and 4.

Initially, the interfaces between the different layers are considered to be horizontal, so that the appropriate conditions are $U_j = V_j = 0$ and $h_j =$ constant.

## B. DEFINITION OF LAYERS

Before running the model we must decide how many layers are required to obtain a reasonable simulation of observed events. Here we shall make use of the S87 data, in which descriptions are given of upwelling events in March 1983 and February 1984. The cross-shelf temperature structure during the former event (Figure 3), which mirrors the density structure, and the coincident current observations at the locations donated by A18, A110, and B24, suggest the presence of at least three layers:

1. A surface mixed layer of O(20 m), in which an offshore flow occurs; as observed at A18

2. An intermediate layer, also of O(20 m), in which an onshore conpensatory flow occurs; as observed at B24
3. A bottom layer, in which only slight cross-shore motion occurs; as represented by A110

The layers are therefore characterized by distinctive flow patterns in the cross-shore direction, the experimental data for A18 and B24 during the March 1983 event indicating a flow in the upper two layers of O (10 to 20 cm $s^{-1}$). In the alongshore direction, however, all three layers showed a similarly directed motion of O (30 to 60 cm $s^{-1}$), consistent with the wind direction towards the northwest. The above pattern was repeated in the February 1984 event, though no data were available from the A18 site on this occasion. Thus, a three-layer model would seem to be an appropriate choice.

## C. MODEL AREA

A second concern lies in the spatial definition of the area to be modeled. Infrared satellite images demonstrate that in the presence of a stable southeasterly wind field, cooler upwelled water lies over the continental shelf between Portland and Cape Jaffa, and in an extended band to the northwest of this "active" region (presumably due to alongshore advection). Cooler water is also seen off the southwest coast of Kangaroo Island (Figure 4).

Ideally, we would like to model the area encompassing all of these features, while using three layers. However, such a scheme places considerable strain upon the available computing facilities. As a compromise, we propose first to model with three layers the major upwelling region between Portland and Cape Jaffa. This region includes the S87 mooring sites. Dependent upon the success of the three-layer model, it is then proposed to test the feasibility of downgrading to a two-layer model covering the larger area, and ultimately to see whether or not the localized feature off Kangaroo Island can be reproduced.

The rectangular area covered in the three-layer model is delineated by ABCD in Figure 1, and is 525 km long and 300 km wide. The coastal boundary AD is aligned in a 144°T/324°T direction so as to be parallel to the coast and shelf edge in the vicinity of the S87 mooring sites. Model grid lines are $x$- and $y$-directed and are parallel to the sides of the rectangle; the grid has square elements with a 7.5 km side, i.e., $\Delta x = \Delta y = 7.5$ km. Thus, there are 70 elements in the $x$ direction and 40 elements in the $y$ direction. For the two-layer model, the rectangular area is extended in the $x$ and $y$ directions by 25 and 30 elements, respectively, so as to include Kangaroo Island to the northwest. The total area of the two-layer model is then 712.5 km by 525 km (AEFG in Figure 1).

## D. BOTTOM TOPOGRAPHY

The bottom topography and coastline used in the model (Figure 5) have been derived from the 1:300,000 series of British Admiralty Charts spanning the region. The most important feature of the bottom topography is the broadening of the shelf inshore of the 100 m isobath to the northwest of Cape Jaffa. Elsewhere the shelf is relatively narrow (30 to 60 km) and uniform in slope. Seaward of the shelf edge (at $\approx$ 180 m) the depth increases rapidly to 2000 m within 20 to 50 km.

In the three-layer model, the maximum layer thicknesses are 20, 20, and 260 m, and for the two-layer model, 20 and 280 m. Where the actual depth is greater than the deep reference level of 300 m, the topography is assumed flat, and an interfacial rather than bottom stress is invoked. The minimum depth is taken to be 50 m, so that the second interface in the three-layer model is initially everywhere at least 10 m above the bottom topography.

## E. TIME STEP

The time step used in the model is constrained by the Courant-Friedrichs-Lewy (CFL) linear stability criterion:

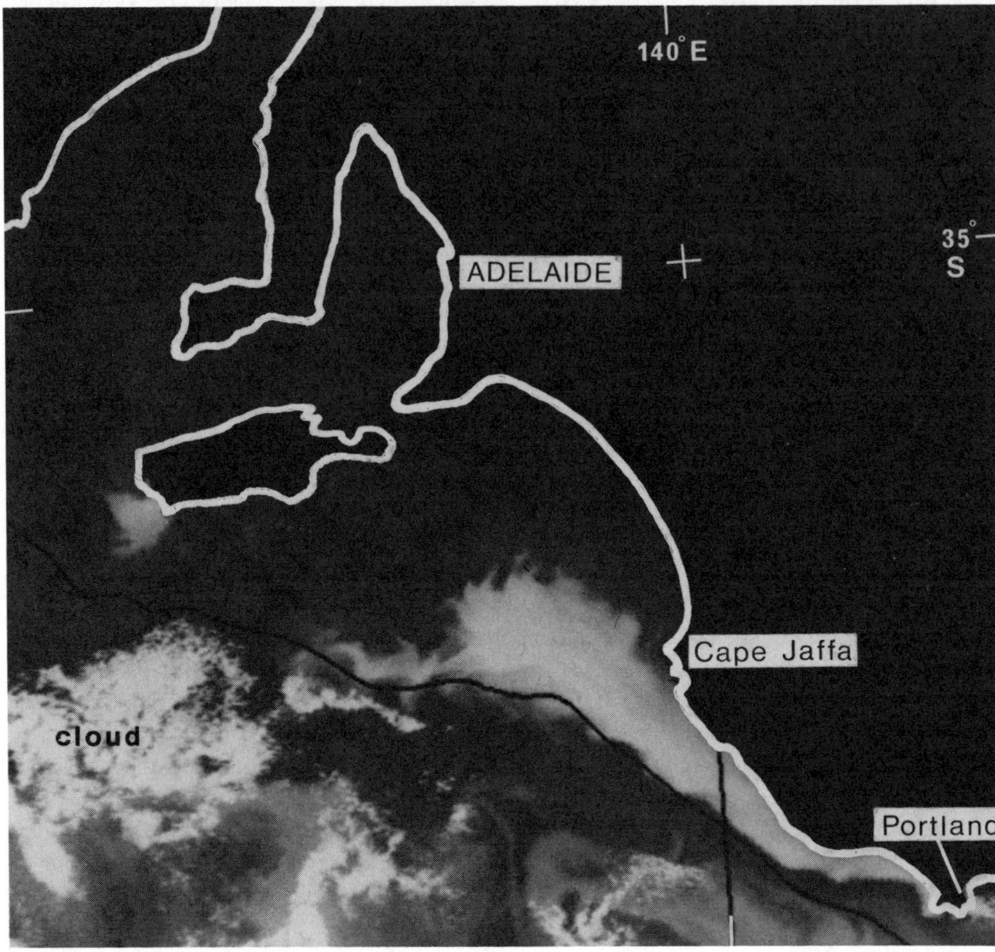


FIGURE 4. Infrared sea-surface temperature image, February 14, 1985. Cooler (light shading) upwelled water lies over the shelf between Portland and Cape Jaffa, with advection of cooler water to the northwest; cooler water is also evident off the southwest coast of Kangaroo Island. The 200-m isobath is shown.

$$\Delta t \leq min(\Delta x, \Delta y)/[g \cdot max(h_1 + h_2 + h_3)]^{1/2}$$

We have chosen $\Delta t = 90\ s$, about two thirds of that permissible using the CFL criterion.

## F. WIND FORCING

Comparisons between summertime wind data at the sites utilized in S87 (Enterprise Station and Cape Jaffa in consecutive years) and that at Neptune Island to the northwest reveal broad agreement. For simplicity, and in the absence of any offshore wind data, we thus assume a uniform wind stress throughout the model region. Since our interest lies primarily in the near-coastal zone, the assumption is that the bulk of the model area serves simply to produce realistic boundary conditions for that zone. This assumption creates some anxiety in view of the limitation of the offshore spatial scale by comparison with the work of others practising a similar philosophy.[8] Nevertheless, in hindsight, the agreement between model output and experimental data eventually serves to quell such anxieties.

Time series of wind observations have been low-pass filtered using the Lancz6 operator of Thompson so as to eliminate diurnal and shorter period signals,[17] and the stress subse-

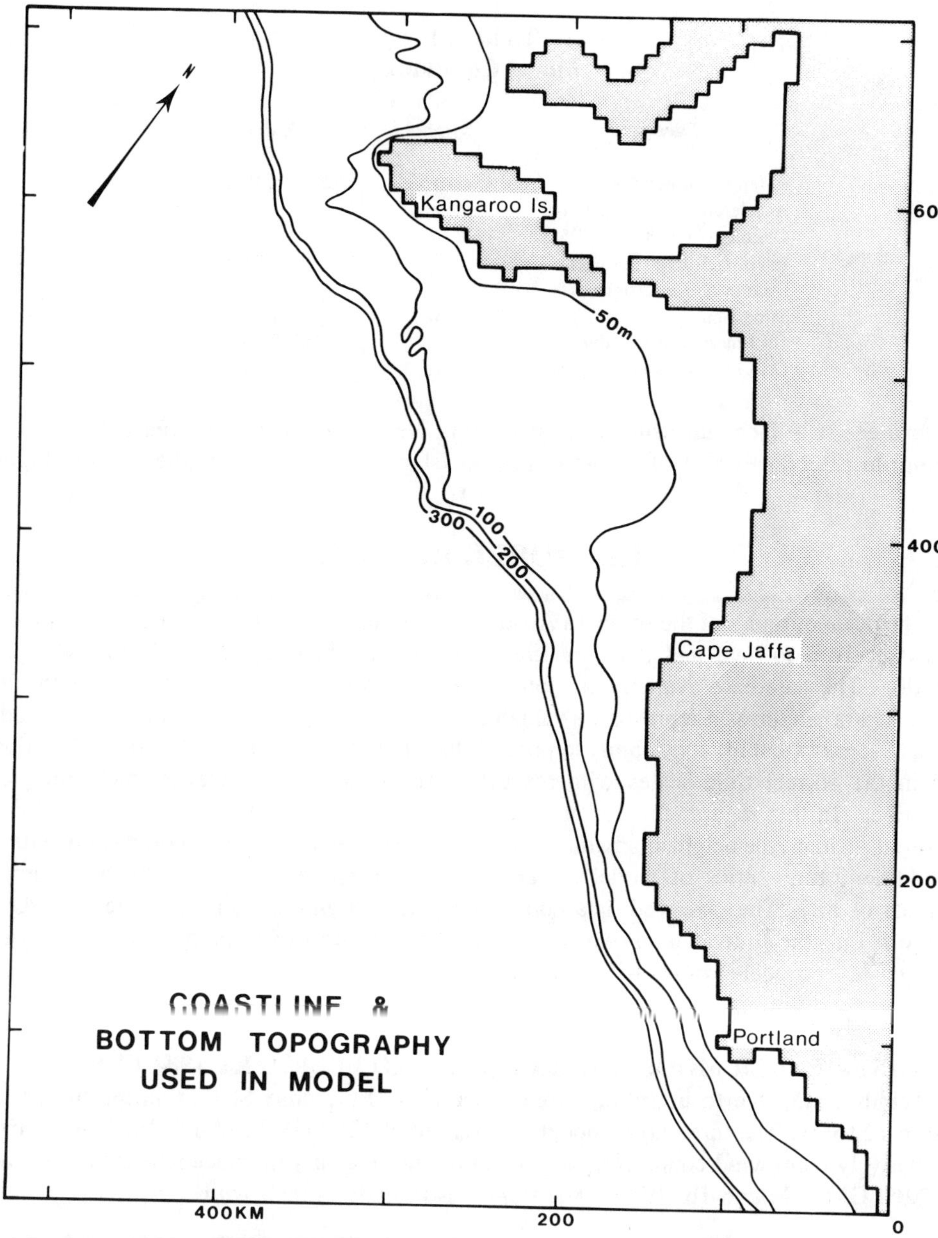


FIGURE 5. Coastline and bottom topography used in model, based upon 1:300,000 scale British Admiralty Chart nos. 348, 349, 1008, 1094, 1762, and 3359. The maximum depth in the model is taken to be 300 m, and the minimum 50 m. Isobaths deeper than 300 m are not shown. The area shown corresponds to the rectangle delineated by AEFG in Figure 1.

quently calculated. The three-layer model is run using real wind data and bottom topogaphy for the events described in S87, i.e., March 1983 and February 1984. The model is also run with a flat bottom (of 300 m) for the former event so as to ascertain the effects of bottom topography.

The two-layer model is first run for the March 1983 event with the three-layer area. The level of agreement between the two model, in the surface-layer currents and first-interface movement, provides justification for the extension of the two-layer model to the

**TABLE 1**
**Model Constants**

| Parameter | Symbol | Value |
|---|---|---|
| Reference water density | $\rho$ | $1.026 \times 10^3$ kg m$^{-3}$ |
| Acceleration due to gravity | $g$ | 9.8 m s$^{-2}$ |
| Coriolis parameter at 37.5°S | $f$ | $8.88 \times 10^{-5}$ s$^{-1}$ |
| Interfacial drag coefficient | $C_I$ | $5 \times 10^{-5}$ |
| Bottom drag coefficient | $C_B$ | $10^{-3}$ |
| Time step | $\Delta t$ | 90 s |
| Grid increment in $x$ direction | $\Delta x$ | 7.5 km |
| Grid increment in $y$ direction | $\Delta y$ | 7.5 km |

larger area. It is then run using wind data from Neptune Island in February 1985, in an attempt to reproduce the feature off Kangaroo Island observed in satellite images (Figure 4).

# III. MODEL RESULTS

The oceanography of the southern Australian shelf is conspicuous by the sparse attention it has received to date and the very limited facilities which are available to rectify the situation. The data base available to "prove" this model, therefore, has serious shortfalls, but in Australian terms it represents a valuable resource. Acceptance of the three-layer model relies to some extent upon its ability to produce the upwelling features in the correct locations and in the correct time scales with respect to the onset of the relevant meteorological conditions. In this regard, the upward movement of the layer interfaces from their initial positions — interface height anomalies — will be used to identify areas disposed to upwelling. In addition, time series of model currents will be compared directly with the observed currents of S87. The observed data come from sites 15 and 37 km along the line PQ in Figure 1, i.e., the equivalent of two and five model grid elements from the coastal boundary. The first-layer model currents will be compared with the observations at A18 (refer to Figure 3), the second layer with those at B24, and the third layer with those at A110.

## A. THREE-LAYER MODEL: MARCH 1983 AND FEBRUARY 1984 EVENTS

Winds conducive to upwelling were observed at Enterprise Station during the period March 12 to 21. The three-layer model was started at 1200 GMT, March 10, 1983, a time of relatively calm wind conditions, and then run until the first interface reached the surface (at 2200 GMT, March 16, 1983). The fluid was initially at rest, with

$$h_1 = 20 \text{ m}, \qquad \rho_1 = 1025.6 \text{ kg m}^{-3}$$

$$h_2 = 20 \text{ m}, \qquad \rho_2 = 1026.2 \text{ kg m}^{-3}$$

$$h_3 = 260 \text{ m}, \qquad \rho_3 = 1026.6 \text{ kg m}^{-3}$$

These density values are based on data presented in S87 and Schahinger.[19] The value of model parameters used in this run, and subsequent runs, unless otherwise stated, are given in Table 1.

The development of the model currents in each of the three layers during the period March 10 to 17 is compared with the observed currents in Figure 6, along with the wind stress components input to the model and the interface height anomalies (IHAs). Here the observed hourly currents have been low-pass filtered with the Lancz6 operator of Thompson

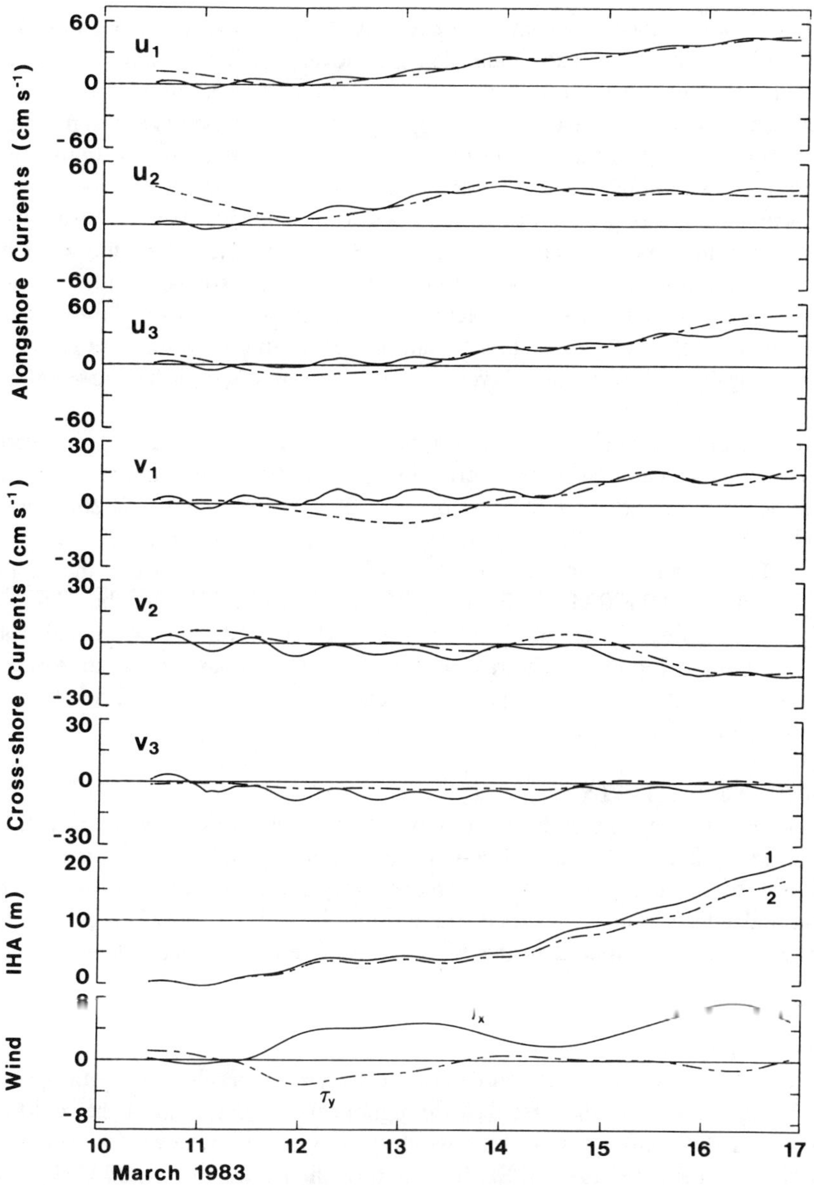


FIGURE 6. March 1983 event: time series of wind stress components input to the three-layer model (units: $10^{-2} * N\ m^{-2}$, predicted IHAs at the coast (point P in Figure 1), and the observed and predicted currents in each of the three layers at points along the line PQ shown in Figure 3. The observed currents have been low-pass filtered with the Lancz6 operator of Thompson,[17] and are shown by the dashed curves in the upper six panels.

and are shown by the dashed curves in the upper six panels of Figure 6.[17] Inertial oscillations of O(3 to 4 cm $s^{-1}$) are evident in the predicted currents (solid curves), a product of the impulsive nature of the model start. The oscillations, which are also apparent in the IHAs, decay slightly during the course of the upwelling event, but are of no immediate concern in the ensuing studies.

There is generally good agreement between the observed and predicted currents in each layer, especially during the latter stages of the event, with the cross-shore flow pattern as expected, i.e., an offshore flow in the upper layer ($v_1 > 0$) and an onshore flow in the intermediate layer ($v_2 < 0$). The model does, however, predict a slight onshore motion in the bottom layer where little was observed. The event can be split into two periods, reflecting the difference in the character of the driving force and the associated current and interface response. From the start of the model until March 13, 14 the wind had a significant onshore component ($\tau_y < 0$) though it was still predominantly alongshore; the upward movement of the interfaces during this period was slow (<5 m from their undisturbed positions) and the cross-shore current response relatively slight (<5 cm $s^{-1}$). Thereafter, the wind stress was dominated by the alongshore component ($\tau_x > 0$) and the response of the system was far more pronounced; the interfaces approached the surface at a rate of ≈5 m/d, equivalent to a vertical velocity of $\approx 6 \times 10^{-5}$ m $s^{-1}$, and the cross-shore motion was O (10 to 15 cm $s^{-1}$). The alongshore flow in each layer was in the same direction as the wind and O (30 to 45 cm $s^{-1}$).

Similar features are apparent in the results obtained for the February 1984 event (Figure 7). In this case, the wind stress time series input to the three-layer model were from Cape Jaffa and the layer densities taken to be $\rho_1 = 1025.7$ kg $m^{-3}$, $\rho_2 = 1026.1$ kg $m^{-3}$, and $\rho_3 = 1026.6$ kg $m^{-3}$. Initial layer thicknesses were again $h_1 = 20$ m, $h_2 = 20$ m, and $h_3 = 260$ m. The model was started at 0000 GMT, February 5, 1984 and the first interface reached the surface at 1100 GMT, February 10, 1984, about 4 d after the advent of winds favorable to upwelling. In contrast to the March 1983 event, the wind had a small but persistent offshore component. The rate of upwelling at the coast was again $\approx 6 \times 10^{-5}$ m $s^{-1}$, the same as that predicted off parts of Oregon by Peffley and O'Brien[18] using a two-layer $\beta$-plane numerical model.

## B. MODEL SENSITIVITY

In this section, the sensitivity of the model to the following parameters will be tested: (1) bottom topography, (2) wind direction, and (3) interfacial stress. Since the simulated March 1983 and February 1984 events displayed very similar features, insofar as time series of current and interface are concerned, the ensuing investigations will focus on the former event. The parameters adopted in the March 1983 event will be presumed to apply until specified otherwise.

### 1. Bottom Topography

The role of bottom topography in determining areas favorable to upwelling is examined by considering contours of the first IHA throughout the model area for the real-topography case and also a flat-bottom case (with a depth of 300 m everywhere). Contours of the first IHA are shown in Figures 8A and 8B for day 6 of the event, i.e., 1200 GMT, March 16, 1983, 10 h prior to the surfacing of the first interface. Shading is used in Figure 8 to highlight areas where the upward interface movement is greater than 15 m.

In the flat-bottom case, the anomaly contours simply follow the coastline at regularly spaced intervals in a zone 20 to 30 km wide. The contour pattern is quite different in the real-topography case, with the most pronounced interface movement occurring next to the coast between $x = 150$ and 300 km. A rapid reduction in upwelling occurs further to the northwest, coincident with the broadening of the shelf. Figures 8A and 8B demonstrate that in the model region, it is bottom topography rather than coastline curvature that is important in causing localized upwelling.

The effect of bottom topography upon currents over the shelf is also significant, as shown in Figure 9. Here the current and first IHA profiles along the cross-shore section PQ on day 6 of the March 1983 event are displayed for both the flat-bottom and real-topography cases. Features evident in the flat-bottom case will be discussed first and include the following:

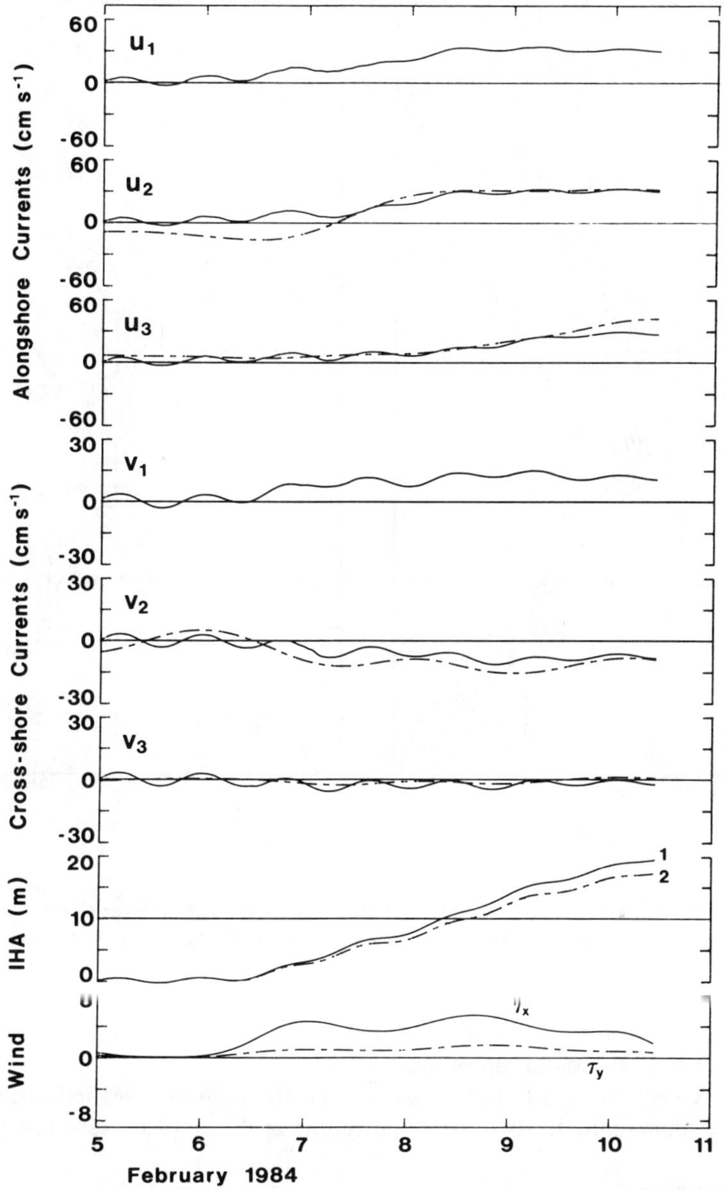


FIGURE 7. As in Figure 6 but for the February 1984 event.

1. Upwelling occurs in a zone abutting the coast of O(20 to 30 km).
2. The alongshore flow ($u_i$) is essentially barotropic in the intermediate and bottom layers, though considerably smaller in magnitude than that in the upper layer (O[25 cm $s^{-1}$] cf. O[33 cm $s^{-1}$]) with the direction in each layer consistent with Ekman principles.
3. An alongshore jet (>60 cm $s^{-1}$) occurs in the upper layer near the coast.
4. The cross-shore flow pattern is as described earlier.

In the real-topography case, several differences emerge:

1. Over the shelf the magnitude of the current components in each layer increases markedly; their qualitative relation to one another, however, remains the same.

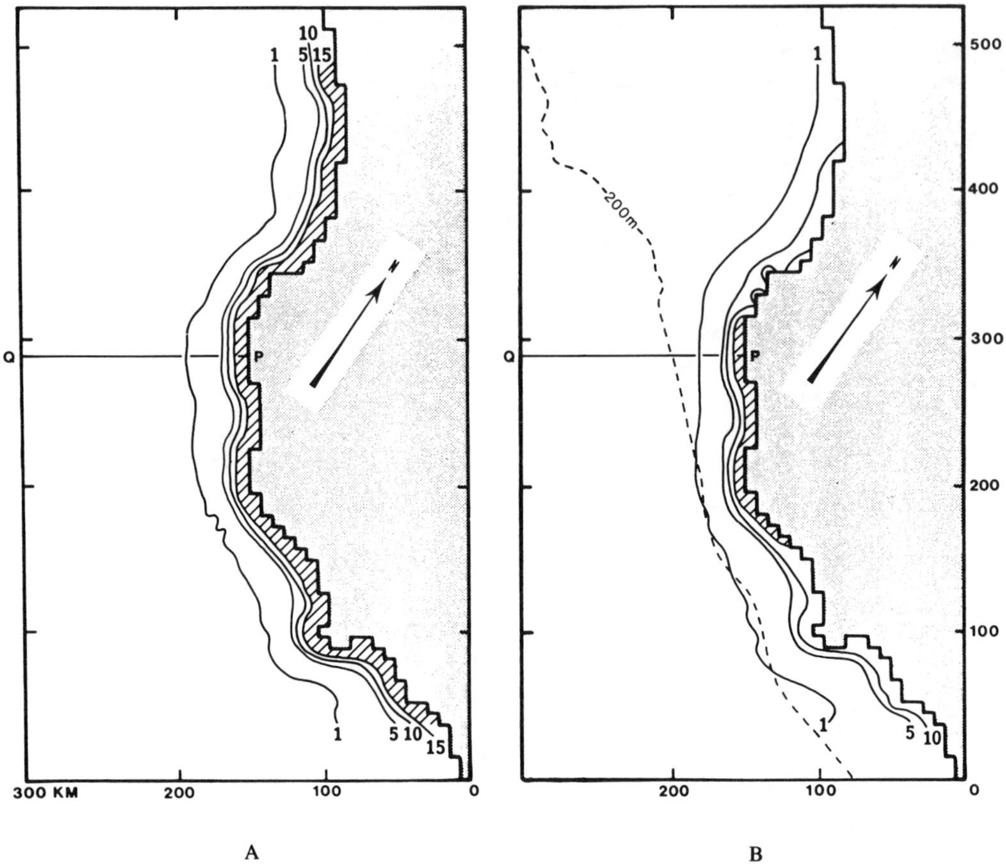


FIGURE 8. First IHA (m) at day 6 of the March 1983 event using the three-layer model; (A) case with a flat bottom of 300 m, (B) case with real topography. The anomaly is greater than 15 m in the shaded areas.

2. The near-shore alongshore jet in the upper layer increases to ≈70 cm $s^{-1}$. (We have no data to confirm or deny the existence of such a jet, though similar features are observed in other coastal upwelling systems.)[18]
3. The IHA near the coast is less than for the flat-bottom case, indicating that in this instance topography tends to delay upwelling at the coast by several hours.

## 2. Wind Direction

The importance of wind direction was tested by driving the three-layer model several times with the same wind strength while using a different angle from the ''upwelling-favorable'' alongshore wind direction in each run, i.e., angles from the positive *x* direction. Real topography was adopted and we have used an artificial wind stress time series in which the magnitude increases smoothly from $0 \rightarrow 0.05$ N $m^{-2}$ in 3 d and then remains constant, a pattern similar to that observed during the February 1984 upwelling event. The rate of upwelling is measured by the time taken for the first interface to reach the surface at the coast (at the point P in Figure 8).

The times for a number of angles are shown in Figure 10 and reveal that the maximum rate of upwelling occurs when the wind is ≈6° offshore of the positive alongshore direction, consistent with previous modeling exercises.[4] This preference for slightly ''offshore'' winds helps to explain the faster rate of upwelling predicted in the February 1984 event as compared to the March 1983 event (another contributing factor is the reduced density difference between the first and second layers in the February 1984 event).

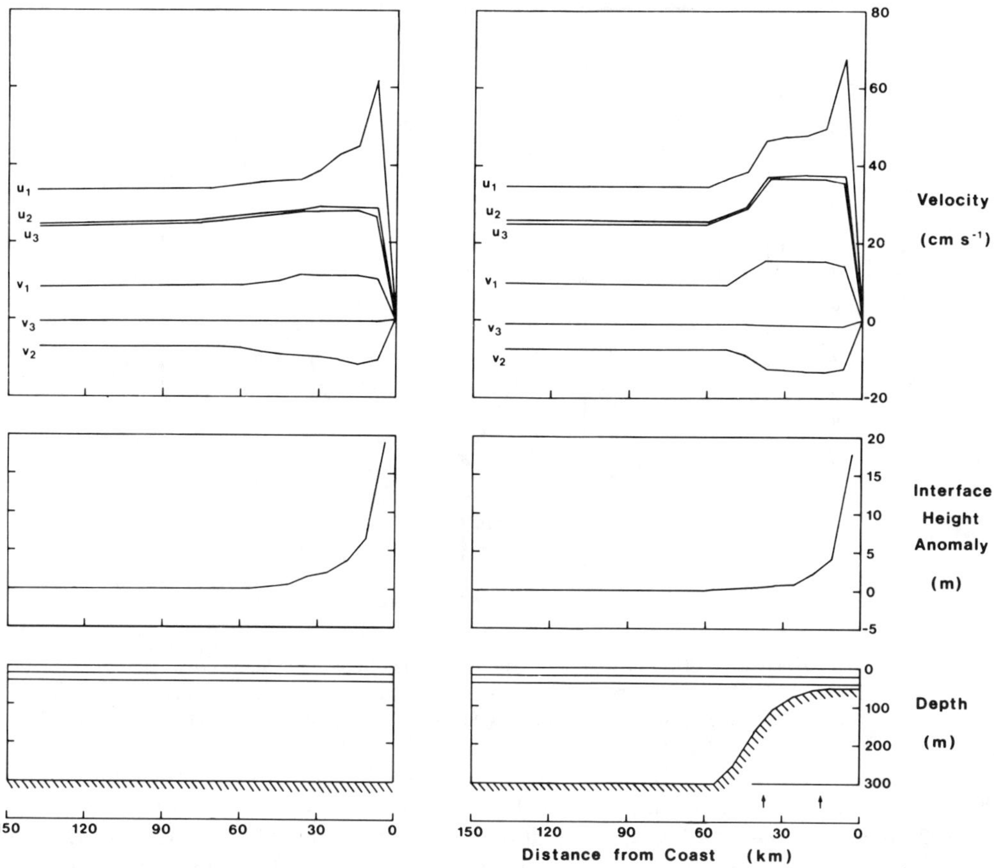


A B

FIGURE 9. March 1983 event, day 6 of the three-layer model run: profiles of the first IHA and the predicted current components in each of the layers along the [illegible] line. (A) case with a flat bottom of [illegible] m, (B) case with real topography. The two small arrows indicate the distances from shore of the current meter mooring sites described in S87.

### 3. Interfacial Stress

The value of $C_I$ employed in other studies[4,8] has been usually in the range $3 \times 10^{-4}$ to $10^{-5}$. In our model runs to date, a standard value of $5 \times 10^{-5}$ has been adopted. To examine the effects of $C_I$ on the model solution, the three-layer model was run again with values of $C_I = 5 \times 10^{-4}$ and $C_I = 5 \times 10^{-6}$, once more using an artificial alongshore wind stress and real topography. The results showed only slight differences from the standard case, with a decrease of less than 1 cm s$^{-1}$ in the upper layer alongshore currents (from a level of ≈40 cm s$^{-1}$) in the "large" $C_I$ run, and the reverse occurring in the "small" $C_I$ run, as one would intuitively expect. Similarly, a greater interfacial stress is associated with marginally smaller offshore currents in the upper layer and a consequent deceleration in the vertical motion of the interface near the coast. The effect upon currents in the lower layers is negligible.

## C. TWO-LAYER MODEL: FEBRUARY 1985 EVENT

The two-layer model was run initially for the March 1983 event over the three-layer model area, i.e., ABCD in Figure 1, with real topography and wind stress. Initial layer

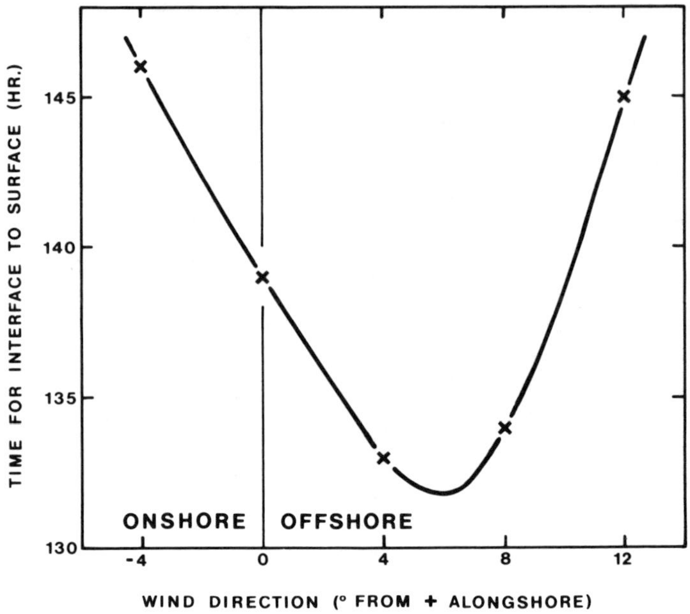


FIGURE 10. Effect of wind direction on the rate of upwelling at the coast. The time taken for the first interface to surface is shown for a number of wind directions varying about the "upwelling-favorable" alongshore direction (+ *x* direction in Figure 1).

thicknesses were $h_1$ = 20 m and $h_2$ = 280 m, with densities $\rho_1$ = 1025.6 kg $m^{-3}$ and $\rho_2$ = 1026.6 kg $m^3$. The interface surfaced at the coast within an hour of the first interface in the three-layer case and the development of both interface movement and upper layer currents in the two cases were almost identical.

Given this high level of agreement and the demonstrated ability of the three-layer model to reproduce the currents observed during upwelling events, it was felt to be a reasonable step to extend the two-layer model to the larger area shown in Figure 5. Reference has previously been made to the fact that the restriction of computing facilities precludes a three-layer treatment for the larger area. The two-layer model will be run for the period prior to February 14, 1985, i.e., the time of the satellite image to which reference was made earlier.

The weather pattern over southern Australia during February 1985 was dominated by a series of high pressure centers which moved steadily across the Great Australian Bight, interspaced by weak cold fronts. The passage of one such cold front on February 5 and the rapid easterly movement of the following high resulted in a strong southeasterly airstream over the model area. It is this episode in particular which is presumed to be responsible for the upwelling event illustrated in the image.

The two-layer model was started at 0500 GMT, February 2, 1985, a day prior to the onset of moderately favorable winds at Neptune Island, with layer thicknesses and densities as for the previous run. The upward movement of the interface at the coast (again considering the point P in Figure 8) commenced late on February 3 and rose at the rate of 3.5 to 4 m/d until about 1200 GMT, February 5. At that point the interface began to rise rather more quickly in response to strong southeasterly winds and continued to rise at a rate of 8 to 9 m/d until surfacing at 0000 GMT, February 7.

Contours of the IHA at day 4.5 of the event, i.e., 1700 GMT, February 6, 1985, are shown in Figure 11. The pattern is very similar to that predicted for the March 1983 (and February 1984) event using the smaller three-layer area (Figure 8B), with the region between

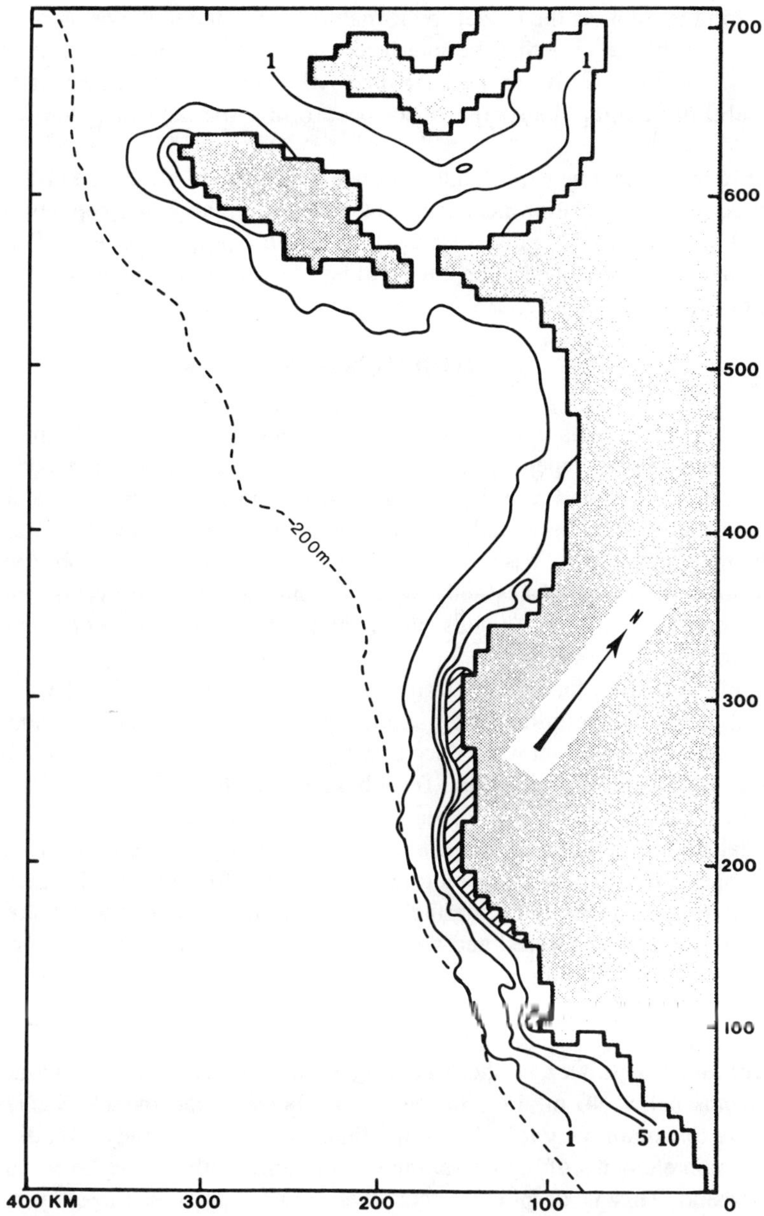


FIGURE 11. IHA (m) at day 4.5 of the February 1985 event using the two-layer model. The anomaly is greater than 15 m in the shaded areas.

Portland and Cape Jaffa being the most propitious to upwelling. A notable feature of Figure 11 is the emergence of the southwest corner of Kangaroo Island as another preferred site.

To digress slightly, it is anticipated that any sea-surface features generated during this event, viz., areas of reduced temperature, would be maintained (if not extended) in the interval between the interface surfacing and the time of the satellite pass due to the persistence of favorable winds. The winds observed at Neptune island during that period, February 7 to 14, 1985, were relatively light and from the southeast quarter, apart from a short-lived southwesterly airstream associated with the passage of a weak cold front on February 13.

Nevertheless, it is somewhat difficult to reconcile the IHA pattern in Figure 11 with the features evident in Figure 4, as the model, by its nature, is limited to the *onset* of upwelling, i.e., the period up until the time that the interface reaches the surface. Satellite images, in contrast, tend to be more sensitive to the later stages of "mature" upwelling. In this sense, an image is useful in defining the spatial scales occurring some days after the surfacing of the interface.

Hence, the model cannot be expected to simulate the observed advection of the interface offshore to an equilibrium position near the shelf edge.[2] Similarly, the apparent advection of cooler upwelled water to the northwest of the "active" region (between Portland and Cape Jaffa) cannot be reproduced. The model does, however, pinpoint the "active" or "source" regions quite effectively.

## IV. DISCUSSION

A layered three-dimensional numerical model has been used to study the physical processes occurring during time-dependent upwelling along the South Australian coast. Guided by observations of current and water properties during upwelling events,[2] a three-layer representation was adopted. The wind-forced model, while having some limitations due to the available computing facilities, has proved to be successful in reproducing observed current time series in the correct time scales and magnitude during two such events. In addition, the model has been used to identify sites favorable to upwelling, in accord with features evident in infrared satellite images.

Variations in the bottom topography (rather than coastline irregularities) were found to be the dominant factor in determining sites conducive to upwelling, and the maximum rate of upwelling at the coast was shown to occur with a wind direction $\approx 6°$ offshore of the upwelling-favorable alongshore direction. Tests have shown that the three-layer-model solution is in large measure independent of the choice of interfacial stress. Indeed, variation of the interfacial drag coefficient through two orders of magnitude produced only barely perceptible changes in the surface-layer currents and virtually none in the lower layers. Perhaps linked to this insensitivity to interfacial stress are other features which are common to the model runs we have processed and which may well be general. For example, experimental indications are that during upwelling events the active surface layer associated with off-shore currents is approximately 20 m thick, and that compensatory onshore currents seem to be restricted to mid-depths (20 to 40 m). The circulation produced by the three-layer model conforms to this cross-shore pattern, though a slight onshore motion is also predicted in depths below 40 m, i.e., in the bottom layer of the model. Such a vertical profile may serve to explain why it is that in the final exercise, a two-layer model managed to simulate the main elements of the upwelling event, namely, the upper layer currents and interface movement. This is despite the fact that a more detailed three-layer model is suggested by observations of the vertical structure of currents and water properties. Note, however, that the two-layer model cannot reproduce the observed concentration of onshore motion at mid-depths. The *general* direction of the cross-shore motion below the upper layer is predicted in both models to be onshore and it appears that the behavior of the first interface is insensitive to a coarse integration of this motion. Because of the agreement between the predictions of the interface movement by both models, it was possible to model a relatively large area with the less time-consuming two-layer model in a successful attempt to identify areas disposed to upwelling.

The present models focus upon the processes which occur during the onset of upwelling. However, the model run terminates when the first interface reaches the surface, typically after 4 to 5 d, and hence precludes the study of subsequent features. To examine the later phases of upwelling, a more detailed model is required in which consideration is given to interfacial mixing and thermodynamics. This is beyond the scope of the present paper but it is anticipated that such matters will be addressed as this work proceeds.

The significance of the upwelling mechanism, in its seasonal influence on water properties of the regional shelf seas, is considerable both in contributing to our understanding of the marine dynamics of these coastal waters and also in its role of provision of nutrients to the local rock lobster fishery. At the present time, the incidence of upwelling events is unknown. The operation of this model in response to wind fields, contemporary and historic, will assist to remove this lacuna in our knowledge.

## ACKNOWLEDGMENTS

Thanks are due to Carl Nilsson and Paul Tildesley of the Division of Oceanography, CSIRO Marine Laboratories, Hobart, for the provision of the satellite image shown in Figure 4. In addition, our appreciation is extended to Gail Jackson for draughting many of the figures. One of us (C. W.) has been supported during this research by a Flinders University Research Scholarship.

## REFERENCES


1. **Hahn, S. D.,** Physical Structure of the Waters of the South Australian Continental Shelf, Ph.D. thesis, Flinders University of South Australia, Bedford Park, 1986.
2. **Schahinger, R. B.,** Structure of coastal upwelling events observed off the south-east coast of South Australia during February 1983 — April 1984, *Aust. J. Mar. Freshwater Res.*, 38, 439, 1987.
3. **Lewis, R. K.,** The southern rock lobster, in *Natural History of the South East,* Tyler, M. J., Twidale, C. R., Ling, J. K., and Holmes, J. W., Eds., Royal Society of South Australia, Adelaide, 1983, chap. 17.
4. **O'Brien, J. J. and Hurlburt, H. E.,** A numerical model of coastal upwelling, *J. Phys. Oceanogr.*, 2, 14, 1972.
5. **Hurlburt, H. E. and Thompson, J. D.,** Coastal upwelling on a β-plane, *J Phys. Oceanogr.*, 3, 16, 1973.
6. **Suginohara, N.,** Onset of coastal upwelling in a two-layer ocean by wind stress with longshore variation, *J. Oceanogr. Soc. Jpn.*, 30, 23, 1974.
7. **Kishi, M. J. and Suginohara, N.,** Effects of longshore variation of coastal geometry and bottom topography on coastal upwelling in a two-layer model, *J. Oceanogr. Soc. Jpn.*, 31, 48, 1975.
8. **Preller, R. and O'Brien, J. J.,** The influence of bottom topography on upwelling off Peru, *J. Phys. Oceanogr.*, 10, 1377, 1980.
9. **Crépon, M. and Richez, C.,** Transient upwelling generated by two-dimensional atmospheric forcing and variability in the coastline, *J. Phys. Oceanogr.*, 12, 1437, 1982.
10. **Crépon, M., Richez, C., and Chartier, M.,** Effects of coastline geometry on upwellings, *J. Phys. Oceanogr.*, 14, 1365, 1984.
11. **Hua, B.-L. and Thomasset, F.,** A numerical study of the effects of coastline geometry on wind-induced upwelling in the Gulf of Lions, *J. Phys. Oceanogr.*, 13, 678, 1983.
12. **O'Brien, J. J. and Heburn, G. W.,** The state-of-the-art in coastal ocean modelling: a numerical model of coastal upwelling off Peru — including mixed layer dynamics, in *Coastal Oceanograpy,* Gade, H., Edwards, A., and Svendsen, H., Eds., Plenum Press, New York, 1983, 113.
13. **Heaps, N. S.,** A mechanism for local upwelling along the European continental slope, *Oceanol. Acta,* 3, 449, 1980.
14. **Heaps, N. S.,** Development of a three-layered spectral model for the motion of a stratified sea. I. Basic equations, in *Physical Oceanography of Coastal and Shelf Seas,* Johns, B., Ed., Elsevier, Amsterdam, 1983, chap. 9.
15. **Heaps, N.S. and Jones, J. E.,** Development of a three-layered spectral model for the motion of a stratified sea. II. Experiments with a rectangular basin representing the Celtic Sea, in *Physical Oceanography of Coastal and Shelf Seas.*, Johns, B., Ed., Elsevier, Amsterdam, 1983, chap. 10.
16. **Cai Wenju and Lennon, G. W.,** Upwelling in the Taiwan Strait in response to wind stress, ocean circulation and topography, *Estuarine Coastal Shelf Sci.*, 26, 15, 1988.
17. **Thompson, R. O. R. Y.,** Low-pass filters to suppress inertial and tidal frequencies, *J. Phys. Oceanogr.*, 13, 1077, 1983.
18. **Peffley, M. B. and O'Brien, J. J.,** A three-dimensional simulation of coastal upwelling off Oregon, *J. Phys. Oceanogr.*, 6, 164, 1976.
19. **Schahinger, R. B.,** unpublished data.

Chapter 6

# ON THE ATMOSPHERICALLY INDUCED VARIABILITY OF THE CIRCULATION OF THE NORTHWEST EUROPEAN SHELF SEA AND RELATED PHENOMENA — A MODEL EXPERIMENT

**Jan O. Backhaus**

## TABLE OF CONTENTS

# I. INTRODUCTION

An important component for the understanding of the shelf sea system as a whole is its circulation since it is the transport medium for a major portion of the biomass, for pollutants entering the sea by different routes, and for heat and salt. The latter interact with the circulation. The decisive role of the circulation in regard to oceanographic and to related environmental phenomena explains its importance within resource management.

Our benefits from shelf seas are manifold. However, quite often there is a conflict between the different uses. Certainly marine pollution in shelf seas counteracts the extraction of protein from the sea to feed the rapidly growing world population. Presently about 90% of the world protein gain from the sea stems from shelf regions. The careful resource management of the shelf seas will be an important future task, especially for developing countries, in order to avoid irreparable damage of the ecosystem and simultaneously to increase the efficiency of gaining goods from the sea.

The northwest European shelf sea is a characteristic example for the conflict situation typical for shelf seas. A number of highly industrialized nations are using its resource potential in manifold ways. It is an appreciated recreational area, a dumping and incineration site for industrial, municipal, and radioactive wastes, a fertile fisheries and fish-farming region, a site for extracting mineral oil and gas from the sea bed, and it has heavily frequented shipping routes which considerably contribute to marine pollution. Presently there is a management for the international share of fisheries catches within the seas of the European Communities (EC), but a well-coordinated resource management considering all relevant components of the marine system does not exist. However, the increasing concern about the status of the EC sea and of the North Sea in particular led to the definition of environmental research programs in which the circulation always occupies a central role.

Besides ecological, environmental, and resource-political questions, a number of still-open oceanographic questions exist which concern the long-term circulation in the northwest European shelf sea and in particular its variability. They gave reason to this investigation.

Since the theory behind this investigation does not consider any essentially new items, the emphasis will be placed on a discussion of the results and of the conclusions rather than on a description of the theory.

# II. STATE OF RESEARCH

Due to a long tradition of research on the northwest European shelf sea it is characterized by a fairly high temporal as well as spatial observational data density. However, up to now a comprehensive and quantitative description of the circulation and its variability does not exist. The available data is still far too scarce and too scattered in time and in space to allow a compilation of a quantitative estimate of the circulation. In particular, large-scale synoptic surveys of oceanographic parameters and long-term records of current observations (covering several months) are still missing. Although the available technology allows the deployment of self-recording current meters over several months, it is unrealistic to expect long-term records in the future as the destructive effect of the intensive fishery prevents the maintenance of such systems.

In the field of modeling the data situation is similar. Model simulations of the long-term circulation, covering periods of several months to years are likewise missing. The present prognostic and mainly barotropic simulations only cover short episodes, typically storm-surge events. The average simulation time of these time-dependent episode simulations is of about the same order as the time scales of synoptic atmospheric disturbances (3 to 5 d).

Estimates for longer time scales of the circulation have been obtained by time-independent diagnostic simulations. These are based upon the concept of prescribing a suitable long-term mean of the atmospheric forcing (a monthly, seasonal, or climatological mean). The resulting circulation subject to stationary forcing is then assumed to represent the respective mean. This approach has the advantage of yielding a fairly decent estimate of the mean circulation with comparatively small computing effort; its disadvantage is that it does not yield any information about fluctuations of the circulation within the averaging period.

As a consequence of the historical development of models and model applications for the northwest-European shelf sea, where up to now preference was given to episode and diagnostic simulations, time-dependent long-term simulations, forced by a transient atmosphere are missing. However, in view of the miserable data situation (which is even worse for other shelf seas) model simulations are presently the only means which allow an estimate of the low-frequency variability of the circulation for time scales between (say) 10 d and 10 years.

A likely reason for this obvious gap in model applications might be the quality of the numerical schemes used within the commonly applied shelf sea models. These did not allow an integration with suitable computing economy and numerical stability over the time spans mentioned above. On the other hand, even the integration of a very uneconomic model over 10 years is still much less expensive than a shipborne field exercise covering just a few months.

## III. THE STRATEGY OF THE MODEL EXPERIMENT

The obvious information deficit concerning low-frequency fluctuations of the circulation and of related phenomena as, for example, the dispersion of biomass and of pollutants on the northwest European shelf provides the main motivation for this investigation.

In this context it is convenient to define low-frequency fluctuations of the flow by time scales which are well above the average frictional response time of the circulation under atmospheric forcing. For the area of interest this is a time span of about 1 to 5 d, depending on the respective local energy dissipation rate induced by bottom friction, which increases with decreasing water depth. Thus, the pattern of the frictional response time reflects the topography of the basin. This dissipation time scale is about the same as that of synoptic atmospheric disturbances (e. g., moving cyclones). Hence, one can assume that a single storm will not contribute significantly to the low-frequency circulation.

For the analysis of the variability of the circulation it is convenient to define a separation of the time scales in relation to the annual signal. Fluctuations with frequencies lower (i.e., time scales larger) than the synoptic frequency band but higher than the annual frequency signal will be defined as "subsynoptic" fluctuations.

The northwest European shelf sea lies in an area of extreme atmospheric activity which shows a pronounced annual signal. Superimposed on the annual cycle there are weather events of stochastic character covering various space and time scales. These events contribute significantly to the considerable observed variability of the atmosphere.

The concept of this investigation is based upon the assumption that the variability of the circulation in a shallow sea at mid-latitudes is characterized to a high degree by the activity of the atmosphere and in particular by the transfer of atmospheric momentum. Due to lacking data, the thermodynamic energy transfer can not be considered directly. However, the influence of the thermal stratification of the sea on the momentum transfer within the water column will be considered according to mean conditions.

A systematic model experiment for the determination of estimates of the variability of the circulation for time scales below as well as above the annual cycle was conducted. In the course of this experiment a nonlinear three-dimensional numerical shelf sea circulation model was integrated over a period of 14 years.

In the following sections some results of the experiment, quantitative estimates of the atmospherically induced low-frequency variability of the circulation and its ratio to the mean flow, will be presented. However, prior to this the essential prerequisites for this investigation are described: the atmospheric forcing for the model, the circulation model itself, and the treatment of the baroclinic fields in the experiment. Furthermore, some details on the compression and handling of the model data are given and, finally, an attempt for a model validation is made. In the last section the results of the circulation model are applied to phenomena related to the circulation. The simulated flow field was used in a transport-dispersion model to estimate the fate of fish eggs and larvae and of pollutants.

## A. THE ATMOSPHERIC FORCING

The data for the meteorological forcing for the model experiment was provided by the Norwegian meteorological office at Oslo. It is a 28-year long time-series (1955 to 1982) of six-hourly air pressure fields at sea level, defined on a 150 × 150 km grid. The data cover the northeastern North Atlantic, the adjacent European shelf sea and parts of the continental coastal regions (Figure 2). The temporal and spatial scales of the atmospheric fields relevant to ocean modeling (air pressure, wind stress) were compiled in an atlas for the region covered by the data.[1] The wind stress was derived from the geostrophic wind according to the nonlinear relationship of Luthardt and Hasse.[2,3]

The Norwegian data was indeed an important prerequisite for the long-term experiment, because the preparation of the meteorological forcing data for oceanographic modeling usually is a rather tedious task which could be avoided due to the well-prepared data.

## B. THE CIRCULATION MODEL

The numerical scheme of the circulation model was developed especially for the purpose of long-term simulations with emphasis on good stability properties. It consists of a new composition (new in relation to commonly applied shelf sea models) of known numerical algorithms in finite differencing.

A detailed description of the scheme would not fit into the frame of this publication and has already been published elsewhere.[4] Here the emphasis will be placed on the physical processes to be simulated and on some mainly practical considerations which together led to the concept of the three-dimensional model.

In a shallow shelf sea the vertical transfer of atmospheric momentum leads to a pronounced current profile, which is induced by the proximity and even the overlap of the surface and bottom Ekman layers and by the presence of lateral boundaries. In the case of atmospheric forcing the latter may cause large scale sea level elevations (Windstau) that drive a quasi-geostrophic compensating flow. Thus, the currents at depth, beneath the directly wind-driven regime may have an entirely different flow pattern. In contrast to the deep ocean, these processes are confined to a very short water column and they are subject to nonlinear interactions. As a whole they require a three-dimensional model.

The vertical momentum transfer and the resulting (barotropic) current profile and thus the circulation are strongly influenced by nonlinear interactions between the wind- and the tidally induced flow. The key point for this interaction is the nonlinear bottom friction. A pilot study investigating these effects showed that the neglection of the tidal oscillations of the flow will lead to a considerable overestimation of the long-term wind-induced flow.[5] The same applies for the density-driven flow, as well as for its superposition with the wind-induced flow. This is a result of the lack of energy dissipation in the absence of the oscillatory tidal currents. On the other hand, for the northwest European shelf sea, the long-term effect of the tidal residual flow itself is almost negligible compared to the magnitude of the wind- and the density-induced flow.

This pilot study was carried out by applying the circulation model in a diagnostic case study. The three main components contributing to the large-scale circulation, the tide, the wind, and the stratification were at first simulated separately, subsequently combined pairwise, and finally superimposed together as depicted in Figure 1.

This pilot experiment led to the conclusion that the nonlinear dissipative effect of the tidal oscillations on the wind- and density-driven circulation may not be neglected in a simulation of the low-frequency flow, even though, for practical reasons, it would be desirable to omit the tide. This is one of the major differences of shelf sea modeling in contrast to deep-ocean simulations. There, indeed, the nonlinear effects of the tides are very small and thus negligible.

The effect of the thermo-haline stratification on the vertical momentum transfer within a shelf sea is another mechanism important for the large-scale circulation. During winter, wide parts of the northwest European shelf sea are vertically well mixed due to tidal- and storm-induced turbulence and due to convection by surface cooling. However, beyond the shelf and in the Norwegian Trench a thermo-haline stratification is present throughout the year. During summer, wide areas are thermally stratified, except for the shallow near-coastal regions and the Dogger bank where tidal mixing dominates. In the model, the effects of the seasonally varying stratification on the momentum transfer are accounted for by a simple parameterization based upon the Richardson-number principle. This allows us to estimate spatial and temporal changes of the "eddy-viscosity" momentum transfer coefficient due to varying winds and/or due to neutral or stable stratification.

The discretized topography of the model area is shown in Figure 2. The spherical model grid has a resolution of 1/5 and 1/3 of a degree in meridional and longitudinal direction, respectively. The average grid distance is about 20 km. In accordance to the seasonally varying vertical stratification 8 (winter) or 13 (summer) computational levels resolve the vertical domain. The space between adjacent computational levels shall be called a model layer. The thicknesses of the 7, respectively, 12 model layers are of the order of 10 m in the upper 100 m of the water column and of the order of 100 m in the deeper parts. There are about 4,500 "wet" grid points in the first model layer and in total there are either about 21,000 or 32,000 wet grid points, respectively, depending on the season. The thickness of the respective lowest model layer may be reduced in order to account for topographic changes within the layer. This implies that in the very shallow coastal regions the model may "degenerate" to a vertically integrated two-dimensional model as described by Backhaus.[6]

The time step of the model was set to 20 min and later in the course of the experiment to 40 min. There were no differences in the estimates of the long-term flow caused by the coarser time step. Even the rather coarse resolution of the tidal flow which led to phase errors of about 1 to 3 time steps (a huge error margin for episode simulations) did not have a significant influence on the nonlinear tidal effects, the retention of which being desired.

The circulation model used in the experiment is based upon the nonlinear "primitive" shallow-water equations in polar coordinates. It has a free surface and at the interfaces between water and land the usual no-flux boundary conditions are applied. At the open boundaries sea level elevations are prescribed which are linearily superimposed from three components. These are tidal elevations for the dominant semi-diurnal lunar tide M2 and elevations due to the inverse barometric effect of air pressure variations. Both these elevations are time dependent, whereas the third component is constant in time (see also Section III. C). This third boundary condition reflects the sea level elevations of the flow being in geostrophic balance with the respective density field prescribed for a summer or a winter season. This elevation is obtained from a line integral (along the boundary) of the vertically integrated baroclinic pressure gradients parallel to the boundary. It is equivalent to the depth mean geostrophic flow normal to the boundary obtained by the assumption that the bottom itself is a "layer of no motion". There remains an integration constant which determines the absolute sea level at the open boundary. This constant was derived empirically with the

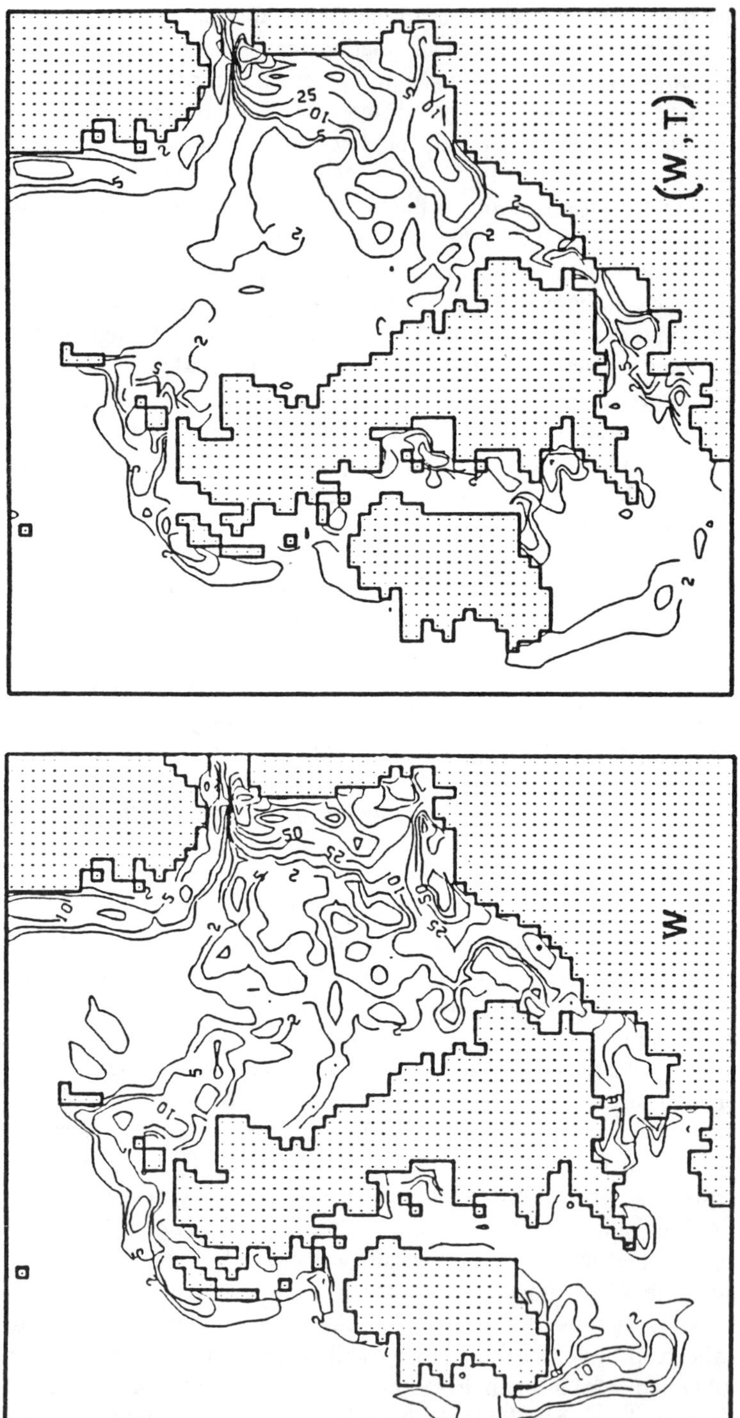


A

FIGURE 1. The influence of the tide (T) on the kinetic energy (cm/s)**2 of the (depth mean) mean circulation due to winds (W) and/or stratification (S) for mean climatological winter conditions. In the respective left figures the tide was excluded. Cases: (A) Wind-induced circulation; (B) density-induced circulation; (C) combination of wind- and density-induced flow.

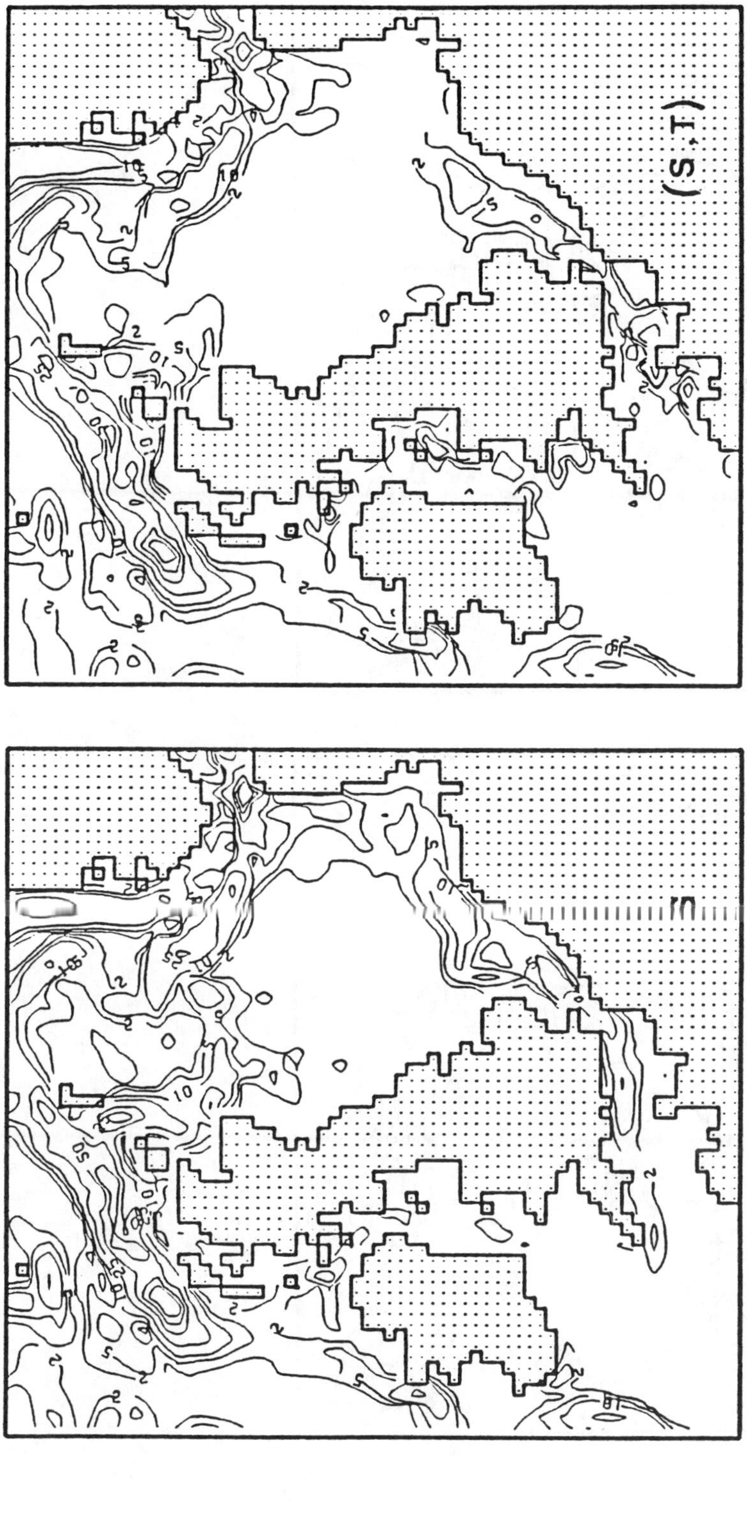


FIGURE 1B.

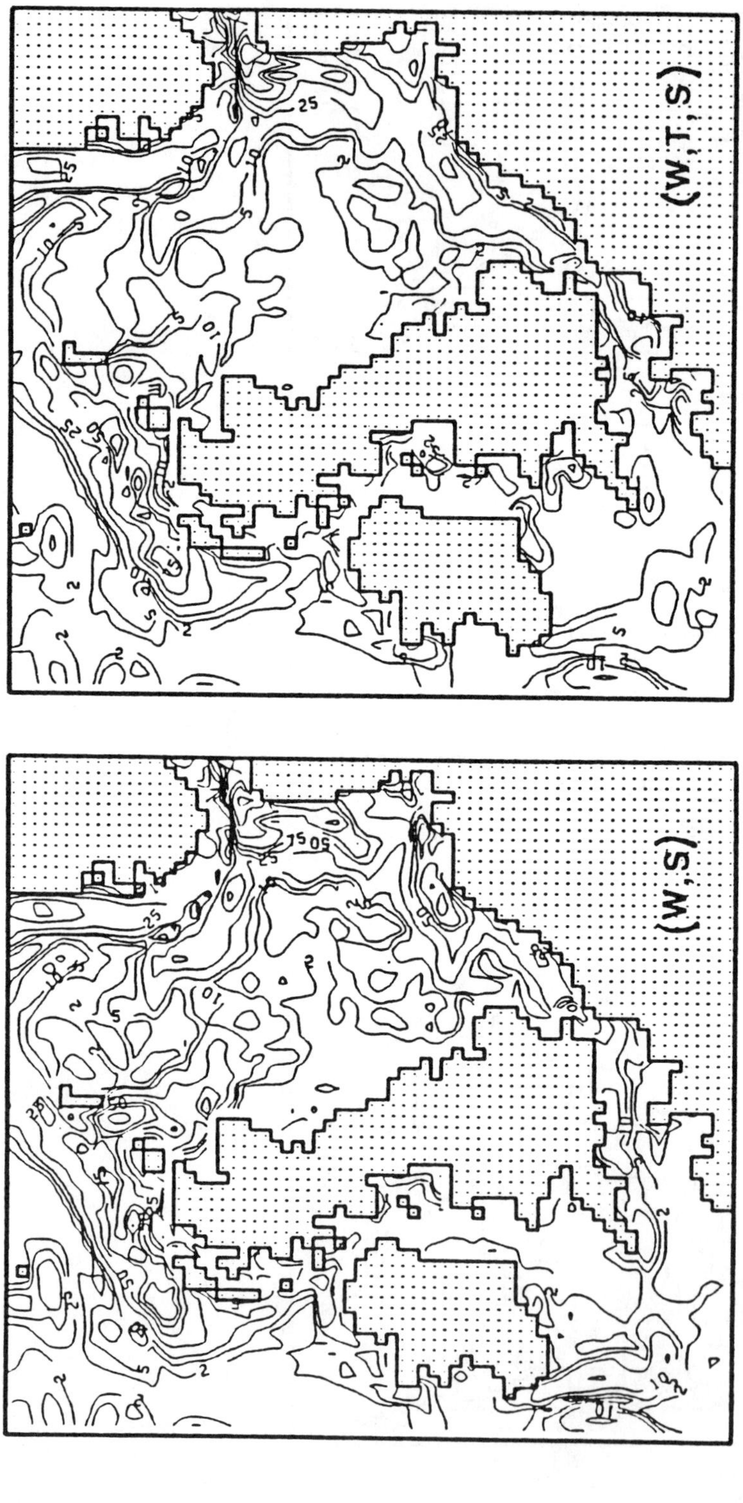


FIGURE 1C.

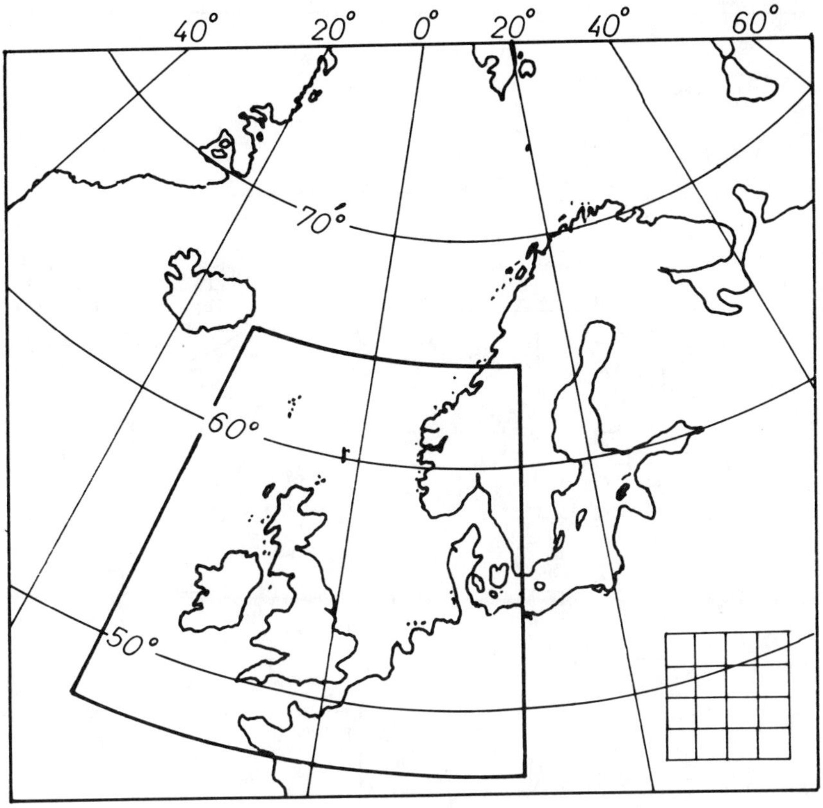


A

FIGURE 2. (A) Top: location of the model domain and region covered by the meteorological grid point data (150-km grid indicated). (B) Bottom: model domain for northwest European continental shelf sea; horizontal grid resolution 1/5 and 1/3 degree in latitude and longitude, respectively. Winter version: 7 layers (21,000 ''wet'' grid points); summer version: 12 layers (32,000 ''wet'' grid points). Location of sections indicated (cf. Figure 6).

help of the model by varying it in a number of trials and by finally adjusting it in a way that the sea levels in the interior of the model domain matched coastal tide gauges with an accuracy of about 20% related to mean sea level.

The open Atlantic boundaries of the model (Figure 2) are placed in deep water beyond the continental margin. The time-dependent forcing of the model at the open boundaries are mainly barotropic elevations, with amplitudes being much smaller in the deep ocean than on the shallow shelf (e.g., tides, surges, etc.). It turned out that placing the boundary in this position induced fewer problems (reflections, etc.) in contrast to locating it on the shelf itself where nonlinearities are much more pronounced (as in many other models). A reflecting boundary condition is applied in the deep ocean. Experiments with nonreflecting boundary conditions on the shelf turned out to lose their advantages whenever almost-stationary and long-lasting (several days to weeks) atmospheric systems occurred. In that case no clear propagation characteristics of the atmospherically induced sea level disturbances near the boundary could be determined. As a result, rather spurious and unrealistic low-frequency sea level fluctuations occurred in the model. These problems were effectively avoided by placing the (clamped) boundary in the deep ocean.

The rather stable semi-implicit numerical scheme was favorable for the choice of the open boundary and for the feasibility of the model experiment as a whole.[4,6] It allowed

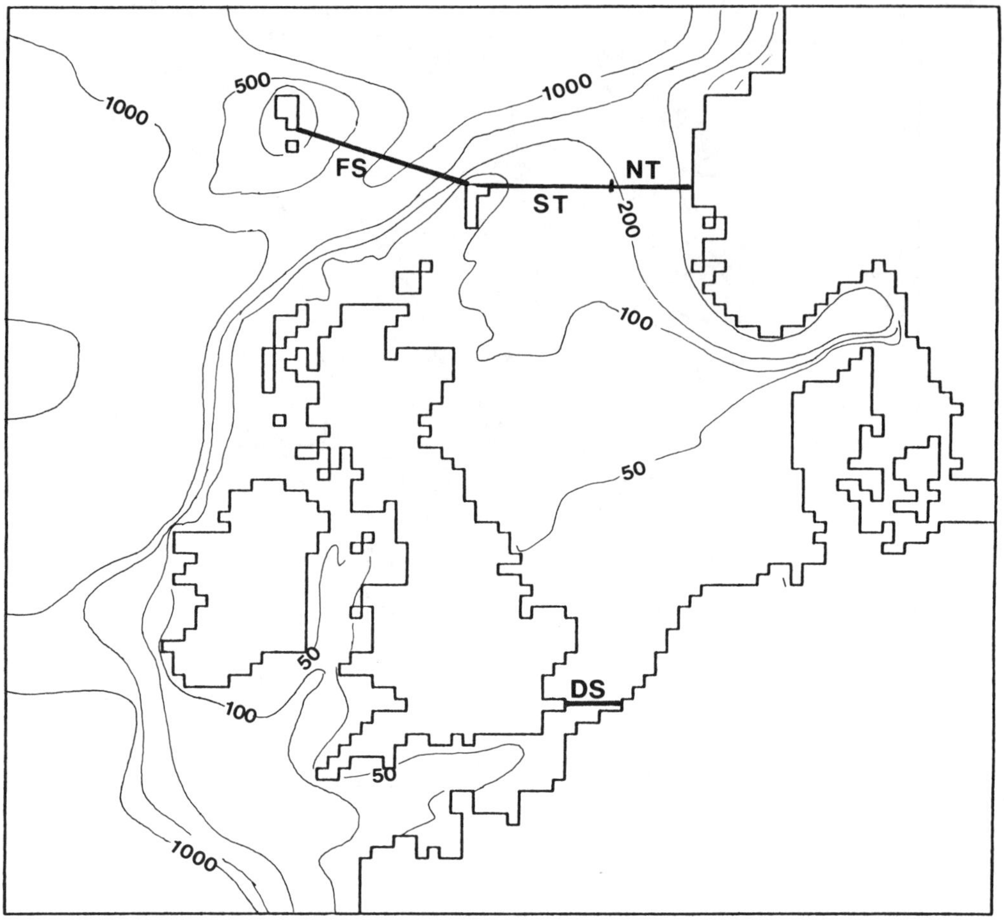


FIGURE 2B.

coverage of the long-term simulations under sometimes extreme atmospheric forcing with a reasonable computing economy and realiability with regard to numerical stability. It is a commonly gained experience from episode modeling, although not often mentioned, that the maintenance of stability for a longer simulation time might be some problem when using a nonlinear model. For this reason in many models the horizontal momentum diffusion term rather serves as an emergency brake to maintain (linear numerical) stability than to parameterize sub-gridscale turbulence. In the experiment the horizontal diffusion coefficient within a model layer was set to a value of 10 m/s times the layer thickness.

## C. THE BAROCLINIC CIRCULATION

Whereas the vertical thermo-haline stratification of the northwest European shelf sea shows a pronounced annual cycle, its lateral haline stratification is present throughout the year. The considerable density gradient is maintained by the Atlantic inflow and by the fresh-water run-off along the coastline. For the winter season this gradient is depicted in Figure 3 by means of the mean surface salinity.

The fresh-water discharges along the continental coastline provide the major contribution due to the large drainage area of the western European hinterland. The maximum run-off occurs in late spring. The influence of the large-scale haline gradient on the circulation can be estimated from Figure 1. For many years it was assumed that the circulation in the North

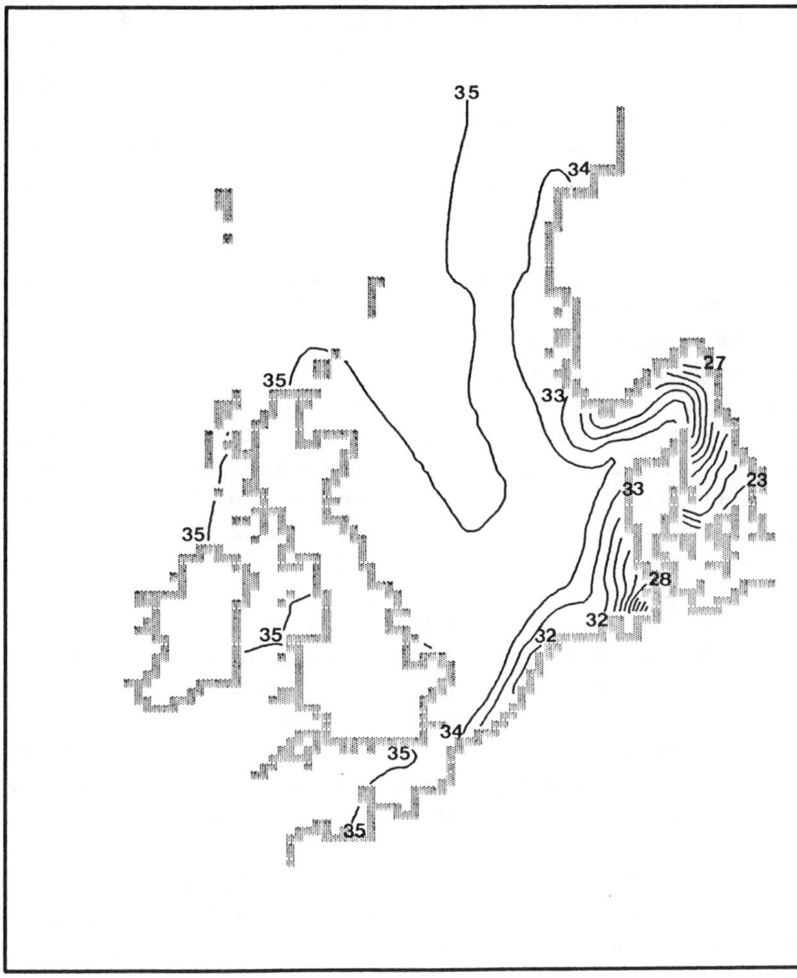


FIGURE 3. Surface salinity in model domain, climatological winter mean

Sea was mainly driven by the winds and by the tides until Backhaus and Maier-Reimer demonstrated the considerable effect of the lateral density distribution.[7] To neglect this effect would lead to a severe underestimation of the circulation, as can be deduced from the display in Figure 1.

There is reason to believe, based on a few large-scale salinity synopses of the North Sea, that weather events with subsynoptic time scales may induce lasting changes of the density structure via the wind-induced circulation. In this way winterly weather events and, due to the atmospheric momentum transfer, also circulation events may leave their trace even until the following summer. Besides the comparatively short frictional dissipation time scale this could be a mechanism which could lend a long term "memory" to a shallow-shelf sea. However, this has as yet not been proven.

In view of this interesting long-term feedback mechanism, it would be desirable to treat the baroclinic fields in a prognostic way in the course of the experiment. However, this was not possible because the necessary lateral and sea-surface boundary conditions could not be obtained due to lack of data. With the exception of the North Sea, there are only very few data available for the rest of the shelf region. Therefore, in order to at least retain the mean effect of the density distribution on the circulation, climatological seasonal means for summer and winter, respectively, were incorporated in the experiment. The data were obtained from charts of climatological means of temperature and salinity for the North Sea region whereas for the remaining western portions of the model domain, data defined on a 1 × 1 degree grid, prepared for general circulation models of the ocean, were used.[8-10] Considerable spatial interpolations were necessary due to the coarse grid of the latter data. Inconsistencies due to interpolation were eliminated by means of the baroclinic model itself. In a prognostic model run a "dynamic" interpolation of the baroclinic fields was carried out. After about 10 to 30 simulation days a first-order balance between the baroclinic fields, topography, and circulation was obtained.

The dynamically balanced seasonal means of temperature and salinity were prescribed in terms of "frozen" baroclinic pressure gradients. This is a severe drawback of the experiment because an adjustment of the baroclinic fields to the actual circulation forced by a transient atmosphere is inhibited. However, at least for meteorologically calm periods the circulation will spin down to a pattern which reflects a long-term seasonal mean of the flow. Hence, the diagnostically simulated baroclinic circulation should be regarded as a seasonally varying background signal on which the atmospherically induced fluctuations are superimposed. This implies that the experiment mainly describes the barotropic variability of the circulation and hence it must be anticipated that the actual variability of the circulation will be underestimated.

### D. DATA COMPRESSION AND HANDLING

The experiment was conducted with two separate model versions of different vertical resolution, one for the summer (May to September) and one for the winter (October to April). A fairly smooth transition between the two models was obtained by starting a respective version with an overlap of 1 month in order to "warm up" the model. The overlapping month was then discarded. A direct initialization from the respective previous run was avoided because of likely interpolation inconsistencies due to the different vertical grid resolutions for summer and winter, respectively.

In the experiment, circulation scenarios over a period of 14 years (1969 to 1982) were simulated; it produced a huge amount of data exceeding by far the usual amount known from episode modeling. Some data compression was necessary. Therefore, while running the model, the output was averaged over two tidal cycles. The result, a daily mean of the three-dimensional flow field and the tidal- and wind-induced variance was stored.

The crude filter removed the poorly resolved linear tidal oscillations but it retained their nonlinear effects. Simultaneously, the filter defines a cutoff frequency for subgrid-scale motions. The Lagrangian excursion of a water particle due to tidal- and wind-induced fluctuations within the averaging period is generally smaller than the grid size of the model (20 km). Therefore, with respect to applications of the flow field in a dispersion model, the time filter also implies a cutoff for spatial scales in a Lagrangian sense. The stored daily variances of the flow will then be used to parameterize the diffusive effect of the filtered out subgrid-scale excursions.

## IV. RESULTS OF THE EXPERIMENT

In view of the information deficit concerning estimates of the variability of the circulation, outlined in Section II, the main focus of the presentation of the experiment results will be placed on the description and quantification of the low-frequency variability of the circulation of the northwest European shelf sea. In order to emphasize the temporal variability, the model raw data were compressed again, this time by a vertical integration. However, the full output was retained for dispersion simulations.

The description begins with the mean annual cycle of the circulation encompassing the shelf region and adjacent slope and ocean margins as previous diagnostic simulations yield no information concerning the circulation induced by the combined action of wind, tide, and stratification.

This is followed by a description of the simulated seasonal and subsynoptic variability. Finally, the experiment results are compared with other estimates of long-term means of the mass transport for the North Sea entrances, with some direct measurements of the flow in the Norwegian trench and with surface elevation records.

## A. THE MEAN ANNUAL CYCLE OF THE CIRCULATION

Climatological seasonal means of the depth mean circulation, averaged over the simulated ensemble of 14 years, are presented in Figure 4. The circulation is displayed by means of the transport streamfunction in Sverdrup units (1 mio m**3/s).

For the entire shelf region the transport rates are well below 1 Sverdrup. The 200-m-depth contour representing the shelf edge coincides with the 1 Sverdrup contour (cf. Figure 2). A considerable amount (about 75%) of the Atlantic water masses entering the North Sea basin in the north through the Orkney-Shetland (Fair Isle) passage sweep across the northern shelf plateau to rejoin the Atlantic via the Norwegian Trench at the eastern boundary of the basin. Only a smaller fraction of the inflowing Atlantic water masses contributes to the exchange and flushing of water masses in the southern parts of the basin.

The pattern of the open cyclonic recirculation cell in the northern North Sea and its seasonal and spatial variations is visualized, for example, by the spatial variations of the 0.6 Sverdrup contour in Figure 4. The mean annual variation of the transports in the southern portions is around 0.2 Sverdrup, whereas in the recirculation cell, in particular in its eastern part, the variations are about 0.5 Sverdrup. There the highest flow rates occur during summer and autumn in contrast to the southern North Sea where the maximum flow occurs during winter. These differences cannot be explained by spatial changes of the mean atmospheric forcing for this time of the year since these are not very pronounced. Hence, during summer and autumn the recirculation cell is mainly driven by the baroclinic pressure gradients (cf. Figure 1). The shallow southern parts of the North Sea, on the contrary, reflect in good agreement the annual cycle of the wind forcing.

The mean flow south and west of the British isles shows comparatively low transport rates in the Channel and in the Celtic and Irish Sea, whereas for the waters around the Hebrides at the west coast of Scotland about the same values as in the southern North Sea are obtained.

Beyond the shelf (cf. Figure 2), in the vicinity of the continental slope, there is an energetic and well-organized flow aligned with the slope; its transport rates range between 2 and 6 Sverdrups. This is the extension of the Gulfstream, feeding the eastern boundary current of the North Atlantic. It follows the shelf edge from the west coast of Ireland over the Faroe-Shetland ridge into the Norwegian Sea. The recirculation within the northern North Sea is partly fed by the water masses of this shelf edge flow.

The atlas compiled from the Norwegian meteorological data allows establishment of links between the mean circulation and the average weather patterns. As an example, in Figure 5 the climatological winter means of the wind stress components, the wind stress curl, and of the air pressure at sea level are displayed. The model region lies within the transition area between the low pressure system reaching from Iceland up to the Barents shelf and the winterly high pressure system over the European continent. This global weather situation leads to a mean southwesterly wind forcing in the region of interest. Its consequences on the flow can be judged from Figure 1. Almost the entire basin of the North Sea and the Baltic are covered by a pattern of positive (cyclonic) wind stress curl, whereas over the western portions of the model domain negative rotation prevails.

An analysis of the circulation of the North Sea in relation to variations of the direction

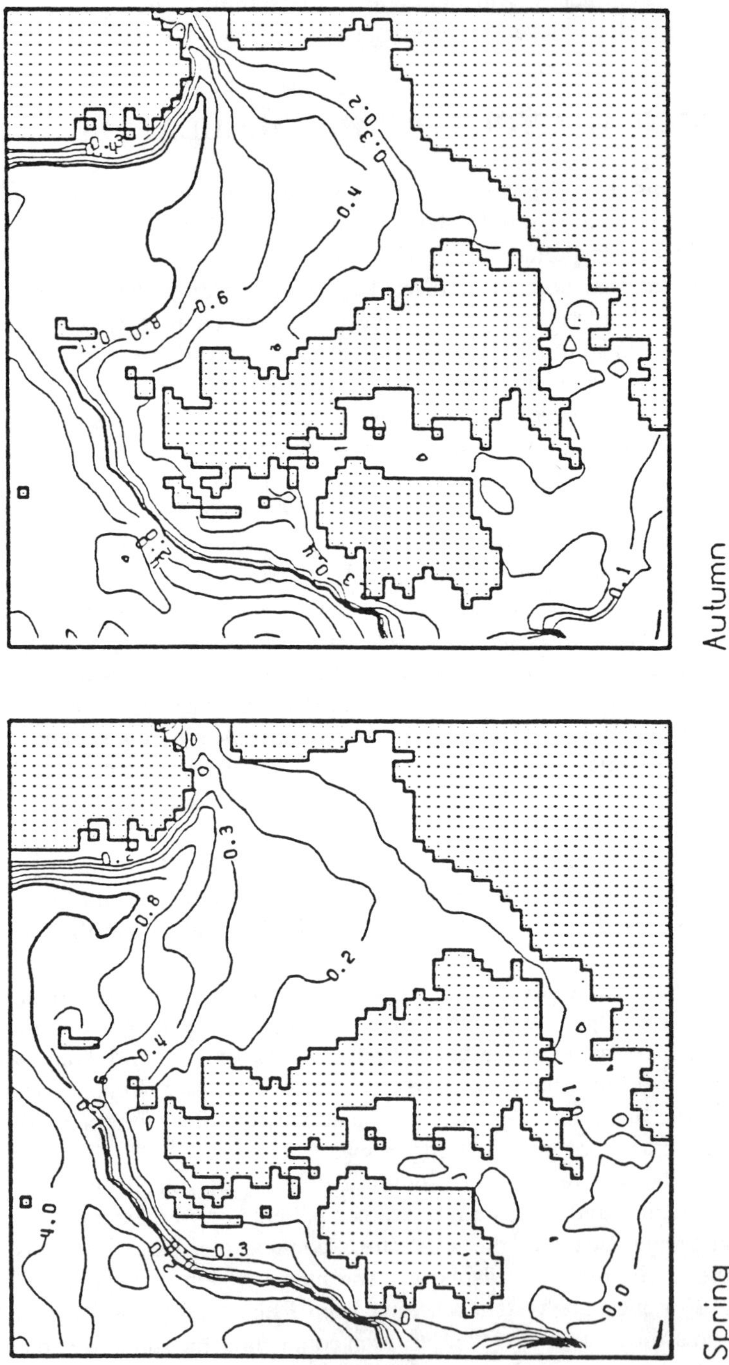
Autumn
Spring

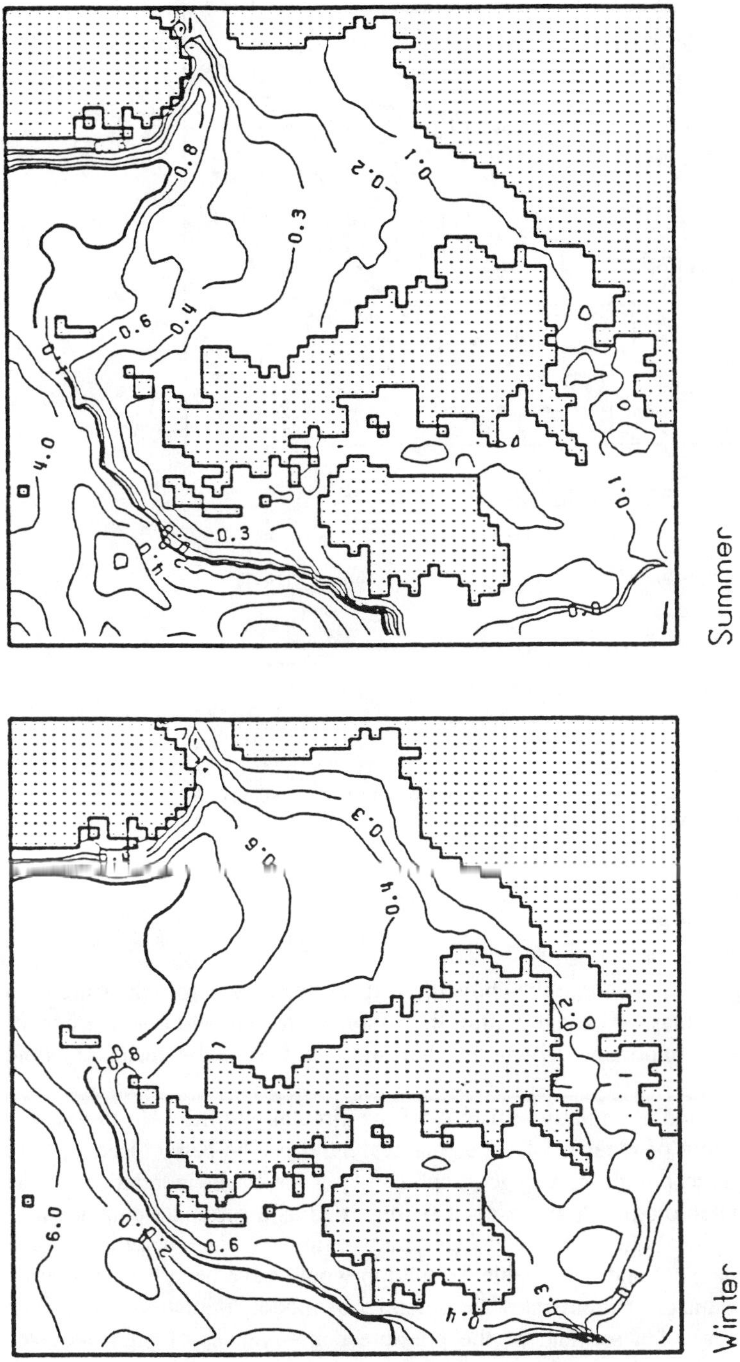


FIGURE 4. Climatological seasonal means of the transport stream function for the northwest European shelf sea in Sverdrup units (10**6 m**3/s).

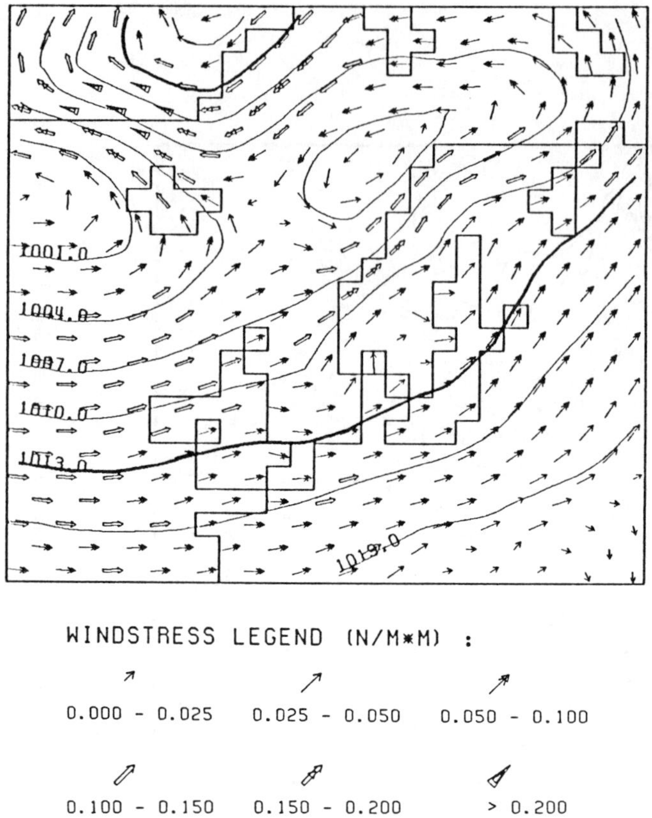


A

FIGURE 5. Climatological winter mean of the atmospheric forcing compiled from Norwegian grid-point data (150-km grid, cf. Figure 2). (A) Arrows indicate wind stress direction and magnitude (N/m**2) and air pressure (hPa) distribution indicated by contour lines; (B) contours of wind stress curl (10** − 1 N/(m**2 * 1000 km), regions with negative curl are shaded.

of the mean wind forcing has shown that the above-described mean meteorological situation leads to an optimum water mass exchange within the basin.[11] The average positive curl of the wind stress is another important driving mechanism that favors the mean cyclonic circulation.

Obviously the favor of the assimilative capacity of the circulation of the North Sea due to the wind-induced flushing or renewal of water masses could not be better. Hence, with regard to the anthropogenic pollution, the geographic tilt of the basin in relation to the direction of the mean meteorological forcing must be regarded as a fortunate coincidence.

It is merely a philosophical question to consider what would have happened to the dump site North Sea if the dynamical and geographical situation would have been less favorable. However, the above conclusion demonstrates the potential of model simulations to provide estimates of the sensitivity of a sea and of the possible consequences of anthropogenic influences on marine ecosystems derived from the knowledge of the physical environment.[12] Further evidence with respect to this scope of model applications is given in Section V.

## B. VARIABILITY OF THE CIRCULATION

The circulation scenarios simulated in the course of the experiment allow for a separation

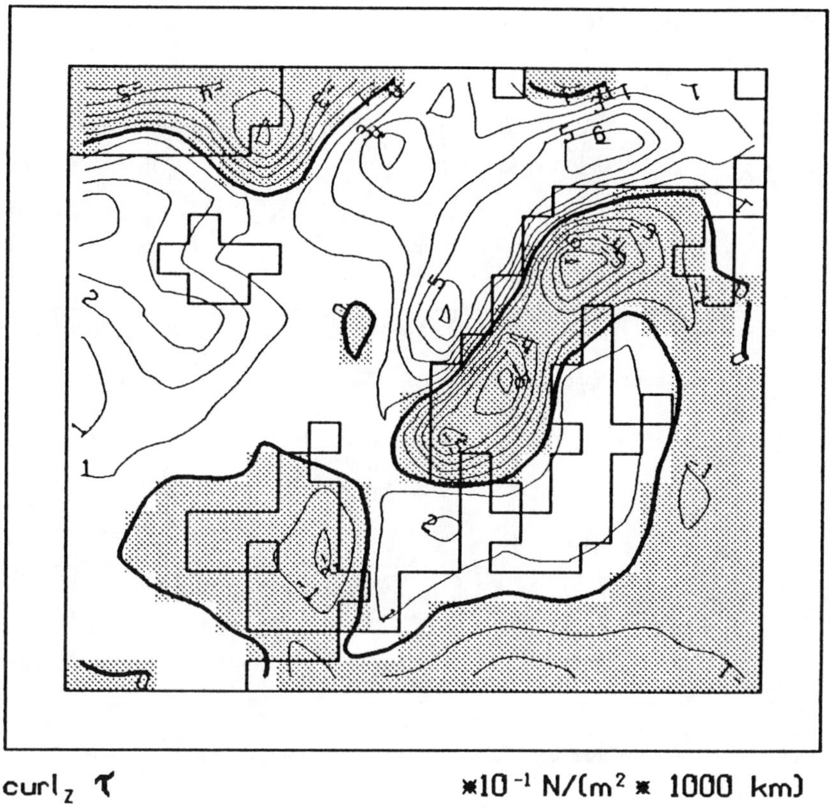


FIGURE 5B.

of time scales in relation to the annual cycle. Again it should be pointed out that the simulated variability will certainly underestimate the actual variability since a feedback between the baroclinic fields and the wind-induced circulation was inhibited. In the following, only some examples of the still ongoing analysis of the model experiment are presented. For (re ) interpretations of historical observational data within the simulation period (1969 to 1982) an atlas of the model results was prepared.[13]

A first impression of fluctuations of the circulation for time scales above and below the annual cycle is depicted in Figure 6. It shows time series of water mass transports through selected sections. In this display a separation for low-frequency (time scales above about 100 d) and for subsynoptic fluctuations (time scales below 100 d but above 6 d) was obtained by suitable low- and bandpass filtering. The low-frequency flow is shown by means of a full line, whereas the subsynoptic fluctuations appear as a shaded variance around the curve. This display provides an estimate of the signal-to-noise ratio of the flow for the given time scales bearing in mind that synoptic fluctuations, contained in the raw data, have been omitted.

In general, a more or less pronounced annual signal is present in each of the time series of Figure 6. However, superimposed on the annual cycle are stochastic and sometimes rather energetic fluctuations of different time scales. They reflect the stochastic character of the forcing atmosphere.

For the section between the Shetland Islands and the Norwegian Trench the annual signal dominates; it clearly exceeds the subsynoptic "noise". This section runs through the center of the recirculation cell, which obviously is rather insensitive to varying atmospheric forcing. The recirculation obtains its energy mainly from the global haline density gradient between

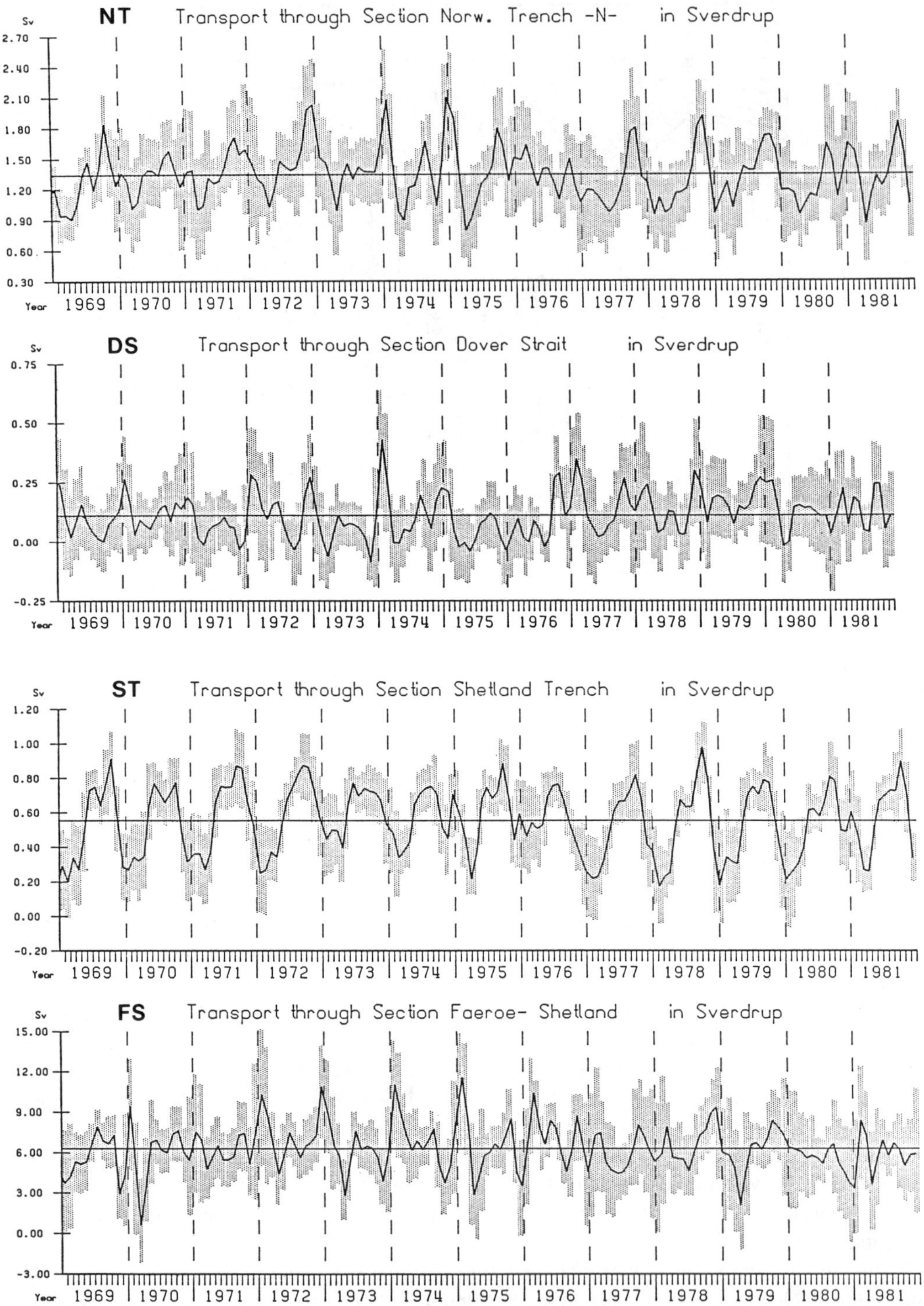


FIGURE 6. Time series of transport in Sverdrup units for selected sections; black curve; filter cut-off about 90 d, positive values correspond to northward flow for the sections: FS = Faroe-Shetland; NT = Norwegian Trench; DS = Dover Strait; and to southward flow for section: ST = Shetland-Norwegian Trench. Locations of sections indicated in Figure 2. Region shaded: subsynoptic fluctuations.

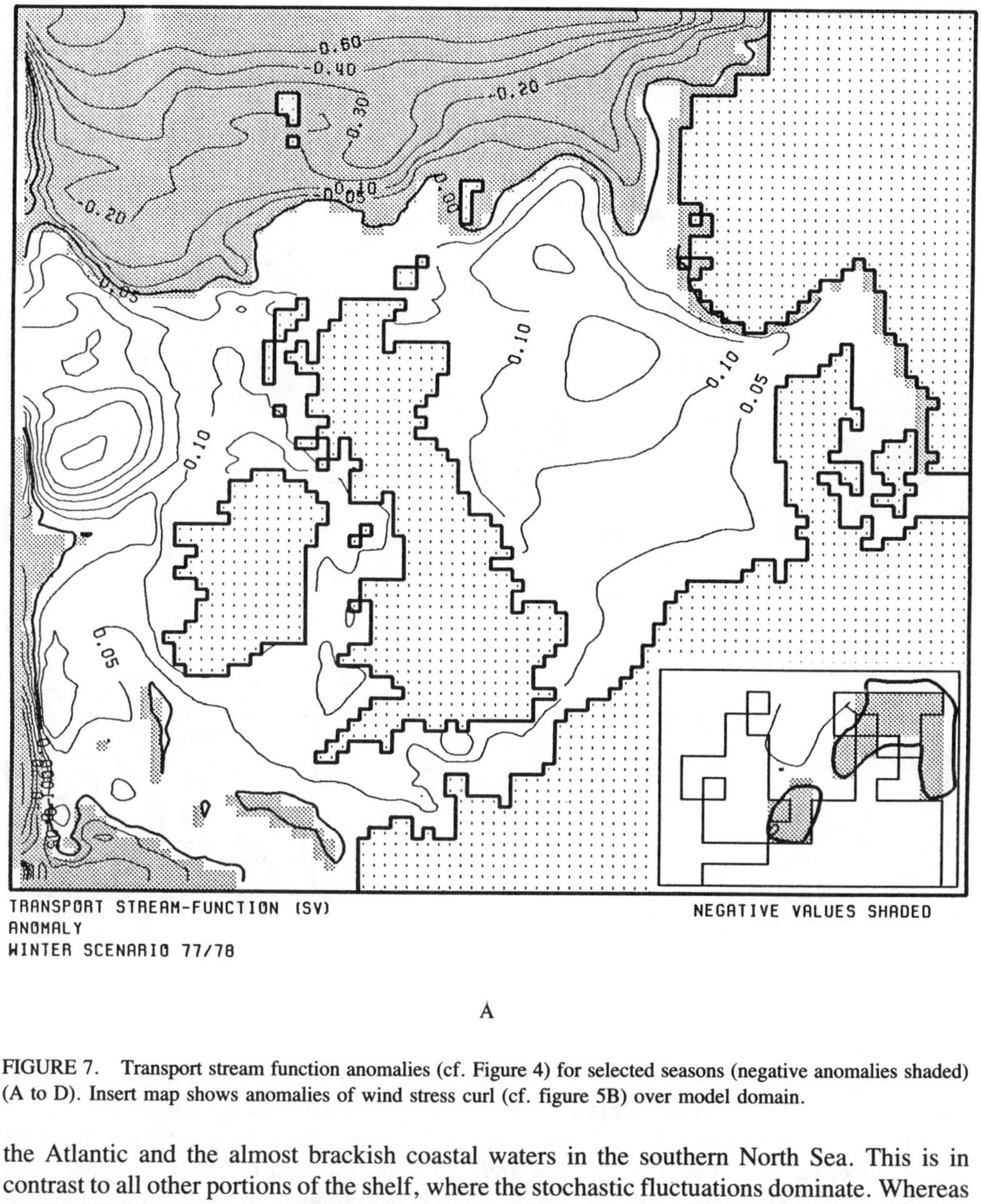


A

FIGURE 7. Transport stream function anomalies (cf. Figure 4) for selected seasons (negative anomalies shaded) (A to D). Insert map shows anomalies of wind stress curl (cf. figure 5B) over model domain.

the Atlantic and the almost brackish coastal waters in the southern North Sea. This is in contrast to all other portions of the shelf, where the stochastic fluctuations dominate. Whereas in general the peak flow rates are reached during winter when the atmospheric momentum input is at maximum in the recirculation cell, the maximum flow rates appear already in late summer or autumn (cf. also Figure 4). This model result is supported by observations.[14]

## C. SEASONAL VARIABILITY

For individual seasons the inter-annual variability of the circulation is visualized and quantified by means of streamfunction anomalies related to the climatological seasonal ensemble means shown in Figure 4. Some examples of the simulated seasonal anomalies of the circulation are given in Figure 7. They demonstrate the considerable spatial as well as temporal variability of the flow. The deviations from the mean are of similar magnitude as the mean itself.

Frequently, the anomaly patterns for the recirculation cell and for the flow in the southern North Sea differ considerably. Thus, the forcing atmosphere may induce a decoupling of the two flow systems.

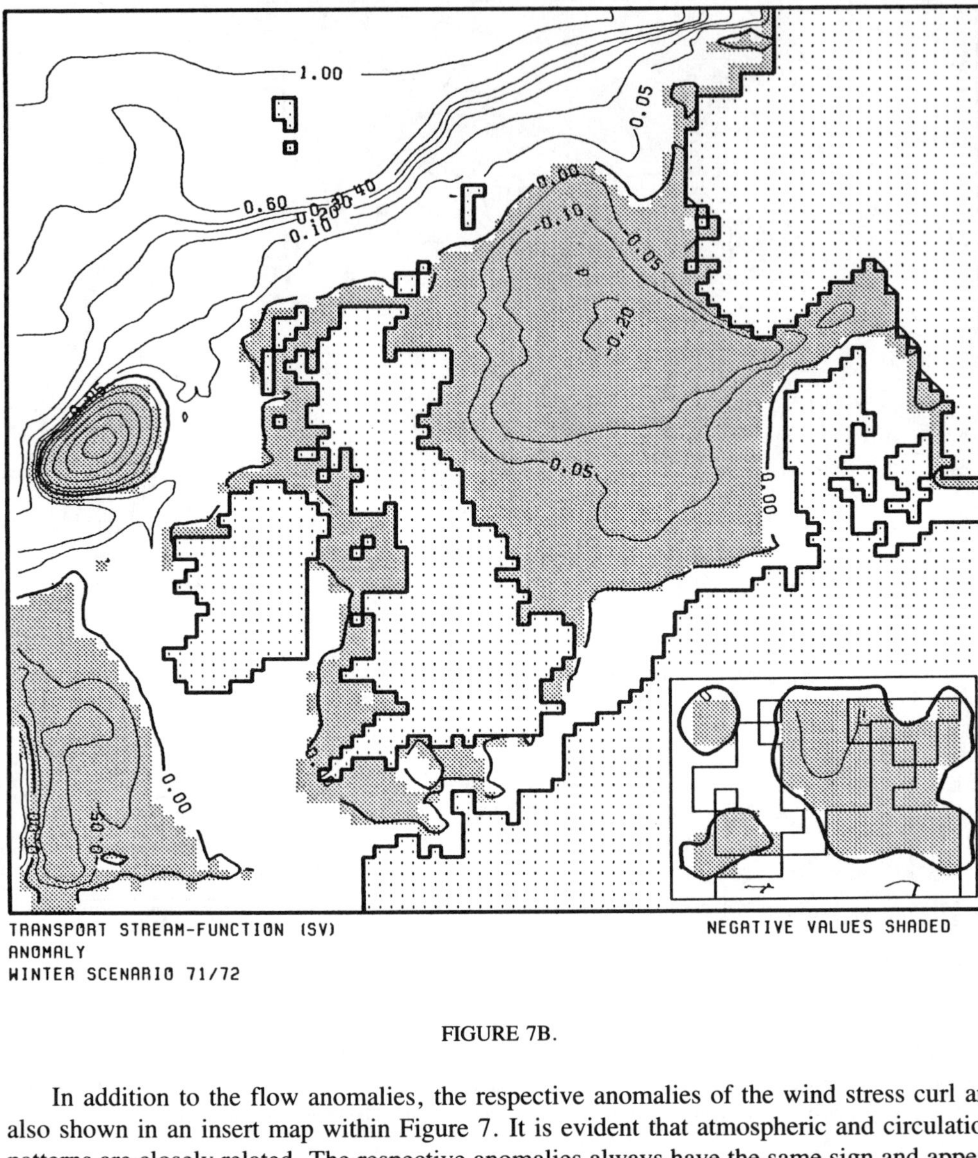


TRANSPORT STREAM-FUNCTION (SV)
ANOMALY
WINTER SCENARIO 71/72

NEGATIVE VALUES SHADED

FIGURE 7B.

In addition to the flow anomalies, the respective anomalies of the wind stress curl are also shown in an insert map within Figure 7. It is evident that atmospheric and circulation patterns are closely related. The respective anomalies always have the same sign and appear in the same regions. Thus, the very low frequent fluctuations of the circulation are rather sensitive to changes of the wind stress curl. A much less pronounced correlation to the other meteorological forcing components was found. This finding can be supported theoretically by a vorticity balance for a stationary case which shows that the large-scale depth mean flow is decisively determined by the curl of the wind and the bottom stress.

## D. VARIABILITY FOR SUBSYNOPTIC TIME SCALES

A much higher correlation of fluctuations of the circulation with wind stress and air pressure anomalies was found for the subsynoptic frequency band. Apart from the average meteorological situation depicted in Figure 5 which characterizes the usual transient westerly activity of the atmosphere, there are some other frequently but stochastically occurring and sometimes rather long-lasting large-scale weather situations which have a considerable influence on the circulation.[15]

Global negative anomalies of the circulation covering the major portions of the shelf

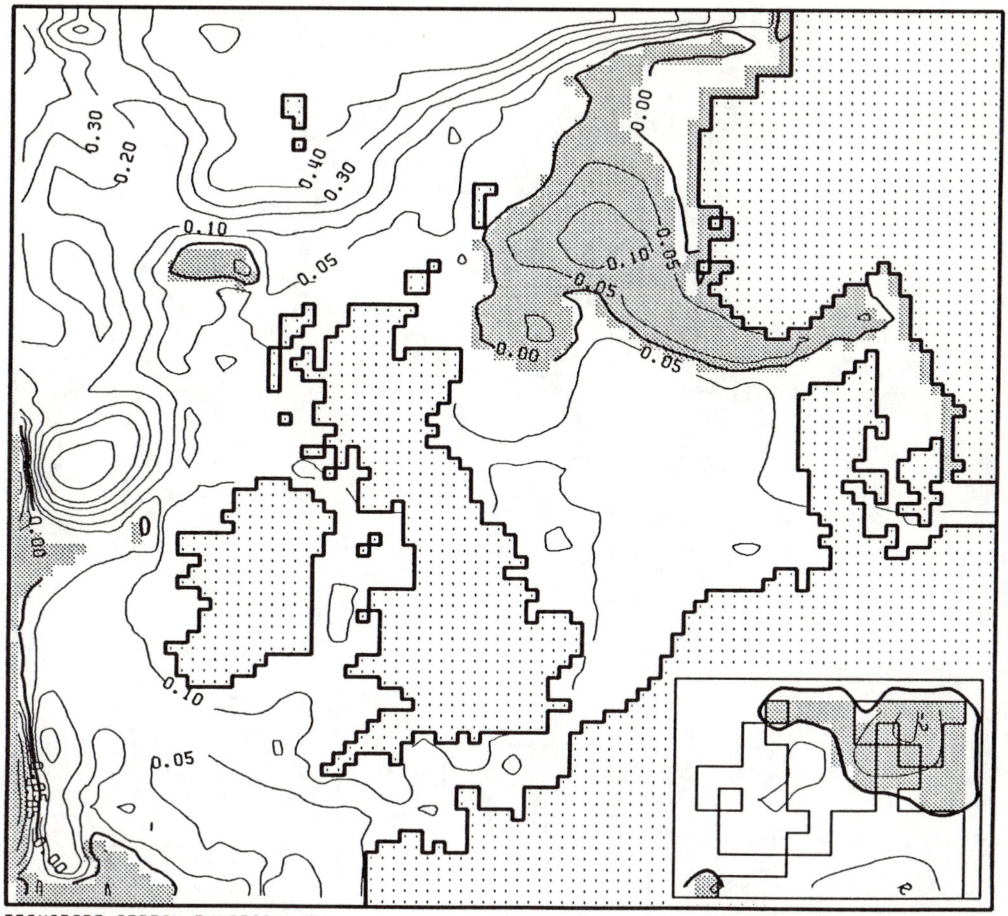


FIGURE 7C.

are induced by large-scale atmospheric-blocking situations which may last for a couple of days up to some months. They appear in conjunction with a positive air pressure anomaly induced by an intensification or by a westward shift of the continental high pressure system. The usually prevailing westerly winds are then replaced by winds of mainly easterly direction. For this anomalous wind forcing, the usually cyclonic circulation may come to a complete rest or it may even reverse entirely.

Another frequently observed weather pattern over northwest Europe are monsoon-like situations where the continental high pressure system is replaced by a persistent low and with the Azores high moving anomalously far north. This so-called European summer monsoon is accompanied by a lot of bad weather with prevailing and fairly strong northwesterly winds. Under this forcing the normally well-organized basin-wide flow within the North Sea, consisting of the northern recirculation and the southern cyclonic flow, breaks up into numerous mesoscale eddies. This situation induces a considerable large-scale lateral mixing of the water masses but a well-defined mean flow does not exist anymore.

The stochastic sequence of the above-described large-scale atmospheric patterns leads to the considerable variability of the circulation for subsynoptic time scales. A closer inspection of these fluctuations, depicted in Figure 6 by the shaded variance, is carried out by means of spectral estimates of transport time series for some of the sections given in the figure.

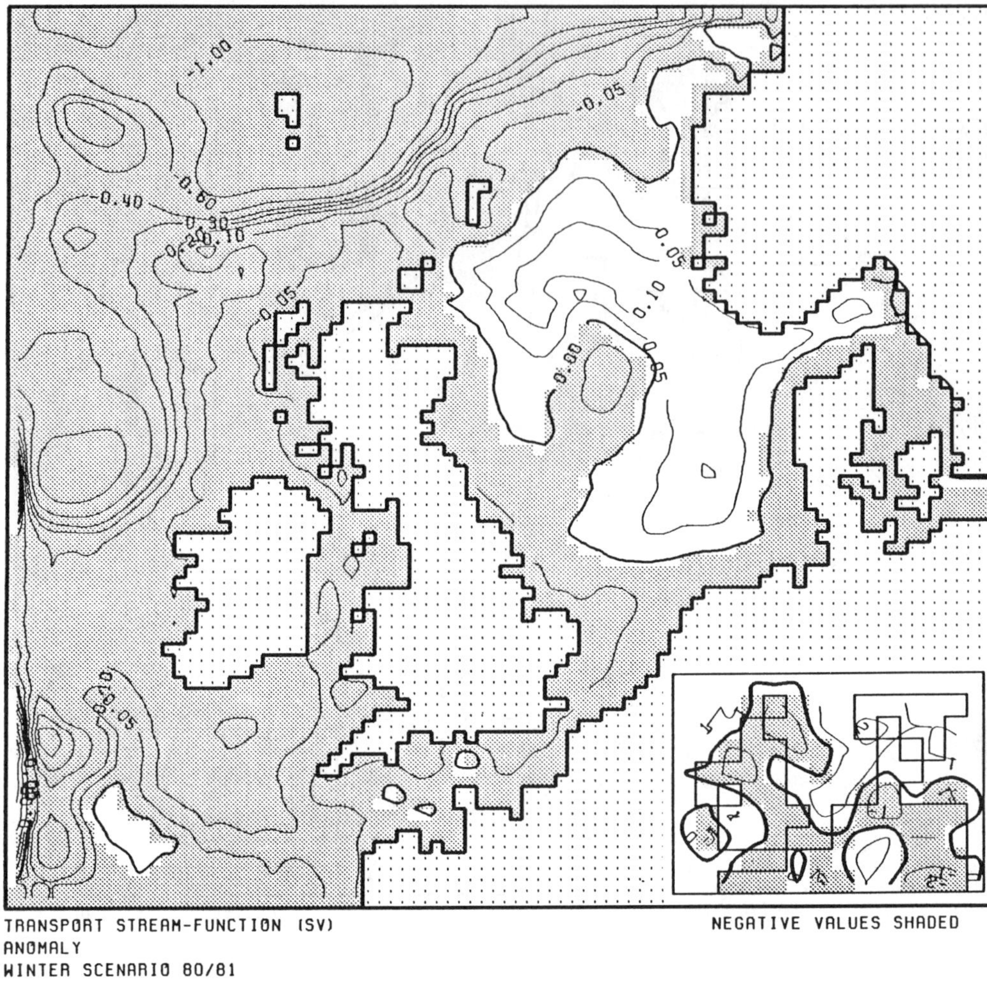


FIGURE 7D.

To provide an impression about the extremes of the variability, the time series will be confined to the winter period (defined here for October to April). The spectral estimates were obtained with the maximum entropy method, which gives a better resolution of the low-frequency part of the spectrum than a Fourier decomposition. The spectra shall only give a qualitative impression about energetic fluctuations of the flow, especially within the subsynoptic frequency band. Therefore, no confidence limits are shown.

For each section a series of spectral estimates for a number of individual winter scenarios is depicted in Figure 8. A comparison with the signals of the forcing atmosphere is given by Figure 9 where a similar series of spectra of the wind stress components for a point in the central North Sea is depicted. These spectra also resolve the synoptic frequency bank and thus allow a comparison of the energy levels of the subsynoptic with the synoptic fluctuations. Clearly the latter have considerably lower energy levels supporting the assumption, mentioned in Section III, that a single storm will not contribute significantly to the low-frequency flow.

For both the atmospheric and the transport spectra the respective ensemble mean spectrum is drawn at the top panel. The stochastic and energetic peaks which appear in the individual spectra are not present in the overall mean. This illustrates the drawback of diagnostic

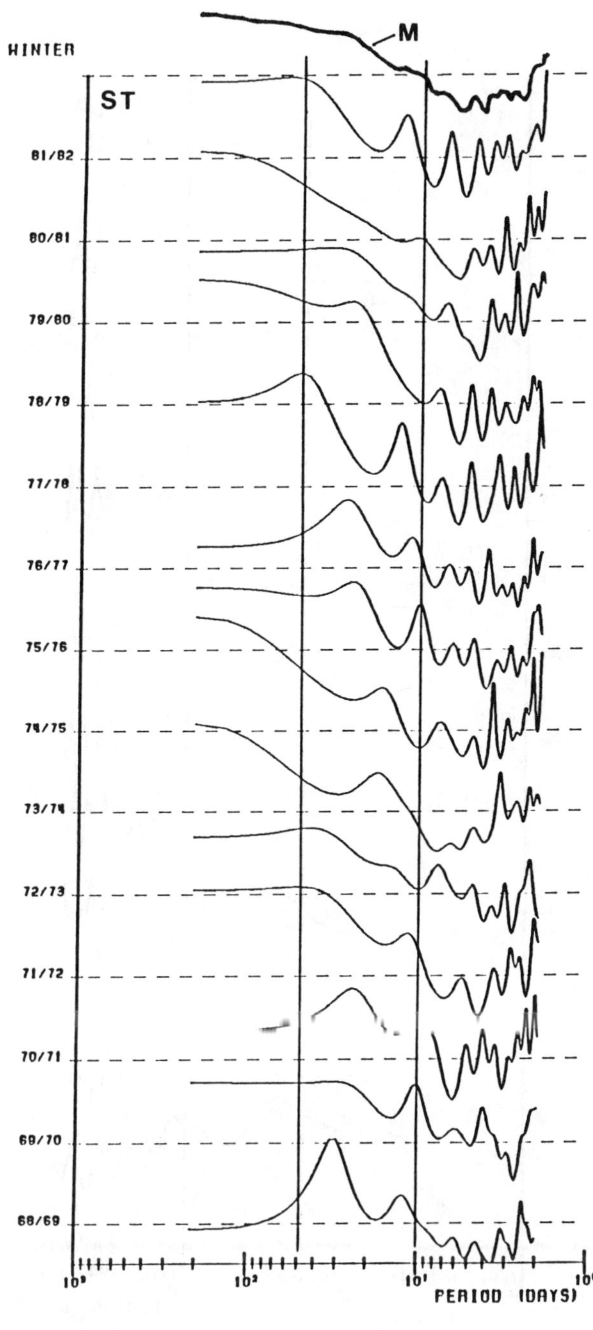


A

FIGURE 8. Spectral estimates of transport through selected sections in the northern North Sea for the winter periods (October to April) 1968/69 to 1981/82. Subsynoptic frequency band indicated by vertical lines. M: mean spectrum. (A) Shetland-Norwegian Trench (ST); (B) Norwegian Trench (NT); (C) Faroe-Shetland (FS). Locations of sections indicated in Figure 2.

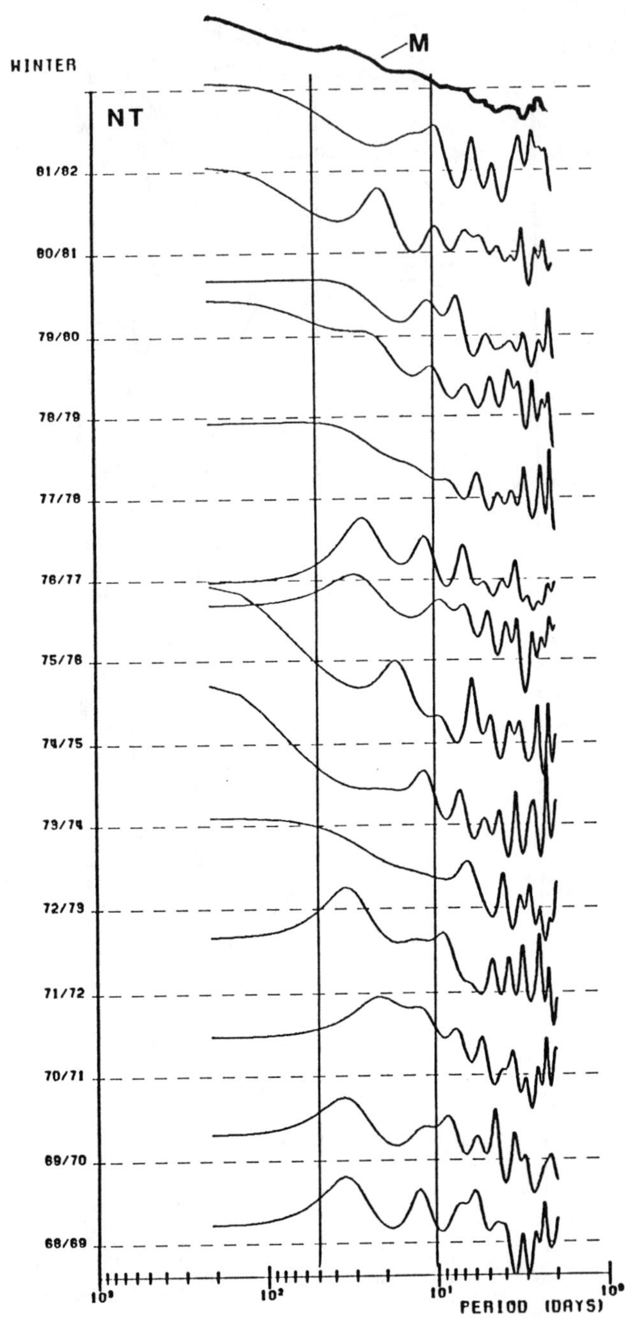


FIGURE 8B .

simulations which due to the time-averaging a priori omit the quantification of subsynoptic fluctuations. The result further demonstrates that well-defined averaging periods cannot be found due to the stochastic character of the forcing atmosphere.

In contrast to fluctuations with inter-annual or seasonal time scales, generally a high correlation between the wind stress and the circulation was found for the subsynoptic frequency band. However, once again a conspicuous exception is the Shetland-Norwegian Trench section where rather low energy values appear for near synoptic frequencies. Ob-

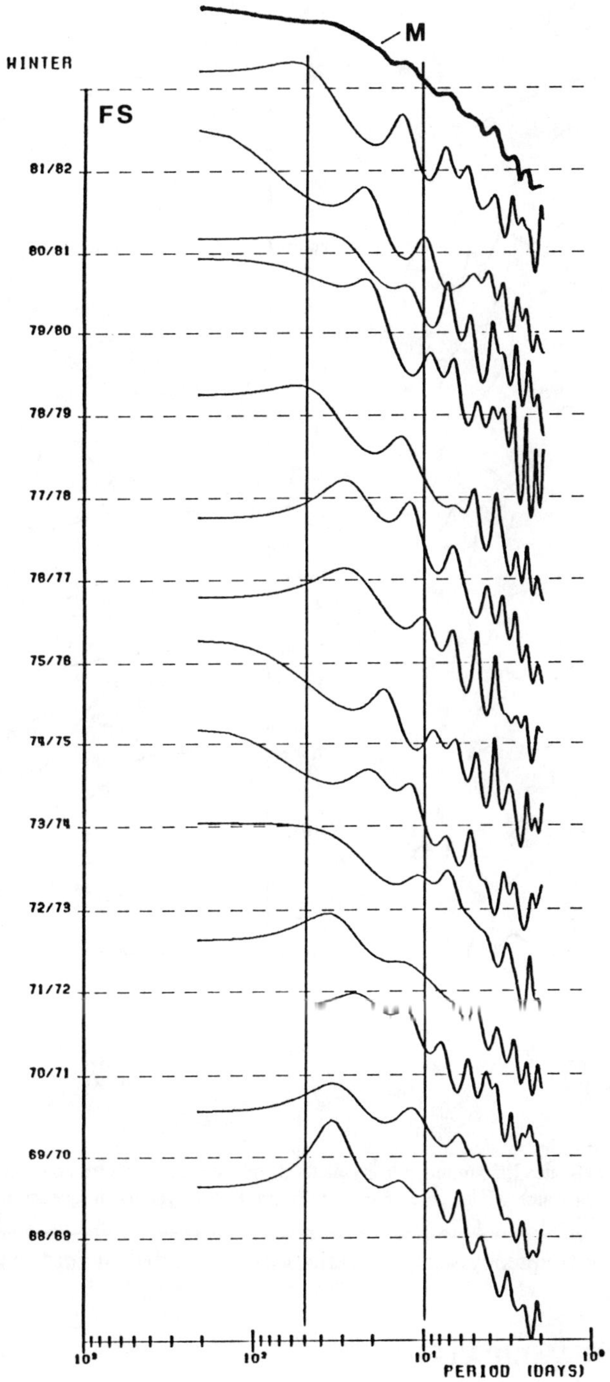


FIGURE 8C .

viously the recirculation cell preferably reacts to an atmospheric forcing with subsynoptic frequency.

The average spatial signal-to-noise ratio of the energy of the subsynoptic fluctuations vs. the energy of the mean flow is depicted in Figure 10. This display shows that for the entire shelf region the subsynoptic fluctuations clearly dominate. This result is different

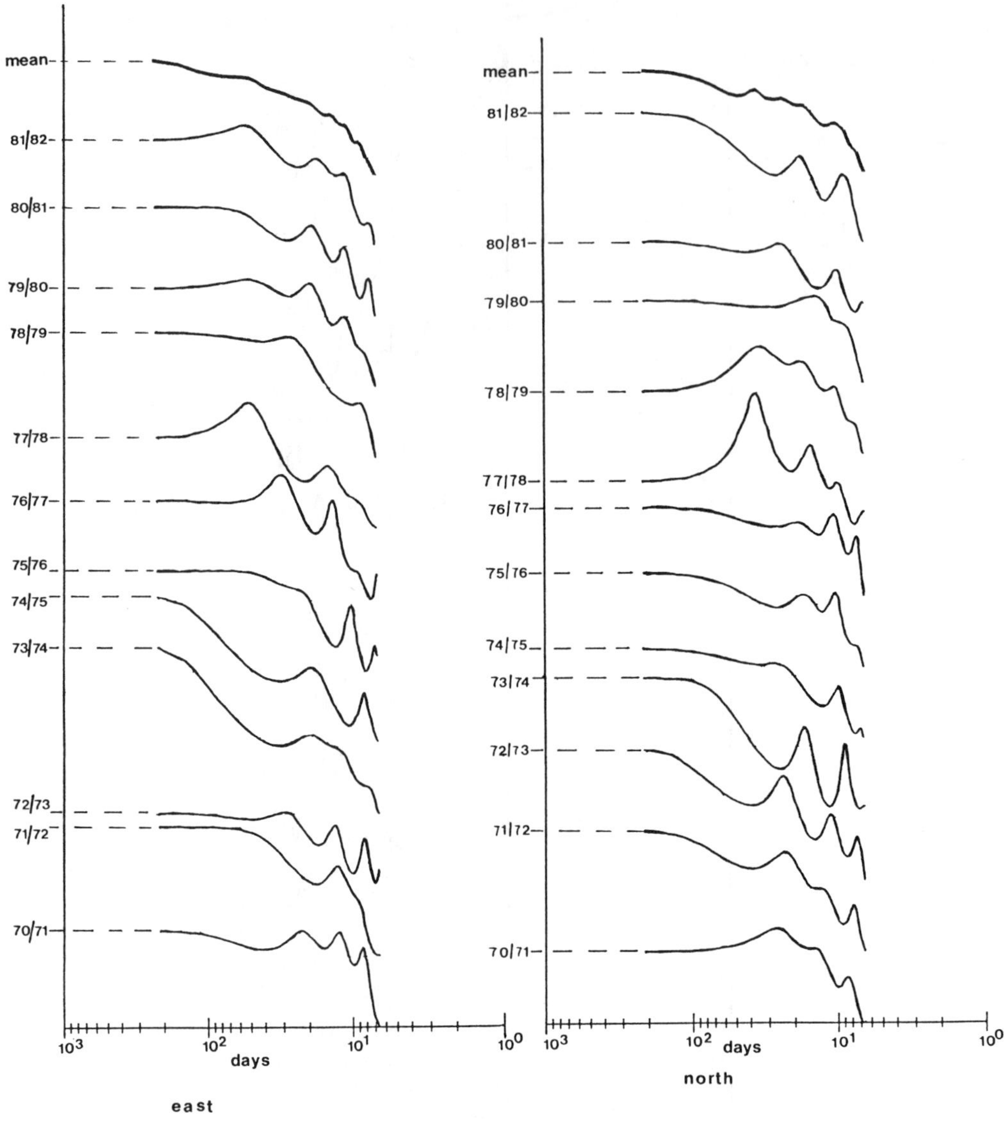


FIGURE 9. Spectral estimates (maximum entropy method) of the east and north components of the wind stress over the central North Sea, same winter ensemble as in Figure 8. Subsynoptic frequency bands indicated.

compared to the low-frequency seasonal variability where the fluctuations were of the same order as the mean.

## E. SAMPLE-TEST VALIDATION

A general and satisfactory validation of the model experiment will never be possible due to the miserable data situation described in Section II. It will rather have the character of a sample test, because only occasional data records of suitable length are available. Many particularly long current records for the northern part of the shelf region have been obtained in the course of field surveys for the oil industry. However, this data is usually classified and hence not available, which is another shelf sea conflict.

Within the International Council for the Exploration of the Sea (I.C.E.S.) estimates for the respective mean in- and outflow rates at the openings of the North Sea to the Atlantic

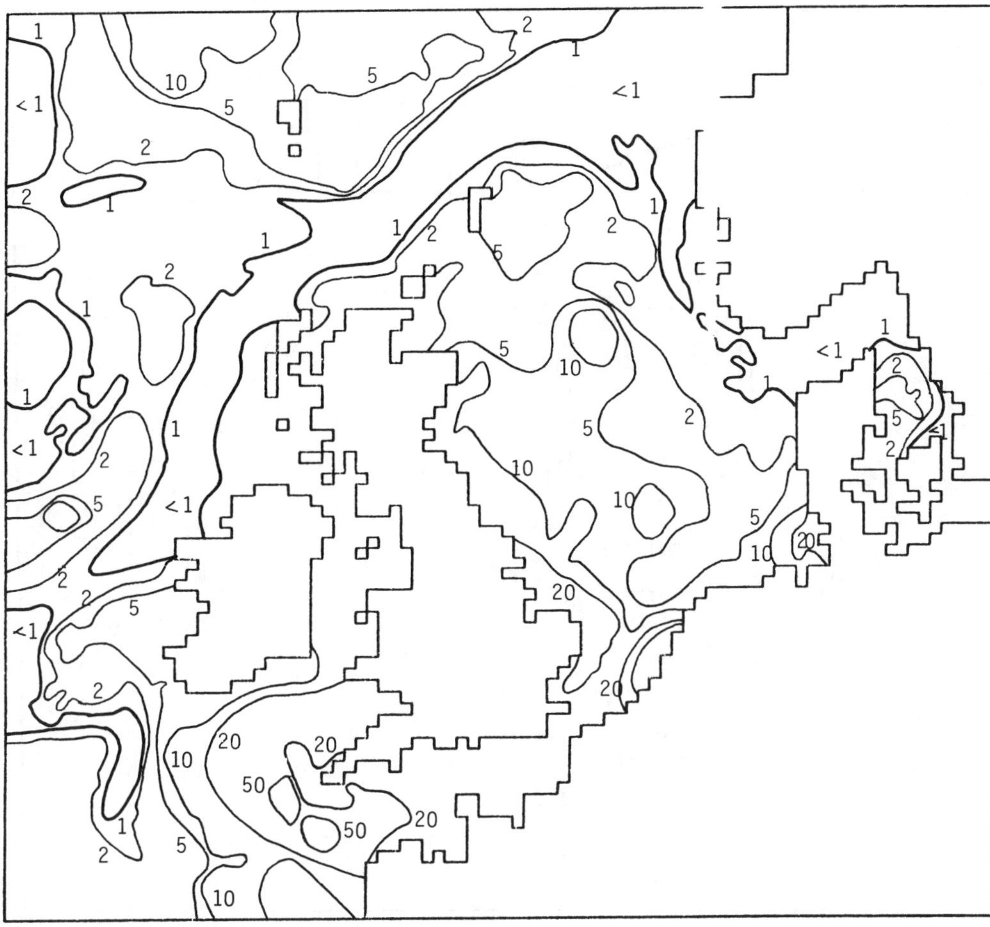

FIGURE 10. The ratio of the energy of the subsynoptic fluctuations of the flow vs. the energy of the mean flow. Average values for the winter period (October to April).

were compiled.[16] In Figure 11 these estimates are compared to the results of the experiment. In general a good agreement is found, bearing in mind that the observational estimates presumably have about the same error margins as the model results.

Within the past 10 years a number of current meter riggs were deployed across the Norwegian Trench where short sections with a comparatively low expenditure of apparatus allow estimation of the temporal changes of the main outflow of the North Sea. For a 2-month period from one of these experiments (NORSEX 1979) transport rates were compiled from the current meter observations which allow for a direct comparison with the results of the experiment.[17] This comparison, presented in Figure 12, shows that the model indeed reproduces a good deal of the observed variability which in particular holds for the low-frequency components of the record. However, deviations occur, especially for short-term events (order of 3 to 5 d). These fluctuations are likely caused by baroclinic eddies and meanders appearing frequently off the coast of Norway.[18] These small-scale features cannot be reproduced by the model because of its too-coarse grid and due to the fixed baroclinic fields.

A final example for the sample-test validation is provided by a spectral comparison of sea levels at coastal tide gauges. A typical result of the coherence and phase of observed and simulated sea-surface elevations for a 7-month winter period is depicted in Figure 13.

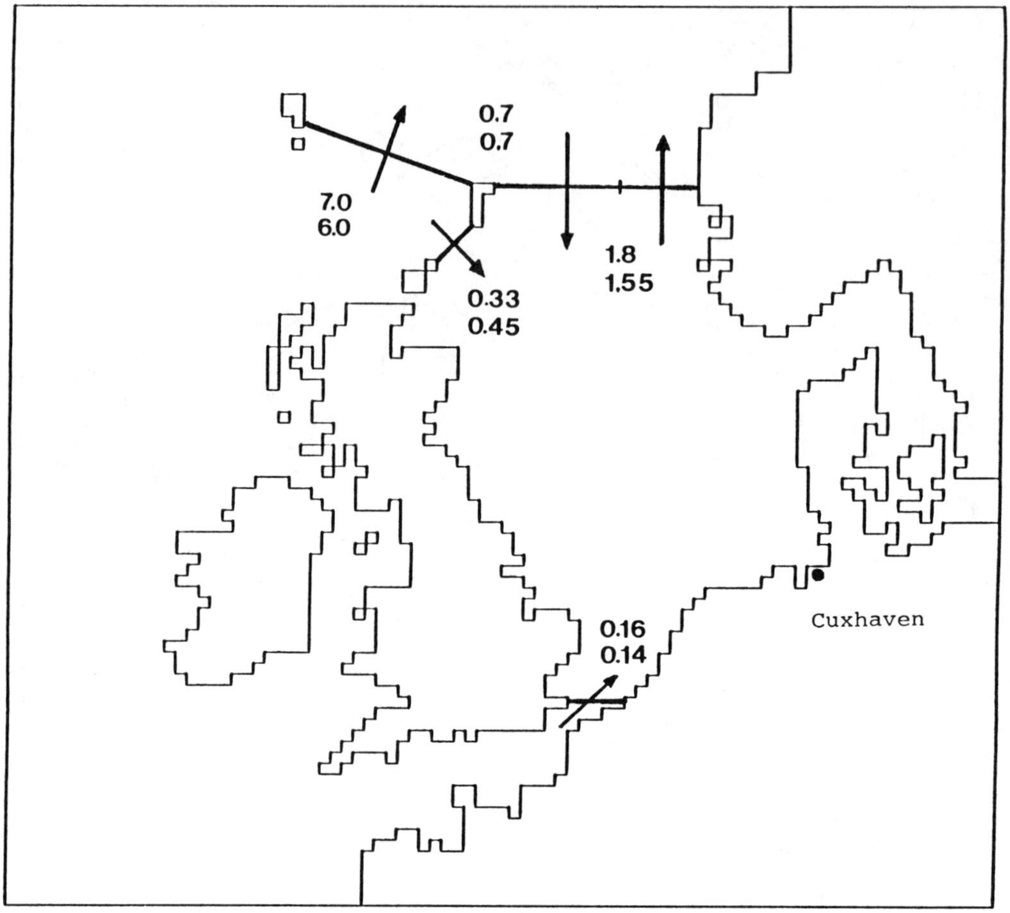


FIGURE 11. Mean transport estimates in Sverdrup units for selected sections as deduced from the model experiment and from observations. Lower number: model estimates.

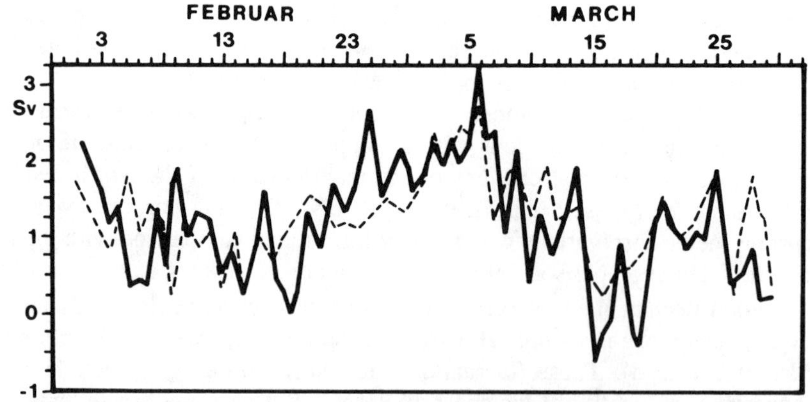


FIGURE 12. Comparison of observed (NORSEX 79) and simulated (dotted) transport rates (Sverdrup units) through a section in the Norwegian Trench for February and March 1979. Observed values were redrawn from Reference 17.

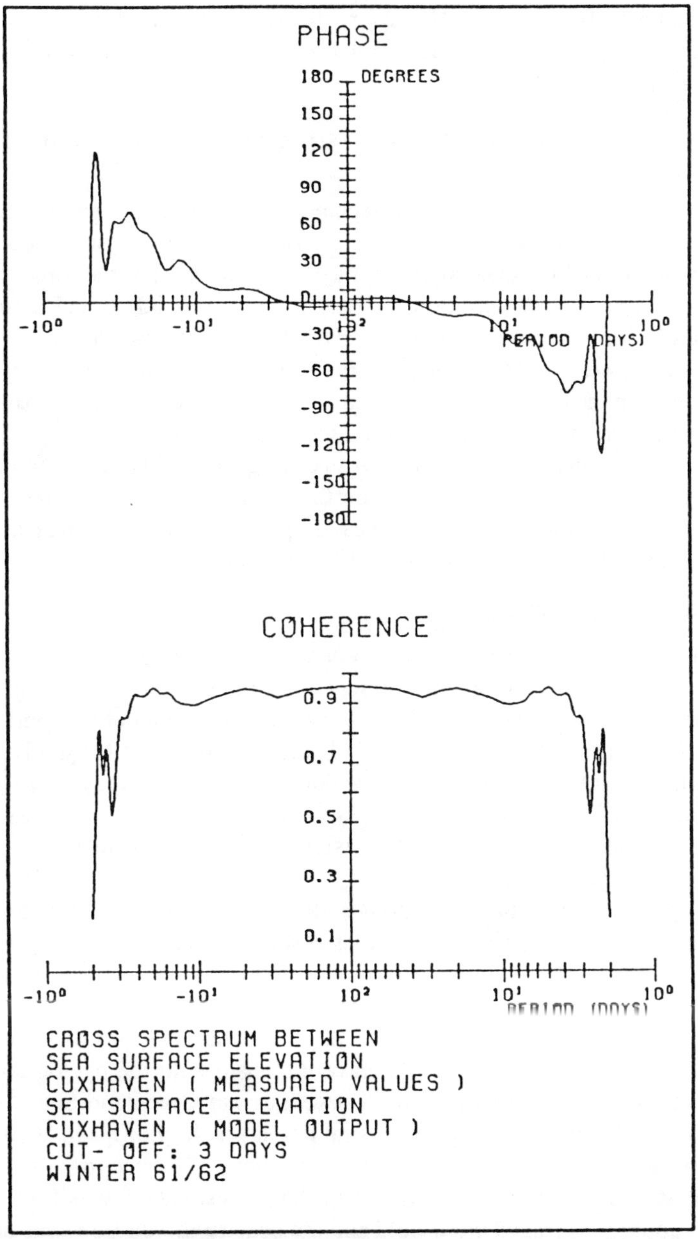


FIGURE 13. Coherence and phase (maximum entropy method) of observed vs. simulated sea-surface elevations at tide gauge Cuxhaven (cf. Figure 2) for a 7-month winter period (1961/62). Data: Deutsches Hydrographisches Institut Hamburg.

A rather high coherence is obtained for the subsynoptic frequency band. For near-synoptic frequencies a somewhat lower coherence is obtained. This might be explained by the lacking spatial resolution of the atmospheric forcing data (150 km). Usually the observational network of the weather services does not resolve small-scale atmospheric fronts and storms within the synoptic frequency band as good as the large-scale low-frequency weather systems. On the other hand, the tides which are badly resolved by the model, and tidal interactions with the wind-induced sea level changes might be another explanation for the deviations found for the high-frequency fluctuations. However, the effects caused by synoptic disturbances are covered reasonably well by episode simulations. They are of minor importance for this investigation.

# V. APPLICATIONS TO RELATED PHENOMENA

It is near at hand to use the circulation scenarios of the model experiment for dispersion simulations in order to gain insight into the fate of particulate or dissolved matter within the sea. In view of the scarce data available for a direct validation of the flow, the comparison of observed and simulated concentration patterns allows for an additional indirect validation of the flow field as long as passive (conservative) tracers are considered.

In the following sections examples of Eulerian and Lagrangian transport-dispersion simulations, based upon the model experiment, are presented. The first example concerns observed passive tracers which were used for a validation of the circulation and the transport model. In the second study active (vertically migrating) tracers are considered which make use of the three dimensionality of the flow on the northwest European shelf. Finally some estimates of the water quality of the North Sea are presented which emphasize the temporal variability of dissolved substances discharged by riverine point sources.

## A. DISPERSION OF PASSIVE TRACERS

The well-documented observations of the concentrations of nuclear wastes, e.g., cesium-137, discharged into the sea by nuclear-reprocessing plants are an excellent (and, in fact, the only presently available) data base for comparisons with model simulations.[19-21] There are three plants within the North-European shelf region: Sellafield (formerly Windscale) in the Irish Sea which discharges the largest amount of wastes, Dunraey at the north coast of Scotland, and a French plant at Cap de la Hague in the Channel. The radioactive wastes released by these plants are almost-perfect passive tracers and, as such, a good opportunity for comparisons with model simulations.

The discharges started in the early sixties. Regular surveys (once or twice a year) of the concentrations of the released radionuclides reveal that the radioactivity stemming from European waters can be traced as far as the Greenland Sea.[22] On the northwest European shelf the radionuclides either travel through the Dover Strait (Channel discharge) or around the north coast of Scotland (Irish Sea and Scottish discharges) into the North Sea. There they merge in the southern North Sea and leave the basin with the cyclonic mean flow via the Norwegian coastal current towards the Norwegian and Greenland Sea.

A previous two-dimensional dispersion simulation for the European shelf region reproduced the basic gross features of the observed cesium-137 concentration patterns.[23] The vertically integrated linear circulation model in that investigation was forced by a homogeneous and 3-monthly-averaged wind field and the density-driven flow was not considered. The transport model was based upon an Eulerian finite difference scheme accounting for the effects of numerical diffusion. Similarily, the three-dimensional dispersion model, some results of which to be presented here, was also based upon an Eulerian finite difference (vector-upstream) scheme, which incorporated an antidiffusive flux-correction algorithm.[24] The dispersion model is supplied by the daily three-dimensional flow fields of the model experiment whereby the diffusion, which exceeded the estimated numerical diffusion, was derived from the daily variance of the flow field.[25]

Originally it was intended to include all relevant European sources of radioactivity in the model domain, in particular the Sellafield plant, which increased the levels of emission considerably in the late 1960s. However, this idea had to be abandoned. It turned out that the simulated advection of radioactivity in the West-British waters on its way from the Irish Sea, around the northern tip of Scotland, towards the North Sea was considerably retarded and too diffusive. As a result the concentrations in the North Sea were too low. Further, in disagreement with observations, an unrealistic diffusive leakage of concentrations from the Irish towards the Celtic Sea was simulated. The latter effect did not occur in the study of Prandle, cited above, but likewise his concentrations were in general too diffusive in the North Sea. His linear vertically integrated circulation model had a comparable (in fact even slightly coarser) grid resolution than the three-dimensional circulation model.

The comparison of both circulation/dispersion models yields the puzzling result that a simple and coarser resolving two-dimensional model forced by a homogeneous and only slowly varying wind field gives a better estimate of the evolution of the cesium-137 concentrations than the dispersion model based upon a more complex three-dimensional nonlinear circulation model forced by a fluctuating and inhomogeneous wind field.

The failure of the transport, respective circulation model for the West British waters is explained by the diffusive effect of a time-varying flow field in a topographically complex and highly dynamical region like the waters at the British west coast. The rectification of the flow forced by a slowly varying and homogeneous wind field is much more pronounced than in the case when a realistic temporal and spatial variability of the wind field is prescribed. In that case small-scale disturbances of the flow are generated by the wind forcing in conjunction with nonlinearity and topography; the accuracy of the model is then crucially determined by its spatial resolution.

The synoptic and subsynoptic fluctuations of the wind field are the primary source for the disturbances. The homogeneity and the applied 3-monthly time averaging of the wind field prescribed in Prandle's study effectively suppressed or even eliminated these fluctuations. The result is a strong rectification of the wind forcing which is reflected by the resulting flow. The modified signal-to-noise ratio of the forcing and of the flow causes a relative enhancement of the rectification. Nevertheless, in reality, fluctuations do occur and they are likely to dominate as can be deduced from the previous section.

Valuable information about erroneous features of the simulated circulation and about their reason was gained from the comparison of observed and simulated concentrations. This demonstrates the potential of dispersion simulations with respect to the indirect validation respective falsification of the simulated flow.

The above-described and explained failure of the circulation model led to the conclusion that the resolution of the model should be chosen in a way that it matches with the spatial scales of the flow induced by the combined effect of topography and a time-varying wind field. Obviously this requirement does not hold for the spatial resolution of the West-British waters by the model. However, as will be shown below, it is still valid for many other regions of the model domain where the topography is less complex (e.g., the open North Sea).

The original intention to include all sources for radioactivity in the model domain had to be discarded because of the described model failure. Consequently, the dispersion model was applied to a reduced domain which essentially encompassed the North Sea region. Therefore, at the northern entrances of the North Sea observed cesium-137 concentrations were prescribed accounting for the effects of the discharges of Sellafield and Dunreay. The model was run for a period of 10 years, starting in the late 1960s when the discharges were drastically increased.

An example of the simulated spatial distribution of cesium-137 caused by the combined effect of the three sources is depicted in Figure 14. An area with high concentration values (above 300 Bq/m**2) at the east coast of Britain was accurately reproduced by the model. In general, the patterns of the cesium-137 concentrations were in good agreement with observations.[26]

A comparison of temporal variations of the concentrations is depicted in Figure 15. The coarse space-time density of the observations (about twice a year with station increments of

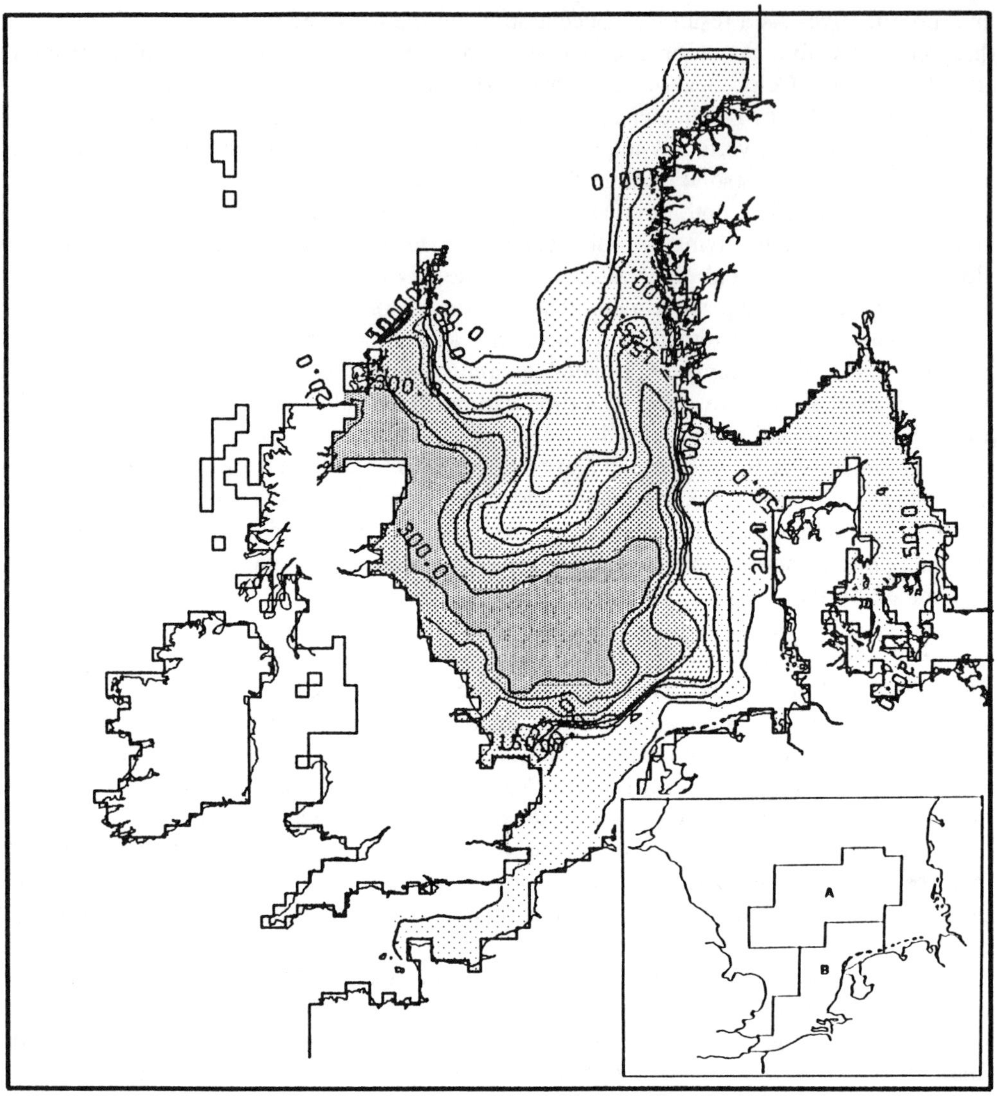


FIGURE 14. Spatial distribution of simulated concentrations (Bq/m**3) of cesium-137 in the North sea for February 1978. Insert Map: locations of subregions A and B, cf. Figure 15.

the order of 50 to 100 km) did not allow compilation of adequate time series for isolated (single) positions. Therefore, time series are shown where the data was collected for certain regions (e.g., the I.C.E.S.-boxes), the dimensions of which are approximately 200 by 300 km. For each box the simulated time series are presented for the box mean and for the extremes found within the box.[26] The observational data points obviously fit well into the concentration range hindcasted by the model. This also applies for the general trend within the 10-year period, bearing in mind that the variance of the concentrations in this display is caused by both temporal and spatial variations of the concentrations.

In general, this comparison was rather satisfying. Together with the findings of section IV.E there are reasons for the assumption that the estimates of the model experiment reproduce the variability of the circulation except for those regions where the topography is inadequately resolved.

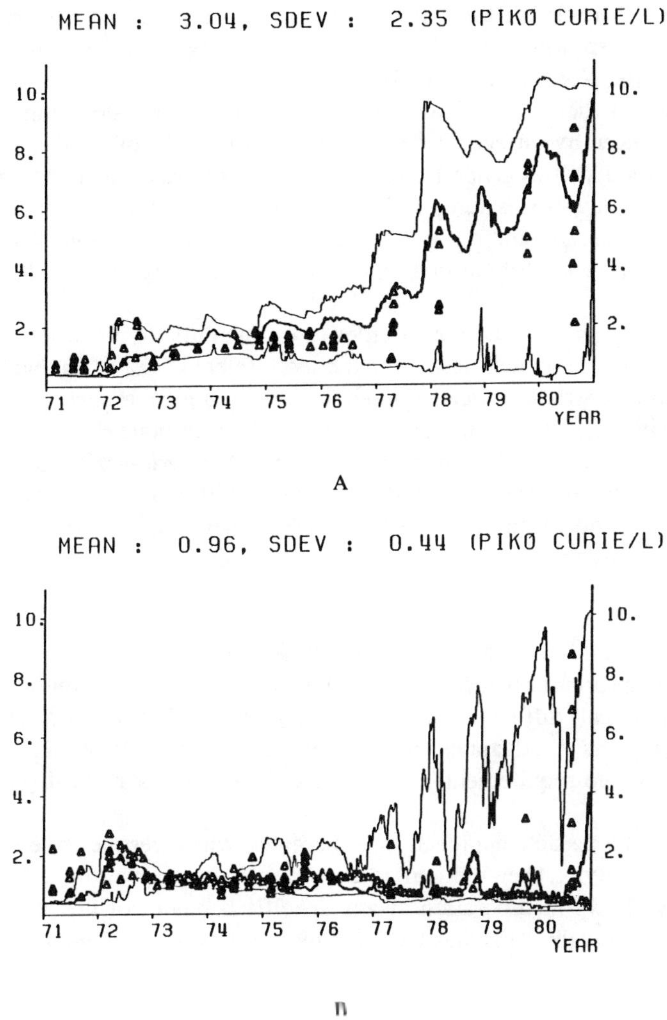


FIGURE 15. Time series of cesium-137 concentrations (pci/l) for two subregions (cf. Figure 14) in the North Sea; (A) Central North Sea, (B) Southern North Sea. The three curves represent the simulated regional mean concentrations (thick line) and the extremes (thin lines), respectively; observed values are indicated by triangles. Data: Deutsches Hydrographisches Institut Hamburg.

## B. CIRCULATION AND FISH RECRUITMENT

In recent years, and in particular within the I.C.E.S., attention has been focused on the influence of the circulation on the recruitment of fish, which in their early life stages (eggs and larvae) drift almost passively with the currents.[27] For certain species this critical time span may last for almost half a year. In this context it is interesting to note that during evolution, fish species obviously have adopted strategies to make use of the circulation in order to retain their genetic information within a certain shelf-sea area. There is reason to believe that similar strategies are also applied by marine organisms on lower trophic levels, like macro zoo-plankton (e.g., copepodes).

It is crucial for the survival of young fish, besides predation and other influences, to arrive at the end of their nearly passively drifting stadium at a region where enough food is available. The coastal and wadden seawaters along the eastern, continental coast of the

North Sea are a rich reservoir for nutrients and as such a preferred nursery area for young fish. The preferred spawning areas of some fish species, however, are the western and northern parts of the North Sea, about 1000 km apart.[27]

The routing and, hence, survival strategies of fish incorporate among other processes the turbulence induced by surface waves and vertical variations of the flow, and/or upstream spawning in relation to the general circulation. It is supposed that the variability of the physical environment (temperature, salinity, circulation, turbulence, light, etc.) besides biological fluctuations and anthropogenic influences (food, predation, diseases, pollution, over-fishing, etc.) may contribute in a critical way to the considerable observed variability of fish stock.

An investigation was carried out in order to estimate the hypothesized influence of the variability of the circulation with respect to the recruitment of fish.[28] In that study the results of the circulation experiment were applied in a Lagrangian trajectory (particle) model. Diffusion was simulated by means of a Monte-Carlo technique.[29,30] The random particle excursions were parameterized from the (deterministic) subgrid-scale variance of the three-dimensional flow computed with the daily means of the flow fields (cf. Section III.D). The space- and time-dependent diffusion coefficient A is then defined by:

$$A = VV / (2\ dt)$$

where VV is the (vector-) variance and dt the time step.

The Lagrangian approach incorporating the Monte-Carlo technique effectively avoids the artificial (numerical) diffusion which is a major drawback of Eulerian (finite difference) methods. However, its disadvantage is the need of huge particle numbers (and thus huge computing and storage requirements) to ensure accuracy and statistical significance of the simulation.

Preference to the particle approach must be given whenever the tracers to be simulated have an "Eigenactivity". Then the Langrangian clearly is superior to the Eulerian approach because it easily allows incorporation of any specific behavior of the particles besides the Monte-Carlo random walk. This was the main reason why a Lagrangian model was applied in the recruitment study.

An example for the crucial depth dependence of vertically migrating herring larvae is depicted in Figure 16, where for a 6-month scenario two tracer (larvae) ensembles (A and B) with different eigenbehavior were released at a specific spawning ground (Orkney). For more details reference is made to the study of Bartsch.[31] In case A the larvae stayed in the near-surface region (about the Ekman-layer), migrating vertically in dependence of the daily light cycle. In the case B it was assumed that they prefer to stay in the near-bottom region. Obviously, their final fate at the end of the larval stadium depends on their preference for certain depth ranges in relation to the vertical structure of the flow. It can be deduced from observations that only the case A where the larvae tracers have covered a distance of 600 to 1000 km in about 6 months realistically describes the behavior of the herring species considered and characterized by their respective spawning area.

However, it is known that other species at the British east coast (e.g., the Buchan herring) stay close by their spawning area. Their strategy seemingly gives preference to the near-bottom flow, the prevailing direction of which being onshore. The model reproduced this phenomenon and helped to understand the respective strategy.

Investigations of this kind provide new insight into the rather complex recruitment procedure and they contributed to the definition of new observational strategies. Presently, field studies are being carried out which focus on the vertical abundance of larvae and on the current profile (i.e., the international Autumn Circulation Experiment, ACE, in the northern North Sea in 1987/88). A part of the experiment strategy is the combination of

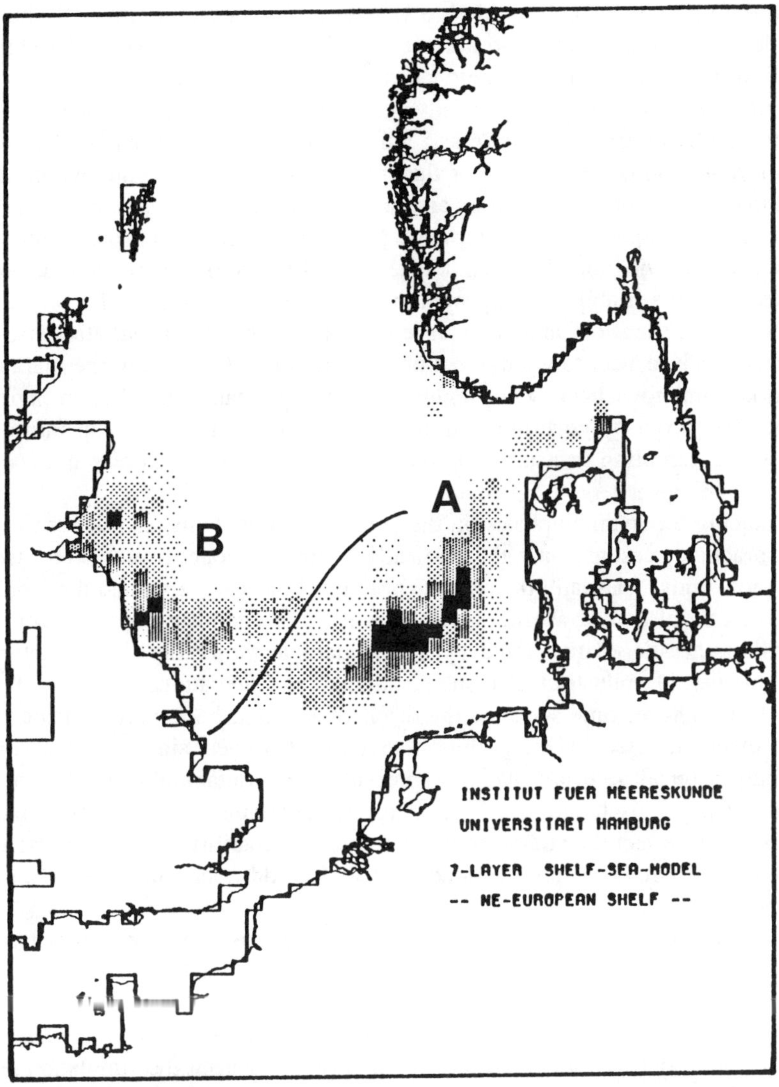


FIGURE 16. Simulated scenario of two active tracer ensembles (resembling fish larvae from Orkney spawning ground); pattern A: light-dependent vertical migration of larvae in upper 25 m of water column. Pattern B: vertical migration of larvae in near-bottom region (range 25 m above bottom). Both patterns drawn for the end of a 6-month dispersion period. Relative larvae concentrations given by shading contours.

field observations and model simulations in order to gain a deeper insight into the complexity of hydro-biological processes. Therefore, the experiment will be accompanied by simulation runs of the circulation and the dispersion model described here.

The development of environmental models where biological and physical processes are equally considered is likely to be a long-lasting future task in view of the complexity of the marine environment.

## C. TEMPORAL VARIABILITY OF DISSOLVED SUBSTANCES

In an interdisciplinary national research program devoted to the fate of pollutants within the North Sea, the results of the model experiment have been used to obtain estimates of the dispersion of riverine pollutants discharges. Previous investigations used a mean flow field within the transport models and hence yielded mean concentrations.[32,33] The results of the experiment, on the contrary, also allow estimation of the influence of the variability of the circulation on the variability of the concentrations induced by riverine pollution.

As pointed out in Section IV, the usually cyclonic circulation may be enhanced, retarded, or even reversed by certain large-scale and long-lasting weather events.[11,15] These changes may cause a reduction or an increase of the concentrations of pollutants within the sea. It is assumed that, in particular, stagnation and reversal events will cause concentration peaks, because it may then happen that a water body passes a contamination source more than just once, each time receiving an additional load. For the fluctuations of concentractions, therefore, an even higher variability is expected than for the circulation itself.

The magnitude, duration, and abundance of peak values of concentrations deserve particular concern with respect to a judgment of the water-quality status of the marine system. These are almost unknown because the regular and internationally agreed-upon environmental monitoring, carried out by governmental agencies, does not allow their quantification due to an inadequate sampling (a usual sampling frequency of these surveys is once or twice per year).

While placing the main emphasis on the temporal fluctuations of dissolved substances, mainly for practical reasons, only the vertically averaged model results were used in this third dispersion study.[34] Usually the riverine pollutants discharges around the North Sea are published in terms of annual averages, although there are good reasons to assume that the effluents will rather be characterized by a considerable variability due to anthropogenic (temporal changes of industrial effluents) and natural fluctuations (seasonal fresh-water runoff, etc.). Therefore, time series of the discharges would be a more realistic and, thus, appropriate means to describe the pollution sources. However, since adequate time series of the discharges for all relevant rivers are presently not available the simulation must rely on constant (average) discharge data.[35] As a result, the simulated variability of the concentrations will solely reflect the atmospherically induced variability of the flow itself since no other (biological or chemical) processes accounting for additional fluctuations were considered.

Again a Lagrangian transport model was used but this time the parameterization of (subgrid-scale) turbulence was not taken into account in order to allow estimation of the diffusive effect of the variable large-scale flow.[34] For each river contributing significantly to the pollution of the North Sea a separate dispersion simulation was conducted for the same period (1969 to 1982) as in the model experiment. During the simulation period one (water-) particle per day was released at the respective river mouth and pursued through time and space until it had left the model domain. Concentrations were computed subsequently from the tracer ensembles by "loading" the particles with an amount of dissolved material scaled to the emission rate of the respective river source. This approach has the advantage that it offers an interesting variety of concepts for a statistical analysis:

1. It allows estimation of the dispersion of a single source but also the superposition of several sources.
2. The scaling of the pollution load can be modified later to simulate temporal discharges once better data is available.
3. The Lagrangian particle ensembles allow an estimate of the probability of stay for different particle "ages" (time span after release).
4. Statistics for the transit time of particles for specified regions can be computed (e.g., duration of stay within a region).
5. Time series of particle densities (and thus concentrations can be compiled at any position within the model domain.
6. The contribution of each individual source to the time series of a superposition of several sources can be determined.

7. An overall temporal mean particle density (or concentration) and its variance can be computed.
8. The particle ensembles allow quantification of the (regional) diffusivity of the variable flow field.

A Eulerian (differential) approach is less versatile and would not be adequate for some of the analysis concepts outlined above. The analysis of the particle dispersion provides further insight into the temporal and spatial variability of the circulation since the tracers were assumed to be passive. Nevertheless, an inspection of the results also gives useful information in regard to environmental pollution since the circulation is the most important transport medium. However, it should be borne in mind that no biological or chemical processes were considered. It is likely that these will account for a variability which is similar as in the figures depicted below.

Only some few examples of the tracer study can be presented here. The overall mean of the riverine pollution load for the North Sea and its standard deviation computed over the entire simulation period is shown in Figure 17. The estimate of the mean load agrees well with other studies.[32,33] The concentrations in Figure 17 represent relative values since a gross sum of contaminents instead of a specific substance was assumed; the values are relative in a sense that the total output of all rivers considered was set to 100%. The picture would not change drastically if the values for specific substances were prescribed. An exponential scaling was used for the shading of the contour intervals.

It is obvious that the fluctuations of the concentrations, depicted by the standard deviation, considerably exceed the mean values everywhere. The continental coastal waters receive the highest loads for both mean and fluctuations due to the discharges of the rivers Rhine, Elbe, and Weser.

For two regions in the southern North Sea time series of concentrations averaged over the region are depicted in Figure 18. The stochastic fluctuations of the concentrations reveal a great number of peak events. Their durations are of the order of several days up to some weeks. Certainly, in regard to environmental risk and/or hazard assessments, these peaks deserve special attention. The likely existence of these fluctuations, which up to now have not been observed due to inadequate sampling, might well be one reason for controversial opinions about the water quality status of the North Sea. It must be assumed that the findings of the present environmental monitoring are heavily biased by the stochastic fluctuations depicted above. A likely consequence is contradicting interpretations.

For the radioactivity released at Cap de la Hague in France a particle dispersion pattern is shown in Figure 19. It allows for a comparison with an investigation where a Lagrangian Monte-Carlo dispersion model was applied to a constant annual mean flow field for the same source.[36] A very similar pattern for the particle distribution emerged. Obviously, the prescribed diffusion coefficient of 25 m*m/s, which was tuned to match the simulated concentrations with observations, exactly accounted for the diffusive effect of the fluctuating circulation depicted by Figure 19. This conclusion is drawn from the fact that the particle cloud in the figure was computed without any explicitly prescribed diffusion. It can further be argued that the diffusive effect of subgrid scale processes (tides, synoptic fluctuations, turbulence, etc.), not accounted for in the above result, is of minor importance for the large-scale dispersion of matter within the sea. Obviously, in that case the variability of the circulation dominates.

## VI. FINAL REMARKS

A model experiment was presented which focused on the estimation of the variability of the circulation on the northwest European shelf for time scales below and above the

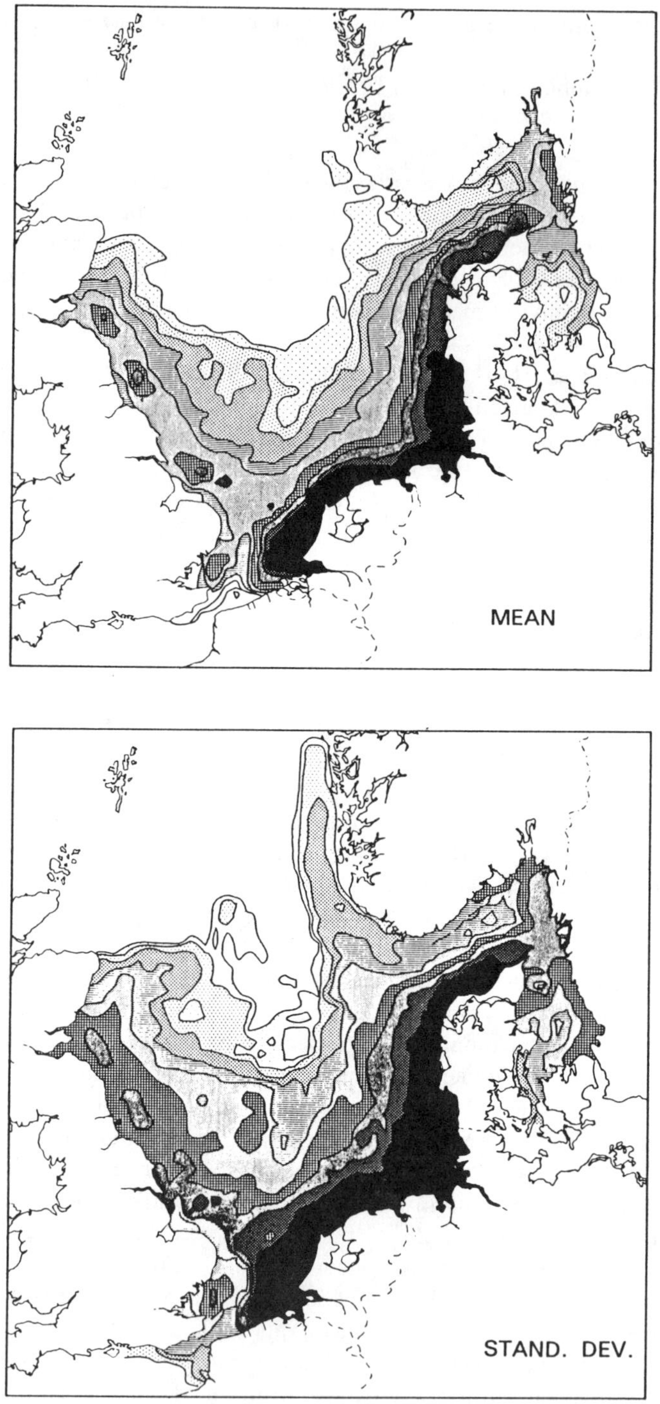


FIGURE 17. Relative pollution load (arbitrary matter) in the North Sea due to riverine discharges, shaded concentration contours drawn in exponential scale (to the base of two). Top: mean concentration; bottom: standard deviation of concentrations compiled over a 14-year period (1969 to 1982). Constant (continous) discharge rates and passive dispersion were assumed, scaled to the average emission rates of the respective rivers (Rhine, Elbe and Werer, Tyne, Firth of Forth, Thames, Humber).

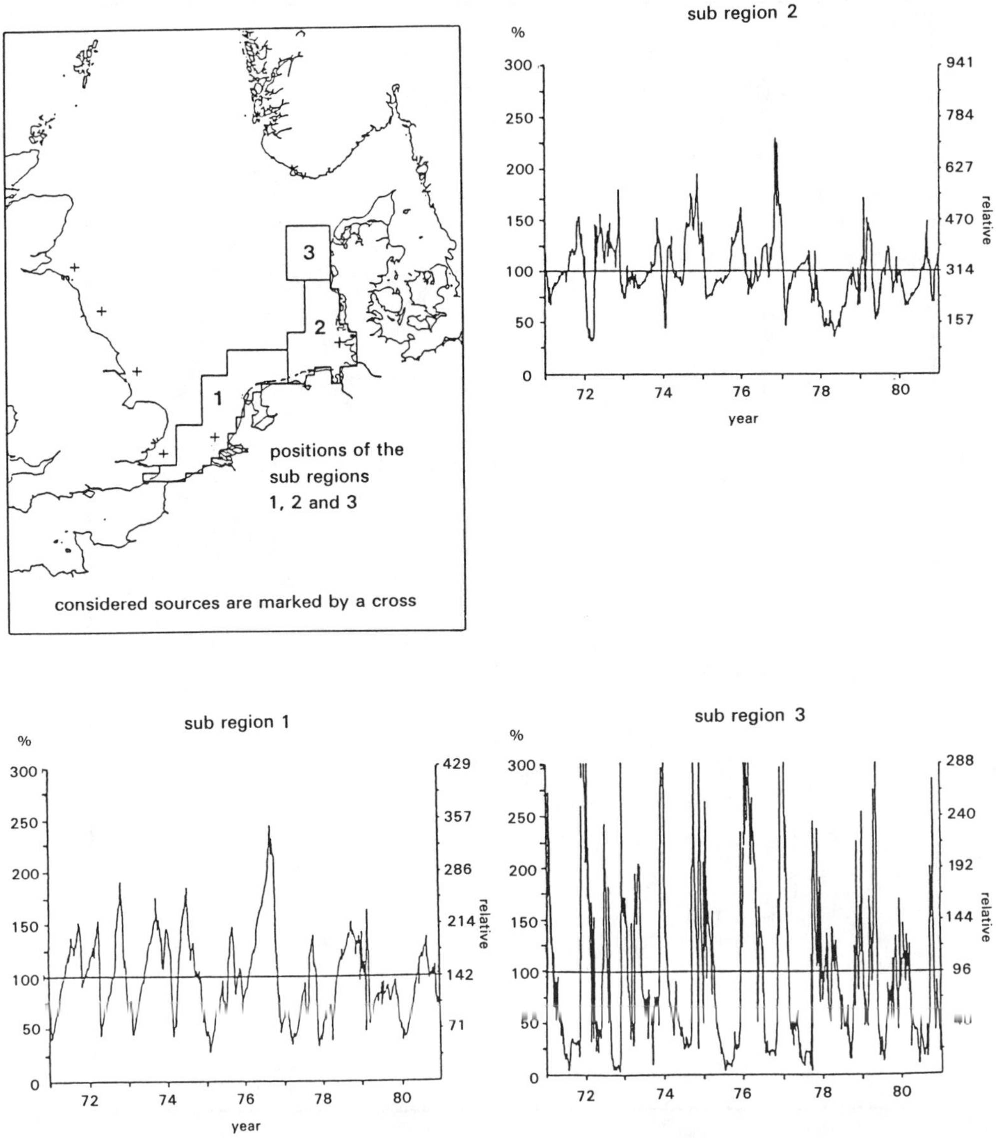


FIGURE 18. Simulated time series of regional averages of (relative) concentrations caused by riverine discharges, shown for three subregions in the North Sea (as indicated) for the period 1971 to 1981. Left axis: concentration percentage, related to estimated regional mean load (100%); right axis: absolute (relative) values.

annual cycle. The results provide information for the subsynoptic frequency band where an information gap existed for observations and model simulations alike. The simulated estimates have shed some new light on fluctuations of the long-term flow which, in a shallow shelf sea, almost directly reflect the transient activity of the atmosphere. Only a sample-test validation of the results was possible due to lacking long-term observations, It is anticipated that the estimated variability still underestimates the real variability of the flow because the baroclinic fields were fixed in time.

The applications of the experiment results to related environmental problems demonstrate the potential of model simulations as a valuable aid for the interpretation of phenomena which are influenced by the circulation. The general concern about the status of the marine environment of the North Sea in regard to pollution is further supported by the likely existence

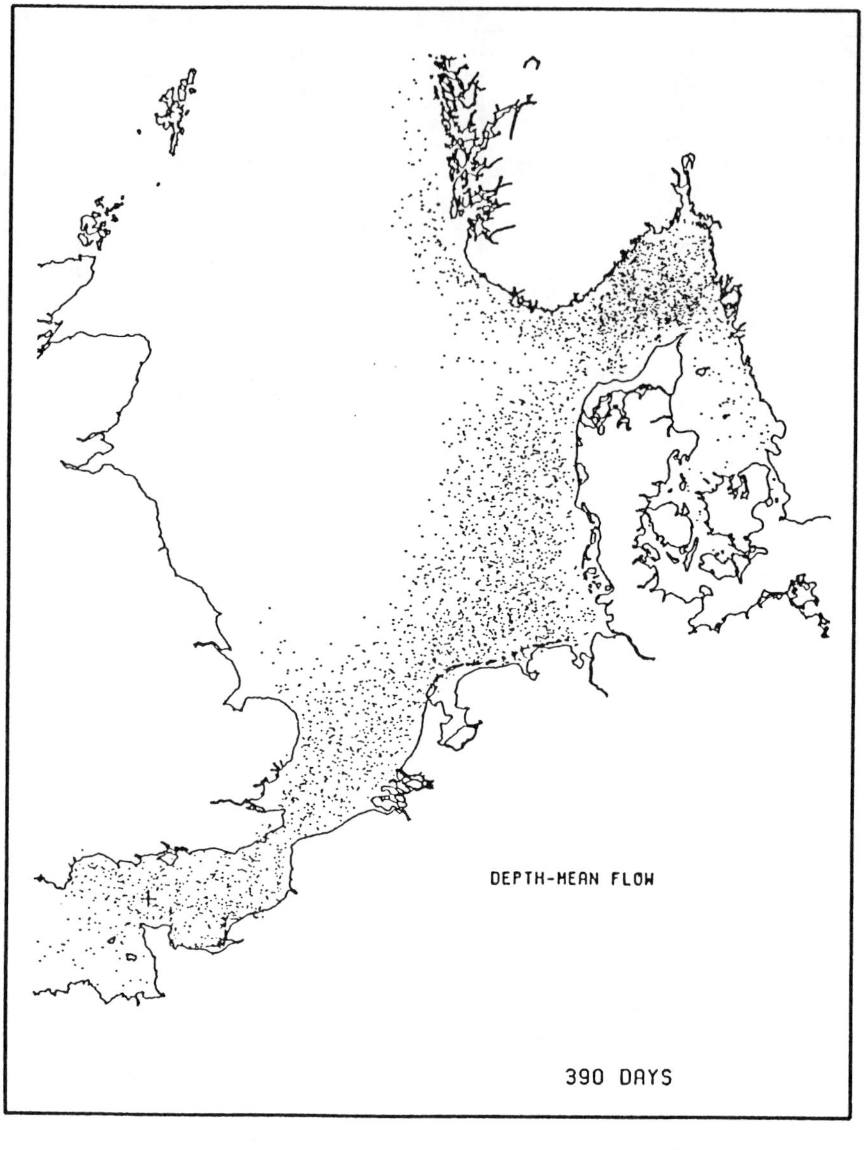


FIGURE 19. Particle pattern for continous discharge (one particle per day) at the nuclear-reprocessing plant at Cap de la Hague for a particle "age" (time after release) of 390 d; period: 1971 to 1981. No diffusion was prescribed.

of considerable fluctuations of pollutant concentrations which may exceed the mean load by orders of magnitude. These results initiated national initiatives for a (re-) definition of the strategy for a marine environmental monitoring aiming at an adequate sampling of the fluctuations of the marine system caused by natural and by human activity.

In the very near future the conflicts due to the different uses of the resource potential of shelf seas will inevitably arise for the developing countries. Their rapidly growing populations will have an increasing demand for protein and for other goods gained from the sea. A careful and optimized resource management will then be a neccessary prerequisite. Model investigations of the kind described here can easily be repeated and/or adopted for other shelf seas. They can effectively be used to provide valuable information for the planning of a resource management.

## ACKNOWLEDGMENTS

The valuable meteorological data provided by the Norwegian Meteorological Office is gratefully acknowledged. Data for the model verification were kindly provided by the Deutsches Hydrographisches Institut Hamburg. During the 8-year period of model development and applications, the author was supported by the Deutsche Forschungsgemeinschaft (DFG-SFB94), the German Federal Ministry for Research and Technology (BMFT-MFU-545), and the Deutsche Wissenschaftliche Kommission Für Meeresforschung (DWK). Without the help of my colleagues J. Bartsch, D. Hainbucher, and T. Pohlmann the work presented here would not have reached its present state.

## REFERENCES

1. **Backhaus, J. Bartsch, J., Quadfasel, D., and Guddal, J.,** Atlas of monthly surface fields of air pressure, wind stress and wind stress curl over the North Eastern North Atlantic Ocean: 1955—1982. Tech. Rep. 3-85, Institut für Meereskunde Hamburg, 1985.
2. **Luthardt, H. and Hasse, L.,** On the relationship between surface and geostrophic wind in the German Bight area, *Beitr. Phys. Atm.*, 54, 222, 1981.
3. **Luthardt, H. and Hasse, L.,** The relationship between pressure field and surface wind in the German Bight area at high wind speeds, in *North Sea Dynamics,* Sündermann, J. and Lenz, W., Eds. Springer-Verlag, Berlin, 1983, 341.
4. **Backhaus, J. O.,** A three-dimensional model for the simulation of shelf sea dynamics, *Dtsch. Hydrogr. Z.*, 38, 165, 1985.
5. **Backhaus, J. O., Pohlmann, T., and Hainbucher, D.,** Regional aspects of the circulation of the North European shelf, International Council for the Exploration of the Sea, C.M. 1986/C:38.
6. **Backhaus, J. O.,** A semi-implicit scheme for the shallow water equations for applications to shelf sea modelling, *Cont. Shelf Res.*, 2(4), 243, 1983.
7. **Backhaus, J. O. and Maier-Reimer, E.,** On seasonal circulation patterns in the North Sea, in *North Sea Dynamics,* Sündermann, J. and Lenz, W., Eds., Springer-Verlag, Berlin 1983, 63.
8. **Goedecke, E., Smed, J., and Tomczak, G.,** Monatskarten des Salzgehaltes der Nordsee, dargestellt für verschiedene Tiefen horizonte, *Dtsch. Hydrogr. Z. Erg. Heft B,* 9, 1967.
9. **Tomczak, G. and Goedecke, E.,** Monatskarten der Temperatur der Nordsee, dargestellt für verschiedene Tiefenhorizonte, *Dtsch. Hydrogr. Z. Erg. Heft B,* 7, 1962.
10. **Levitus, S.,** Climatological atlas of the world ocean, NOAA Professional Paper No. 31, U.S. Government Printing Office, Washington, D.C., 1982.
11. **Backhaus, J. O.,** Estimates of the variability of low frequency currents and flushing times of the North Sea, International Council for the Exploration of the Sea, 1984/C:24.
12. **Koudstaal, R.,** Water quality management plan North Sea: framework for analysis, Ifias research series, Coastal waters No. 1, The Netherlands, 1987.

12a. **Dekker, L., Bower, B. T., and Koudstaal, R.,** Management of toxic materials in an international setting (a case study of cadmium in the North Sea), Ifias research series, Coastal waters No. 2, The Netherlands, 1987.

13. **Hainbucher, D., Backhaus, J., and Pohlmann, T.,** Atlas of climatological and actual seasonal circulation patterns in the North Sea and adjacent shelf regions: 1969—1981, Tech. Rep. 1-86, Institut Für Meereskunde Hamburg, 1986.
14. **Dooley, H. D.,** Seasonal variability in the position and strength of the Fair Isle current, in *North Sea Dynamics,* Sündermann, J. and Lenz, W., Eds., Springer-Verlag, Berlin, 1983, 68.
15. **Backhaus, J. O., Hainbucher, D., Quadfasel, D., and Bartsch, J.,** North Sea circulation in response to varying atmospheric forcing, International Council for the Exploration of the Sea, C. M. 1985/C:29.
16. International Council for the Exploration of the Sea: Flushing times of the North Sea, ICES Cooperative Res. Rep. No. 123, Copenhagen, 1983.
17. **Hackett, B.,** Currents and hydrography off western Norway during NORSEX-79, Rep. 3-82, Norwegian Marine Pollution Research and Monitoring Programme, Norwegian Ministry of the Environment.
18. **Mork, M.,** Circulation Phenomena and frontal dynamics of the Norwegian Current, *Philos. Trans. R. Soc. London Ser. A,* 302, 635, 1981.

19. **Kautsky, H.,** The Caesium 137 content in the water of the North Sea during the years 1969 to 1975, *Dtsch. Hydrogr. Z.*, 29, 217, 1976.
20. **Kautsky, H., Jeffries, D. F., and Steele, A. K.,** Results of the radiological North Sea programme RANOSP 1974 to 1976, *DTSCH. Hydrogr. A.*, 33, 152, 1980.
21. **Kautsky, H.,** Different artificial radio nuclides in the water of the North Sea during the years 1977 to 1981, *Dtsch. Hydrogr. A.*, 38, 193, 1985.
22. **Livingston, H. D., Bowen, V. T., and Kupferman, S. L.,** Radionuclides from Windscale discharges. I. Nonequilibrium tracer experiments in high latitude oceanography, *J. Mar. Res.*, 40, 253, 1982.
23. **Prandle, D.,** A modelling study of the mixing of Caesium 137 in the seas of the European continental shelf, *Philos. Trans., R. Soc. London Ser. A,* 310 407, 1984.
24. **Roache, P.,** On artificial viscosity, *J. Comp. Phys.*, 10(2), 169, 1972.
25. **Maier-Reimer, E.,** On the formation of salt wedges in estuaries, in *Mathematical Modelling of Estuarine Physics,* Sündermann, J. and Holz, K. P., Eds., Springer-Verlag, Berlin, 1980, 91.
26. **Pohlmann, T., Backhaus, J. O., and Hainbucher, D.,** Validation of a three-dimensional dispersion model for Caesium 137 in the North European shelf sea, International Council for the Exploration of the Sea, C.M. 1987/C:34.
27. **Corten, A.,** The recruitment failure of herring in the central and northern North Sea in the years 1974—1978 and the mid 1970's hydrographic anomaly, International Council for the Exploration of the Sea, C.M. 1984/Gen. 12.
28. **Backhaus, J. O. and Bartsch, J.,** Circulation and fish recruitment: North Sea, International Council for the Exploration of the Sea, C.M. 1985/C:30.
29. **Bork, I. and Maier-Reimer, E.,** On the spreading of power plant cooling water in a tidal river, applied to the river Elbe, *Adv. Water Res.* 1(3), 161, 1978.
30. **Maier-Reimer, E.,** Hydrodynamisch-numerische Untersuchungen zu horizontalen Ausbreitungsvorgängen in der Nordsee, Mitteilungen des Institut für Meereskunde Hamburg, No. XXI, 1973.
31. **Bartsch, J.,** Numerical simulation of the advection of vertically migrating herring larvae in the North Sea, Meeresforschung (Reports on Marine Research), Deutsche Wissenschaftliche Kommission für Meeresforschung, Hempel, Sarhage (Eds.), to appear in 1988.
32. **Müller-Navarra, S. and Mittelstaedt, E.,** Schadstoffausbreitung und Schadstoffbelastung in der Nordsee, eine Modellstudie, *Dtsch. Hydrogr. Z. Erg. Heft B,* 18, 52, 1987.
33. **De Ruiter, W. P. M., De Kok, J. M., and Postma, L.,** Transport Atlas of the North Sea, Delft Hydraulics, Rijkswaterstaat, Tidal Waters Division, 1987.
34. **Hainbucher, D., Pohlmann, T., and Backhaus, J. O.,** Transport of conservative passive tracers in the North Sea: first results of a circulation and dispersion model, *Cont. Shelf Res.*, 7(10), 1161, 1987.
35. **Anon.,** Gütezustand der Nordsee, Meereskundliche Beobachtungen und Egebnisse, Carlson, Ed., Deutsches Hydrographisches Institut Hamburg, No. 55, 1984.
36. **Maier-Reimer, E.,** Zum Einfluß eines mittleren Windschubes auf die Restströme der Nordsee, *Dtsch. Hydrogr. Z.*, 28(6), 244, 1975.

Chapter 7

# TIDE AND SURGE MODELS FOR SHALLOW WATER — MORECAMBE BAY REVISITED

**Roger A. Flather and Kevin P. Hubbert**

## TABLE OF CONTENTS

# I. INTRODUCTION

Modeling of tides and storm surges in shallow coastal regions presents a number of special problems, which either do not occur at all or are less extreme in larger scale shelf seas. These problems stem from the prevalence of tidal flats (which are exposed as water levels fall and resubmerged as levels rise again), from the related process of inundation of low-lying coastal land during storm surge events, and, of course, from nonlinear effects in general, which are enhanced with decreasing water depth. The problems have long been recognized and a number of numerical schemes incorporating techniques to deal with them were proposed in the late 1960s and early 1970s, providing a basis for practical models of such regions. Flather and Heaps[1] reviewed some of this early work and presented a simple explicit numerical method including drying and flooding of inter-tidal areas.

The purpose of the present paper is to outline some of the alternative approaches available for representing drying banks with reference to more recent work, thereby indicating the present status of this aspect of shallow-water modeling. Some attempts to refine the technique proposed by Flather and Heaps[1] are also described. In view of the dedication of this volume to Dr. N. S. Heaps, it is particularly appropriate to test these new approaches by applying them experimentally to Morecambe Bay.

# II. MATHEMATICAL FORMULATION

The models considered here are based on depth-averaged equations which may be written in the form

$$\frac{\partial \zeta}{\partial t} + \underset{\sim}{\nabla} \cdot (D\underset{\sim}{q}) = 0 \tag{1}$$

$$\frac{\partial \underset{\sim}{q}}{\partial t} + \underset{\sim}{q} \cdot \nabla \underset{\sim}{q} - f\underset{\sim}{k} \times \underset{\sim}{q} = -g\nabla\zeta - \frac{1}{\rho}\nabla p_a + (\underset{\sim}{\tau}_s - \underset{\sim}{\tau}_b)/\rho D + A\nabla^2 \underset{\sim}{q} \tag{2}$$

where the notation is

| | |
|---|---|
| $\underset{\sim}{q}$ | depth-mean current |
| $\zeta$ | elevation of the sea surface above its undisturbed level |
| t | time |
| f | Coriolis parameter ($= 2\omega\sin\phi$) |
| $\omega$ | angular speed of the Earth's rotation |
| $\phi$ | latitude |
| g | acceleration due to gravity |
| $\underset{\sim}{k}$ | unit vector in the vertical (upwards) |
| $\rho$ | density of sea water, assumed uniform |
| $p_a$ | atmospheric pressure on the sea surface |
| $\underset{\sim}{\tau}_s$ | surface (wind) stress |
| $\underset{\sim}{\tau}_b$ | bottom stress |
| D | total water depth $= h + \zeta$ |
| h | undisturbed water depth |
| A | coefficient of horizontal eddy viscosity |

Equation 1, the continuity equation, expresses conservation of mass and Equation 2, the equation of motion, expresses conservation of momentum.

The mathematical problem is to solve these equations within a prescribed region, subject to appropriate initial and boundary conditions and with the external forces ($p_a$ and $\underset{\sim}{\tau}_s$) defined as functions of position and time. The solution, $\zeta$ and $\underset{\sim}{q}$, gives the surface elevation and depth mean current within the region as functions of position and time. Closure of the problem requires the specification of $\underset{\sim}{\tau}_b$, usually by assuming that it is related to the depth-mean current, $\underset{\sim}{q}$, by a quadratic law

$$\underset{\sim}{\tau}_b = C_b \rho \underset{\sim}{q} |\underset{\sim}{q}| \tag{3}$$

where $C_b$ is a coefficient of bottom stress.

Initial conditions take the form of prescribed distributions of $\zeta$ and $\underset{\sim}{q}$ at time $t = 0$. A number of alternative conditions may be applied on open sea boundaries. Among these are to specify either $\zeta$ or $q_n$ (the component of $\underset{\sim}{q}$ along the outward-directed normal to the boundary) at every point on the open boundary as functions of time. A forced gravity wave radiation condition, of the form

$$q_n = \bar{q}_n + \frac{c}{D}(\zeta - \bar{\zeta}) \tag{4}$$

where $\bar{\zeta}$ and $\bar{q}_n$ are specified functions of time for each boundary point, representing externally generated disturbances entering the model, and $c = (gD)^{1/2}$, is another possibility, intended to permit waves generated within the model to propagate out across the open boundary without being reflected. Conditions for land boundaries are dependent on the treatment of flooding and drying processes and will be discussed in the next section.

Solution of Equations 1 to 3 in practical situations requires a suitable numerical technique. The most popular methods employ finite difference approximations with the variables defined on regular spatial grids and at fixed intervals in time. The equations are such that values of the dependent variables at a given instant may be derived from known values, generally at an earlier time. Repeated solution of the equations for all grid points thus advances the solution through time over the period of interest. The resulting formulations may be explicit, in which new values of the dependent variables are derived directly from known values,[1] or implicit, in which combinations of updated values of the dependent variables are derived directly, with individual variables being separated by further computation (e.g., the "alternating-direction implicit" or ADI scheme of Leendertse and Gritton).[2] Explicit schemes are simpler but have the disadvantage that, for stability, the time step of integration, $\Delta t$, must be such that disturbances do not propagate a distance greater than the grid spacing, $\Delta s$, in time $\Delta t$. This imposes a condition $\Delta t < \Delta t_{crit}$, where $\Delta t_{crit} \sim \Delta s/[2^{1/2}(c + |\underset{\sim}{q}|)]_{max}$, which may limit the time step to a few seconds. Implicit schemes are more complex and require more computational effort per time step but are unconditionally stable, permitting, at least in principle, arbitrarily long time steps. However, in order to obtain accurate solutions in complex regions, relatively short time steps may still be required.[3] Irregular grids have the advantage of allowing the spatial resolution to be increased in regions which are complex or of special interest, and may also allow the schematization to be optimized for explicit schemes in the sense of making $\Delta s/[2^{1/2}(c + |\underset{\sim}{q}|)]$ approximately constant over the integration region. A finite difference method for an irregular grid has been developed by Thacker.[4] However, finite element methods[5] have been more widely adopted for use with nonuniform grids.

## III. FLOODING AND DRYING

A number of alternative techniques have been proposed to deal with the representation of flooding and drying in numerical tide-surge models. In order to facilitate the discussion, we choose to classify these techniques as belonging to one of the following types, characterized by the following two scenarios.

(1) Many estuaries and coastal embayments contain large, shallow, and relatively flat areas separated and interleaved by deeper channels and creeks. When water levels are high, the entire bay may be water covered but, as the level falls, the shallow areas are exposed and, ultimately, the flow is confined to the deeper channels. The dry inter-tidal areas may then occupy a substantial fraction of the total surface area of the inlet and the channels may be only a few tens of meters wide (or less). The nature of such regions can, therefore, change from a two-dimensional open sea area to a complex system of interconnected narrow channels depending on water level. Similar transformations can occur when vulnerable areas are inundated during storm surges; a prime example being the Ganges Delta region in the northern Bay of Bengal.

In contrast to the topographic complexities, the numerical methods used in practical calculations for such regions tend to be comparatively simple, employing regular (usually rectangular) computational grids on which flooding and drying are represented by adding grid boxes which become "wet" as the water level rises and removing grid boxes that become "dry" as levels fall. Alternative sets of rules have been proposed to determine when a grid box becomes wet or dry, with various applied or implied conditions at the land-water interface. Generally a condition of zero normal flow, $q_n = 0$, is applied on the coastal boundary, which is constrained to follow the sides of grid boxes and thus moves in discrete steps. The discontinuous movement of the boundary can generate "noise", which may be reduced by appropriate modification of the drying rules.[3,6] Further examples of this type of approach may be found elsewhere.[1,2,7,8]

(2) Other nearshore regions may be characterized by relatively uniform shelving bathymetry, such that the depth varies smoothly over substantial distances. Similar conditions may apply in idealized problems used as test cases for numerical techniques, with the depth variations being defined analytically. Sielecki and Wurtele[9] argued that, for regions with sloping boundaries, the solution of Equations 1 and 2 should not require any lateral boundary condition but that the fluid should determine its own level according to the motion and bathymetry. They employed three alternative finite difference methods on fixed grids to some idealized problems, verifying the results by comparison with analytical solutions. Their technique involved linear extrapolation of values of elevation or total water depth) and momentum for use in the continuity equation at each grid box containing the land-water interface. Lynch and Gray[5] describe a finite element technique which accounts for the moving land-water interface automatically by continuous deformation of the grid. The boundary conditions applied are

$$D = 0 \quad \text{at} \quad \underset{\sim}{s} = \underset{\sim}{s}_0 + \int_0^T \underset{\sim}{q}^B dt, \qquad \underset{\sim}{q}^B = \underset{\sim}{q} \tag{5}$$

where $\underset{\sim}{s}$ and $\underset{\sim}{s}_o$ denote the position of the boundary at times $t = T$ and $t = 0$, respectively, and $\underset{\sim}{q}^B$ is the velocity of the boundary. At the land-water interface, the velocity of a fluid particle is equal to the velocity with which the interface moves. Lynch[10] discusses the significance of Equation 5. Clearly, a substantial computational overhead can arise because of the grid deformation, and for this reason, Lynch and Gray[5] suggest that the grid be allowed to deform only in the nearshore region. If the grid becomes unacceptably distorted, it is necessary to halt the simulation, redefine the grid, and interpolate the dependent variables to the new nodes before continuing the computation. Perhaps because of these considerations, the method appears to have been used only for idealized problems. A finite difference approach incorporating a continuously moving lateral boundary effectively by means of a deforming grid has been described by Johns et al.,[11] and applied in simulating a storm surge in the Bay of Bengal. In this instance, a transformation is applied to one of the horizontal coordinates such that the distance between the land-water interface and the offshore extremity of the model along this coordinate is mapped into a fixed (unit) interval. A fixed grid in the transformed coordinates then corresponds to a deforming grid in the original coordinates. The boundary conditions (Equation 5) are applied, but differentiation of D with respect to time gives a prognostic equation for $\underset{\sim}{s}$ using which, the interface location can be determined from the change in surface elevation and the slope of the sea bed (assumed constant in the application of the method). Linear extrapolation of $\zeta$ to the boundary is required in the numerical representation of this equation. As with the method of Lynch and Gray,[5] additional computational effort is required because of the grid deformation; in particular, the value of h at each model grid point changes as the grid deforms and so must be recomputed at each time step. A further restriction is that the coordinate transformation must remain valid and well behaved, so that the technique would not be applicable to problems in which the land-sea interface becomes strongly distorted, e.g., for an isolated sandbank surrounded by deeper water.

Although the techniques described under (2) are more satisfactory in that they represent the physical processes of inundation and drying more rigorously, they tend to be much more complex and difficult (or impossible) to apply in regions such as those described in (1). For this reason, it would appear worthwhile to examine further the simple methods mentioned in (1) and to consider whether some of their less acceptable features can be eliminated or at least improved.

## IV. EXPERIMENTS WITH METHODS FOR MODELING INTER-TIDAL AREAS

In this section, we describe some experiments using different approaches and assumptions to represent the processes of drying and flooding. The methods are tested by applying them to compute the $M_2$ tide in Morecambe Bay, a complex embayment on the west coast of England having many of the characteristics described under (1) in the previous section. As a basis for these experiments, we have adopted the Morecambe Bay model developed by Stephens,[12] shown in Figure 1. This model employs east-longitude, $\chi$, and latitude, $\phi$, as coordinates, with resolution 54 grid elements per degree of longitude and 81 per degree of latitude, so that the grid size is 1.21 km in the E-W direction and 1.37 km in the N-S direction. The finite difference scheme is essentially that described by Davies and Flather,[13] which is a version for geographical coordinates of Flather and Heaps[1] Scheme 3. The basic drying procedure is that due to Flather and Heaps.[1]

Since the representation of subgrid-scale features is a key issue, we have digitized the bathymetry in a part of the Bay (the area between Walney Island and the mainland — see Figure 1) at ten times the resolution of the basic model. This detailed depth data made it possible to examine various alternative assumptions relating to the definition of bathymetry and the representation of drying in the basic model and to establish a high resolution (121 m $\times$ 137 m grid) model of this subarea of the Bay, providing a benchmark against which the coarse-grid model results could be assessed. The high-resolution model is shown in Figure 2 and the bathymetry plotted as contours in Figure 3. From Figure 3, it can be seen that the northern part of the area consists of a shallow and narrow channel between Walney Island and the mainland. In the south, there is one main channel (approximately 10 m deep and 100 to 200 m wide) with shallow inter-tidal areas to the north and south. There are also four small islands, one of them (Roa Island) connected to the mainland by an embankment. All of these features are smaller than the grid size of the basic model.

The following experimental schemes have been investigated with the coarse-grid model. In all cases, the alternative assumptions and procedures were applied only within the detailed

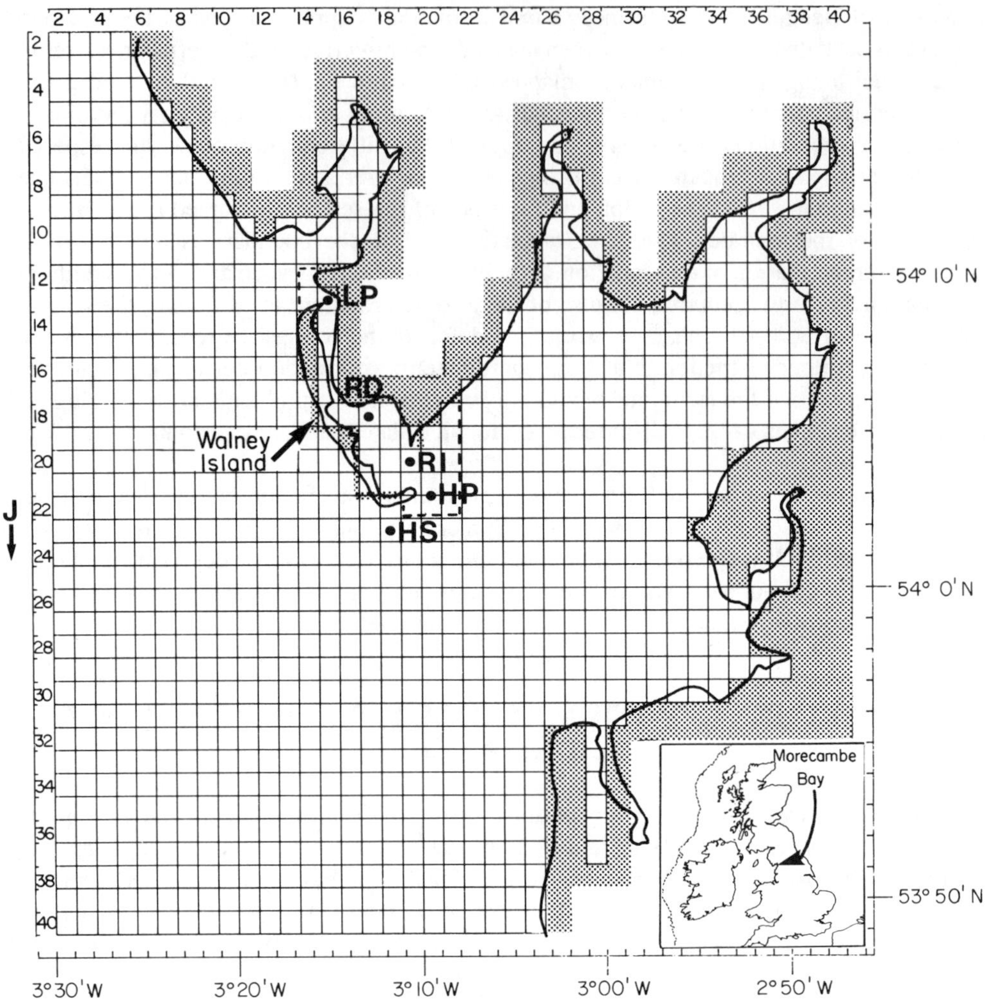


FIGURE 1. Basic model of Morecambe Bay showing the boundary of the study region (broken line), tide gauge positions (●), and location map (inset). LP = Lowsy Point; RD = Ramsden Dock; RI = Roa Island, HP = Hawes Point; HS = Halfway Shoals.

study region between Walney Island and the mainland. In the remaining parts of the Bay the model bathymetry, difference scheme, and drying procedure were as described at the beginning of this section.

1. The depth, defined at the center of each coarse-grid box, within the detailed study region was taken to be

$$
\begin{aligned}
h &= \frac{1}{N}\sum_j (h_j | h_j > h_{land}) && \text{if} \quad 50 \leq N \leq 100 \\
&= h_{land} && \text{if} \quad 0 \leq N < 50
\end{aligned}
\tag{6}
$$

where N is the number of high-resolution grid points within the grid box of the coarse model which are not permanent land points. This corresponds to the basic rule applied when extracting model depth data from bathymetric charts, and can be stated in words as — if the grid box is more than 50% land then define it to be "land"; otherwise define its depth

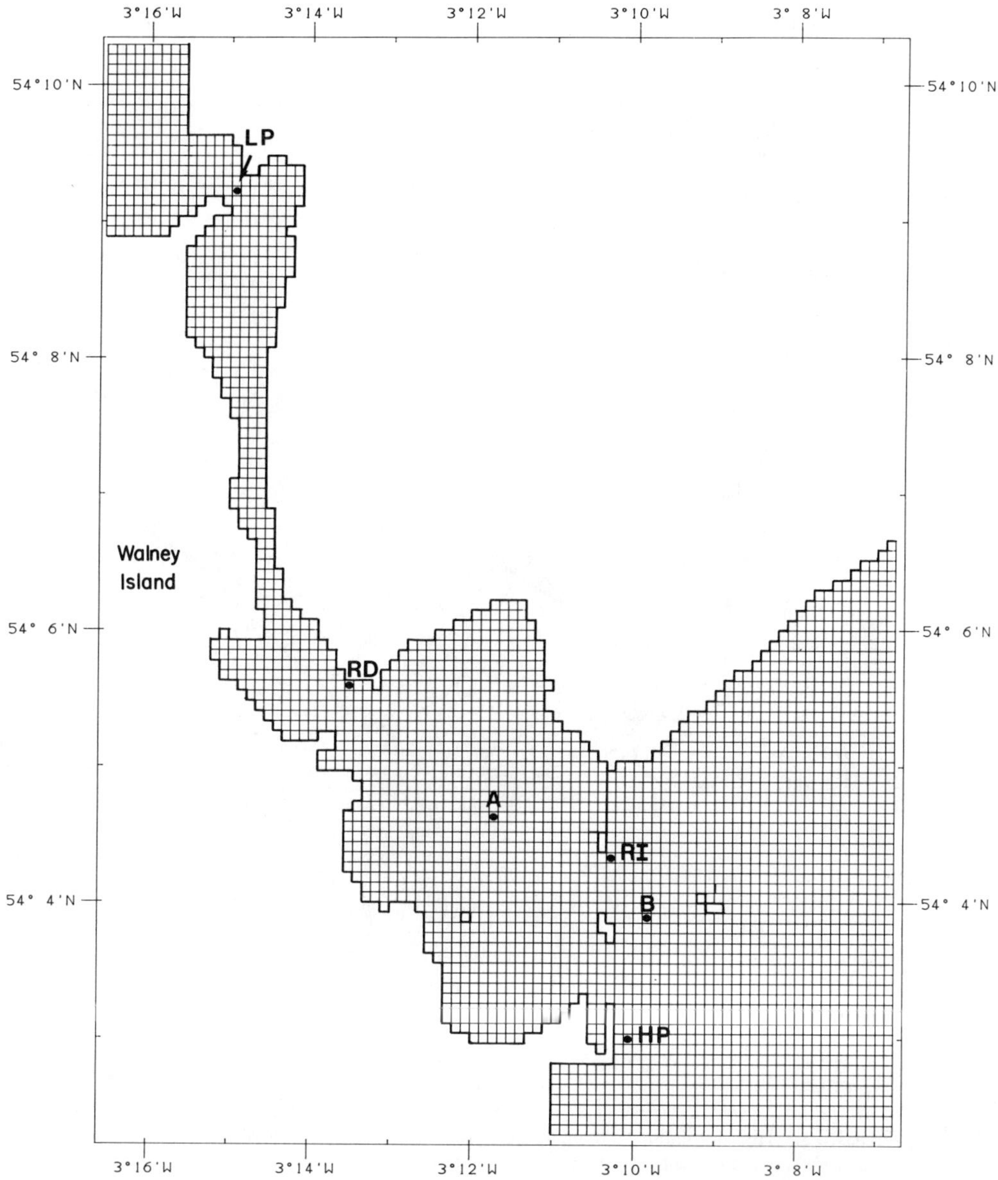


FIGURE 2. High-resolution model of the channel between Walney Island and the mainland with locations of points from which computed results are discussed. A = (column 45, row 78); B = (62,88); RI = (58,82).

to be the mean depth of the "wet" part of the grid box. This definition smoothes out the deep channel entering from the south and blocks the narrow northern channel between Walney Island and the mainland.

2. The depth distribution in 1 was modified by increasing depths along the approximate line of the deep channel to values representative of that channel. This procedure introduces a deep channel one grid element in width extending into the region from the south and restores the northern passage. However, the widths of the channels in the model are, as a result, many times their real widths.

3. Recognizing that these subgrid-scale features would be more realistically represented as one-dimensional (1-D) channels, the basic scheme was modified to make this possible.

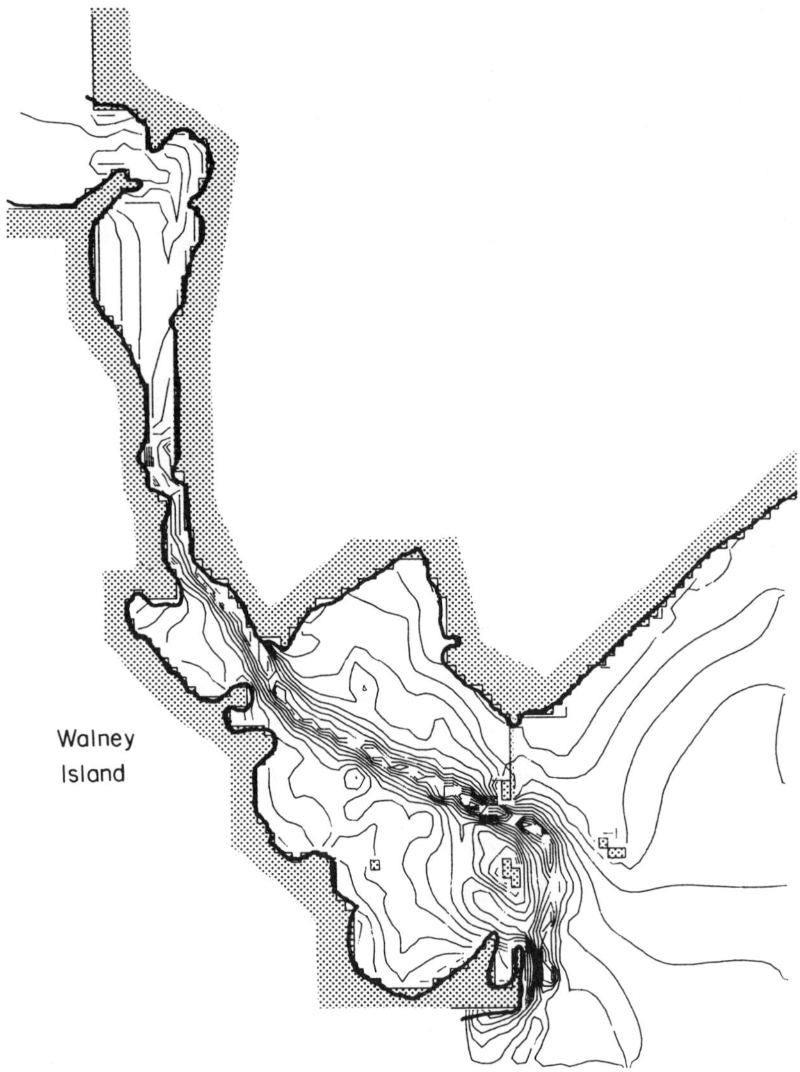


FIGURE 3. High-resolution bathymetry in the region between Walney Island and the mainland. Contours are of depths below mean sea level at 1-m intervals.

Equations of continuity and motion, equivalent to Equations 1 and 2, for one-dimensional flow can be written as

$$\frac{\partial \zeta}{\partial t} + \frac{1}{B}\frac{\partial (BDq)}{\partial s} = 0 \tag{7}$$

$$\frac{\partial q}{\partial t} + q\frac{\partial q}{\partial s} = -g\frac{\partial \zeta}{\partial s} - \frac{1}{\rho}\frac{\partial p_a}{\partial s} + (\tau_s - \tau_b)/\rho D + A\frac{\partial^2 q}{\partial s^2} \tag{8}$$

where B is the breadth of the channel, s is the distance along it, q is the section-averaged current along the channel and $\tau_s$ and $\tau_b$ are along-channel components of surface and bottom stress. If the 1-D channel is aligned with one coordinate axis in the 2-D formulation, then the corresponding equations of motion (Equation 8 and the appropriate component of Equation 2) are compatible and may be solved by the identical finite difference equation. The continuity Equations 7 and 1 can also be made equivalent by including additional factors in the finite difference form of the 2-D equation. With $\chi$ and $\phi$ as coordinates and adopting the grid scheme and notation of Davies and Flather,[13] shown in Figure 4, the finite difference form of the equation of continuity is modified to

$$A_i[\zeta_i(t + \Delta t) - \zeta_i(t)] = -\frac{\Delta t}{R\cos\phi_m}\{[d_i(t)B_iu_i(t) - d_{i-1}(t)B_{i-1}u_{i-1}(t)]/\Delta\chi$$

$$+ [e_{i-n}(t)C_{i-n}\cos\phi_{m-1}v_{i-n}(t) - e_i(t)C_i\cos\phi_{m+1}v_i(t)]/\Delta\phi\} \qquad (9)$$

where

$u_i$ = the E-directed component of depth mean current, $\underset{\sim}{q}$, at u-point i
$v_i$ = the N-directed component of $\underset{\sim}{q}$ at v-point i
$\zeta_i$ = the elevation at $\zeta$-point i
$\Delta t$ = the timestep of integration
$d_i$ = $0.5(D_i + D_{i+1})$, the total water depth at u-point i
$e_i$ = $0.5(D_i + D_{i+n})$, the total water depth at v-point i
$n$ = the number of columns in the rectangular grid array
$D_i$ = $h_i + \zeta_i$ at $\zeta$-point i

and $A_i$, $B_i$, and $C_i$ are the additional factors as follows

$A_i$ = area factor
= wetted area of grid box/total area of grid box
= area wet/$R^2 \cos\phi_m\Delta\chi\Delta\phi$
$B_i$ = breadth factor for u-point i
= breadth over which flow occurs/total breadth of grid box
= flow breadth/$R\Delta\phi$
$C_i$ = breadth factor for v-point i
= flow breadth/$R\cos\phi_{m+1}\Delta\chi$

If $A_i = 1$, $B_i = 1$, $C_i = 1$, then Equation 9 is identical to the 2-D form of continuity given by Davies and Flather.[13] If, for example, $v_{i-n} = 0 = v_i$ in Equation 9, then with arbitrary values of $A_i$, $B_{i-1}$, the equation approximates Equation 7 for 1-D flow in the E-W direction for a channel of arbitrary section. The transformation from 1-D flow confined to a narrow channel at low water levels to 2-D unconfined flow at high water levels can be represented by an appropriate drying procedure which either sets $v_i$, $v_{i-n}$ zero (dry) or computes its value from momentum (wet). In a first implementation of these ideas, we have taken the depth distribution from 2, above, with appropriate *constant* values of $A_i$, $B_i$, $C_i$ derived from the high-resolution bathymetry to represent the correct dimensions of the deep channel entering the region rather than taking those as the grid size in the 2-D scheme. Effectively, this gives a fixed but irregular grid in the region where $A_i$, $B_i$, or $C_i = 1$. Since $A_i$ is assumed constant in Equation 9, the continuity equation can be solved explicitly as in the standard 2-D scheme.

All the above methods employed the drying scheme of Flather and Heaps.[1] In this, tests are performed before each u and v calculation to determine whether the associated (u or v) grid point is either "wet" or "dry". If the point is dry then the velocity at that point is set to zero. If the point is wet then the new velocity value is calculated by solving the appropriate momentum equation. For the u velocity at u-point i and with reference to the notation of Equation 9 and that shown in Figure 4, the procedure is

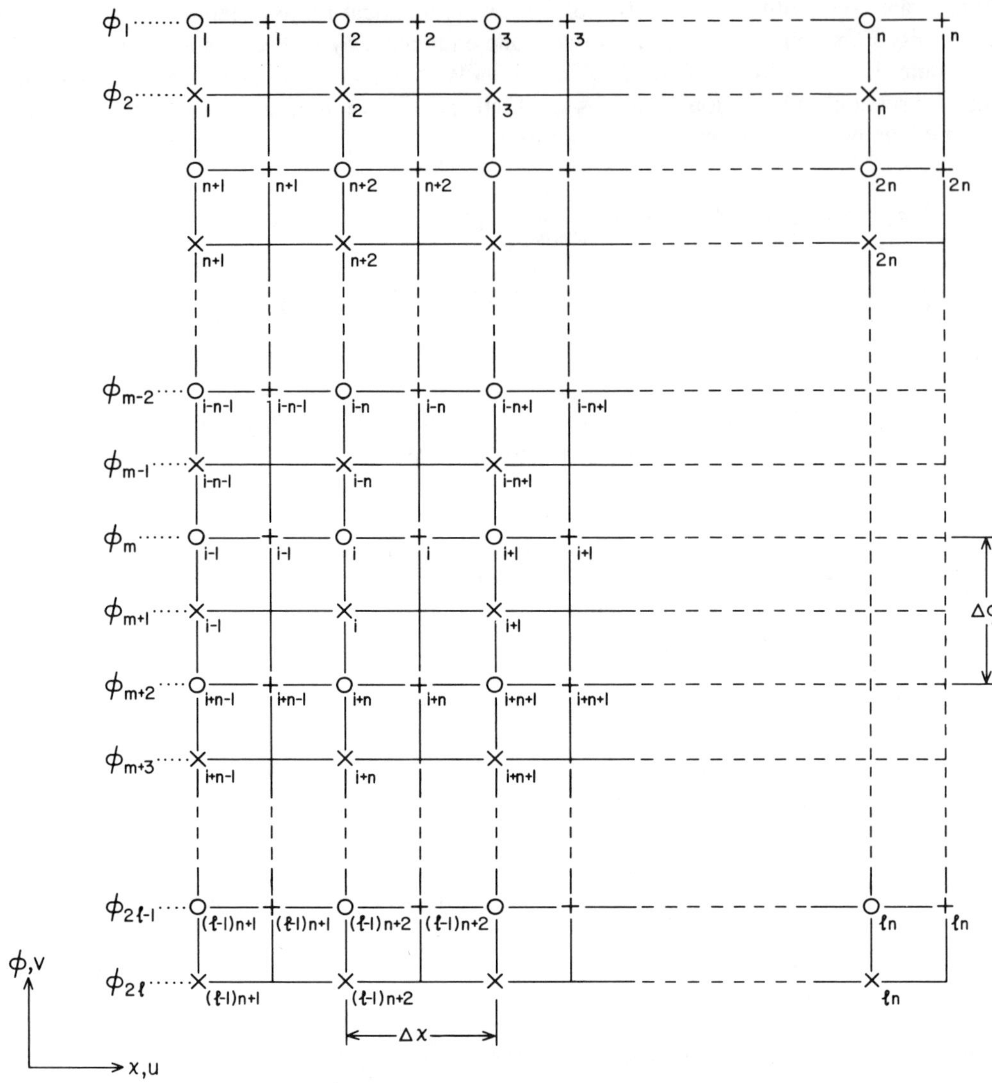

FIGURE 4. Spatial grid for the finite difference scheme used in the model showing the grid point locations ($\circ = \zeta$, $+ = u$, $\times = v$) and the addressing system.

if (i) $D_i$ (t + Δt) > 0 and $D_{i+1}$ (t + Δt) > 0 so that $d_i$ (t + Δt) > 0

or (ii) $D_i$ (t + Δt) > 0 and $D_{i+1}$ (t + Δt) ≤ 0 and $d_i$ (t + Δt) > 0 and $\zeta_i$ (t + Δt) − $\zeta_{i+1}$ (t + Δt) > ε

or (iii) $D_i$ (t + Δt) ≤ 0 and $D_{i+1}$ (t + Δt) > 0 and $d_i$ (t + Δt) > 0 and $\zeta_{i+1}$(t + Δt) − $\zeta_i$ (t + Δt) > ε

then calculate $u_i$ (t + Δt) from the momentum equations, otherwise set $u_i$(t + Δt) = 0.

ε is a critical elevation difference chosen so as to prevent the drying and flooding of a grid box at alternate time steps. A similar procedure follows for a v velocity. In the momentum equations, Equation 2, the surface and bottom stress terms have the total water depth, D, in the denominator. In order to avoid the singularity in these terms as a grid box dries, a minimum value of the total water depth, $D_{crit}$ is employed.

4. The final scheme allowed the area and breadth factors to vary with the changing water level. Thus, for example, as the water level falls, the wetted area within a grid box can decrease progressively until the grid box dries. This is in contrast to the schemes described earlier, under which a grid box is either entirely wet or entirely dry: the transformation from one state to the other being sudden, thus imparting a "shock" or discontinuity to the system. With variable factors, essentially simulating the subgrid motion, such discontinuities should be avoided.

In order to implement this scheme, the breadth factors were combined with the total water depth at the velocity grid points to define new factors, $B_i'$ and $C_i'$, replacing $B_id_i$ and $C_ie_i$ in Equation 9, describing the cross-sectional area of flow across each side of the grid box. The high-resolution bathymetry was used to define these surface and cross-sectional area factors numerically at fixed surface elevation levels, $z_k$. We thus obtained, for each grid element within the study region, values of

$$A_i(z_k), \quad B_i'(z_k), \quad C_i'(z_k) \quad \text{for} \quad z_k = -3.5 + 0.1(k - 1)\text{m}, \qquad 1 \leqslant k \leqslant 71$$

where $0 \leqslant A_i \leqslant 1$ and $B_i'$ = (total water depth at u-point i) × (breadth factor for u-point i)

$$= \frac{1}{10}\sum_{j=1}^{N}(h_j + z_k) \qquad \text{such that} \qquad (h_j + z_k) \geqslant 0$$

where N is the number of high-resolution grid points at the cross-section of the grid box in the coarse model that are wet. In other words, the surface and cross-sectional area factors were defined numerically at intervals of 0.1 m in water level within the range −3.5 to +3.5 m.

With the introduction of the cross-sectional area factors, the finite difference form of the continuity equation is modified to

$$A_i[\zeta_i(t + \Delta t) - \zeta_i(t)] = -\frac{\Delta t}{R\cos\phi_m}\{[B_i'(t)u_i(t) - B_{i-1}'(t)u_{i-1}(t)]/\Delta\chi$$

$$+ [C_{i-n}'(t)\cos\phi_{m-1}v_{i-n}(t) - C_i'(t)\cos\phi_{m+1}v_i(t)]/\Delta\phi\} \tag{10}$$

The required values of $A_i(z)$, $B_i'(z)$, and $C_i'(z)$ for use in Equation 10 were then calculated by linear interpolation between the defined values; for example

$$A_i(z) = a_1A_i(z_k) + a_2A_i(z_{k-1}) \text{ for } z_{k-1} \leqslant z \leqslant z_k \tag{11}$$

where $a_1 = (z - z_{k-1})/\Delta z$, $a_2 = (z_k - z)/\Delta z$ and $\Delta z = z_k - z_{k-1} = 0.1$ m

Since the right-hand side of Equation 10 is defined at the lower time level, t, when all values $\zeta_i$, $u_i$, and $v_i$ are known, there is no difficulty in specifying $B_i'$ and $C_i'$ with z defined as the mean elevation at the associated u or v point at time t. Clearly, alternative methods could be used for the left-hand side of Equation 10, which approximates

$$\int_{z=\zeta_i(t)}^{\zeta_i(t+\Delta t)} A_i(z)dz \tag{12}$$

the change in volume of water in grid box i expressed in terms of the change in water level. The simplest approach is to use Equation 10 with

$$A_i = A_i[\zeta_i(t)] \tag{13}$$

i.e., the area factor defined at the lower time level. Alternatively, since $A_i(z)$ is assumed piecewise linear in z, it can easily be shown that

$$\text{Left-hand side} = A_i\{[\zeta_i(t + \Delta t) + \zeta_i(t)]/2\}[\zeta_i(t + \Delta t) - \zeta_i(t)] \tag{14}$$

provided the elevation remains within one interval,

$$z_{k-1} < \{\zeta_i(t + \Delta t), \quad \zeta_i(t)\} < z_k$$

[Equivalent but more complex expressions result when for example, $\zeta_i(t) < z_k < \zeta_i(t + \Delta t)$]. In this case, the calculation becomes implicit since the value of $A_i$ depends on the elevation, $\zeta_i(t + \Delta t)$, to be determined. However, Equation 10 with Equation 14 can easily be solved by a simple iterative method such as used in some 1-D models.[14] The normal drying condition of Flather and Heaps[1] was also modified to correspond to the representation of subgrid-scale bathymetry. The revised test, applied at individual current points is that the current velocity is zero if the cross-sectional area is zero, so for u-point i,

$$u_i(t + \Delta t) = 0 \qquad \text{if} \qquad B_i'(t + \Delta t) = 0$$

otherwise compute $u_i(t + \Delta t)$ from the momentum equation. As in Flather and Heaps,[1] a minimum value of total water depth, $D_{crit}$, was imposed for the denominator of the surface and bottom stress terms in Equation 2 or Equation 8.

The following model runs were performed:

- C1 coarse grid, Scheme 1 (averaged bathymetry), $\epsilon = 0.1$ m, $D_{crit} = 1.0$ m
- C2 coarse grid, Scheme 2 (deep-water channel), $\epsilon = 0.1$ m, $D_{crit} = 1.0$ m
- C31 coarse grid, Scheme 3 (constant area and breadth factors), $\epsilon = 0.1$ m, $D_{crit} = 1.0$ m
- C32 coarse grid, Scheme 3 (constant area and breadth factors), $\epsilon = 0.01$ m, $D_{crit} = 0.1$ m
- C41 coarse grid, Scheme 4 (time-varying area factors), A evaluated at the lower time step (Equation 13), $D_{crit} = 1.0$ m
- C42 coarse grid, Scheme 4 (time-varying area factors), A evaluated at mid-time step (Equation 14), $D_{crit} = 1.0$ m
- F1 fine grid, Scheme 1 (averaged bathymetry), $\epsilon = 0.1$ m, $D_{crit} = 1.0$ m
- F2 fine grid, Scheme 1 (averaged bathymetry), $\epsilon = 0.01$ m, $D_{crit} = 0.1$ m

Open boundary conditions were obtained as follows. An initial run was performed with the coarse-grid model using an elevation-specified open boundary condition. The elevations were taken from Stephens.[12] The results of this run provided currents normal to the open boundary of the coarse-grid model, which were used with the radiation condition, Equation 4, for all subsequent runs with this model.

A coarse-grid model solution (run C2) was interpolated onto the open boundary of the fine-grid model. The interpolated open boundary elevations and the model open boundary condition were adjusted to account for drying and an initial run was performed with the fine-grid model using an elevation-specified open boundary condition in order to generate currents for use in the radiation condition, Equation 4, used in subsequent runs F1 and F2. The initial run with this model employed the angled-derivative scheme of Davies and Flather[13] to approximate the advection terms in the momentum equations. This led to irregularities in the flow near open boundaries, a problem solved by utilizing upwind differencing of the advective cross-terms. The use of upwind differencing can cause significant numerical diffusion to be present in a solution, manifesting itself in the form of the smearing of velocity gradients and the underprediction of the vorticity in wakes behind islands and promontories. This was not considered to be important in the present work in which the flooding and drying of tidal flats was of primary interest, but may not be acceptable generally.

## V. RESULTS

The results of the various model runs for the $M_2$ constituent of the tide defined on the open boundary are presented in this section. We first consider computed and observed harmonic constants, then examine time series of computed elevation and current and finally, discuss spatial current distributions.

Comparisons of the amplitude and phase of the $M_2$ constituent of the surface tide predicted by the various runs at selected locations (shown in Figures 1 and 2) are given in Table 1, along with equivalent observed values.

The comparison for the coarse-grid model suggests, at first sight, that there is little to choose between the various schemes, except for run C1. This was the run with averaged bathymetry, containing no proper channel through from Hawes Point to Ramsden Dock and beyond. The poor results from this solution, in which the grid point corresponding to Ramsden Dock dries, are, therefore, not surprising. The Hawes Point amplitude predictions from all the coarse-grid model solutions do not compare well with the observations. The tide guage was actually installed at the edge of the deep channel, in a location subject to drying, a situation not readily simulated with coarse resolution. The predictions at Halfway Shoals, which was just outside the study region, show only small variations, suggesting that the effects of the various schemes are localized. Indeed, this is to be expected since the dominant tidal signal in elevation is determined by the response of the whole shelf to the north Atlantic tides. Apart from Lowsy Point, located in a region where there was no detailed bathymetric data, the two fine-grid model runs, F1 and F2, are in close agreement with each other and the observations.

A clearer indication of the performances of the different schemes is obtained from time series of predicted elevation and components of current for various locations within the study region. Model data were stored at every timestep to generate the time series, and so no information was lost and there was no smoothing of the computed results.

A major effect of the alternative bathymetries in the study region is that a grid point may dry out at different levels in the different solutions. This is particularly evident for run C1 with the consequence that the time series of elevations and, particularly, current velocities for this run differ quite markedly from those of the other runs.

Runs C41 and C42 gave almost identical results, suggesting that the surface elevations and hence area and breadth factors, changed sufficiently slowly that the numerical approximation used to calculate the surface area factor did not affect the solution significantly. The results of run C42 will therefore not be discussed further.

The two runs using Scheme 3, C31 and C32, employed different values of the critical elevation difference, $\epsilon$, determining, with the grid size, the minimum elevation gradient required for flow into a dry grid box, and the minimum total water depth, $D_{crit}$. The effect of the lower value of $D_{crit}$ used in run C32 can be seen in Figure 5. This compares the elevation time series for these two runs at grid point (15,13) (Lowsy Point), where the undisturbed water depth is 1.23 m. The elevation between 3 and 7 h (all times are in lunar hours after lunar transit across the meridian of Greenwich) for run C32 is 10 to 15 cm higher than that of run C31 as a result of the smaller $D_{crit}$ and hence higher bottom friction, leading to reduced flow through the channel between Walney Island and the mainland (a slowing

**TABLE 1**
**Comparison between Computed and Observed Amplitudes and Phases of the $M_2$ Tide**

| Location | F1 | F2 | C1 | C2 | C31 | C32 | C41 | C42 | Observed |
|---|---|---|---|---|---|---|---|---|---|
| Lowsy Point | | | | | | | | | |
| (15,13) | 209.3 m | 205.7 m | 207.6 m | 210.7 m | 208.1 m | 198.6 m | 213.0 m | 212.9 m | 235.9 m |
| (16,16) | 335.9° | 335.7° | 332.2° | 334.6° | 332.3° | 332.5° | 331.6° | 331.6° | 335.3° |
| Ramsden Dock | | | | | | | | | |
| (16,18) | 309.2 m | 308.8 m | 152.3 m | 307.8 m | 307.8 m | 307.7 m | 307.7 m | 307.7 m | 307.7 m |
| (29.65) | 332.9° | 332.9° | 337.4° | 333.0° | 333.2° | 333.1° | 330.3° | 330.3° | 330.9° |
| Roa Island | | | | | | | | | |
| (19,20) | 305.9 m | 305.5 m | 282.7 m | 305.6 m | 305.3 m | 305.0 m | 305.0 m | 305.0 m | 305.9 m |
| (58,82) | 330.7° | 330.7° | 329.9° | 330.0° | 330.5° | 330.6° | 329.1° | 329.1° | 328.6° |
| Hawes Point | | | | | | | | | |
| (20,21) | 292.0 m | 291.9 m | 303.1 m | 304.2 m | 304.8 m | 304.1 m | 303.5 m | 303.5 m | 291.8 m |
| (60,100) | 329.5° | 329.4° | 328.1° | 329.3° | 329.7° | 329.7° | 328.4° | 328.4° | 327.3° |
| Halfway Shoals | | | | | | | | | |
| (18,23) | -- | -- | 294.7 m | 295.1 m | 294.6 m | 294.8 m | 294.9 m | 294.9 m | 296.5 m |
| -- | -- | -- | 324.7° | 325.0° | 324.7° | 324.6° | 324.8° | 324.8° | 324.8° |

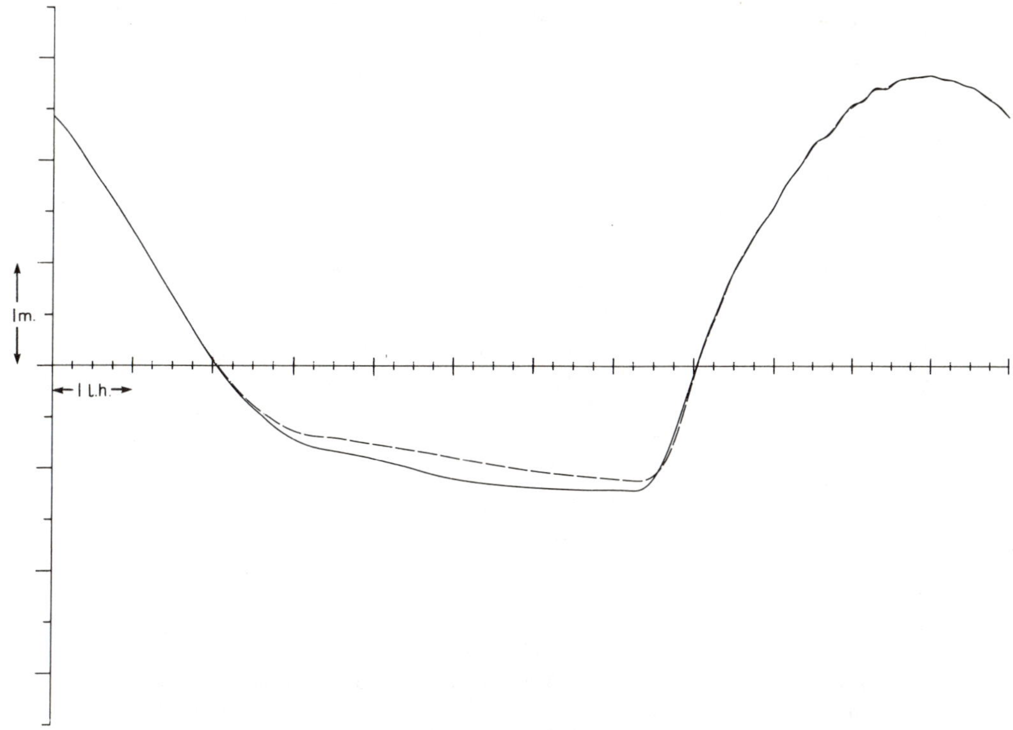


FIGURE 5. Comparison of computed surface elevations at point (I,J) = (15,13) from runs C31 (—) and C32 (- - - -).

down of the tidal propagation) in run C32. Changes in $\epsilon$ and $D_{crit}$ also affect the fine-grid solutions. In addition to the effect on tidal propagation, described above, the value adopted for $\epsilon$ influences the flooding process.

As a consequence of the higher value of $\epsilon$ in run F1, elevation time series at grid points that dry show that these grid points become wet again later in F1 than in F2, with a sudden jump in elevation as the value of $\epsilon$ is exceeded and flow into the grid box resumes. This sudden influx causes oscillations, apparent in the time series, which may be interpreted as seiches set up as the water suddenly flows into a newly wet grid box and is reflected. This is illustrated in Figure 6, which compares the elevation time series for runs F1 and F2 at grid point (58, 82) for a short period of time as the grid point becomes wet again. The elevation and current-velocity time series of F1 are considerably noisier than those of F2, with the grid boxes in F2 drying out and flooding again much more smoothly.

Run C41 produced the smoothest solution of all. Differences between the results of runs C2 and C41 are greatest for lower water levels, with often little or negligible difference at high water. This may be explained by the fact that most grid boxes in the study region are all (or almost all) wet near high water (i.e., the surface area factor is equal to 1.0 in Scheme 4), and consequently Schemes 2 and 4 produce almost the same approximation to the continuity equation. This is illustrated in Figure 7, which compares the elevation time series of runs C2 and C41 at grid point (16,18) (Ramsden Dock). However, low water can be as much as 10 to 15 min earlier and 5 to 10 cm higher for run C41 than for run C2.

A comparison of the elevation at grid point (18,19) of coarse-grid-model run C41 with the elevation at grid point (45,78) of fine-grid model run F2, shown in Figure 8, suggests that run C41 leads run F2 by as much as 5 min over most of the cycle. This phase lead of run C41 over the various other model runs for much of the tidal cycle is confirmed in the

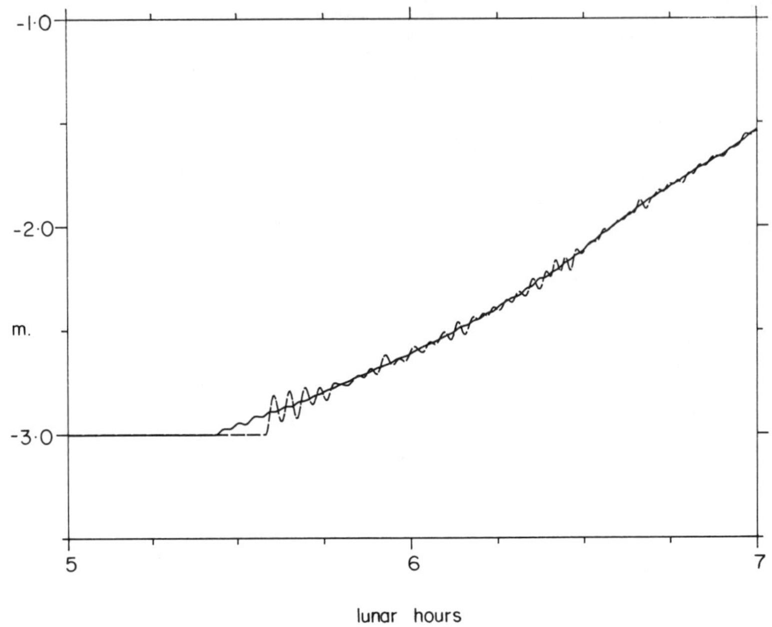


FIGURE 6. Detail showing computed elevations between 5 and 7 lunar hours at point (58,82) - Roa Island, in runs F1 (- - - -) and F2 (—). The differences result from reduced values of the drying parameters $\epsilon$ and $D_{crit}$ in run F2.

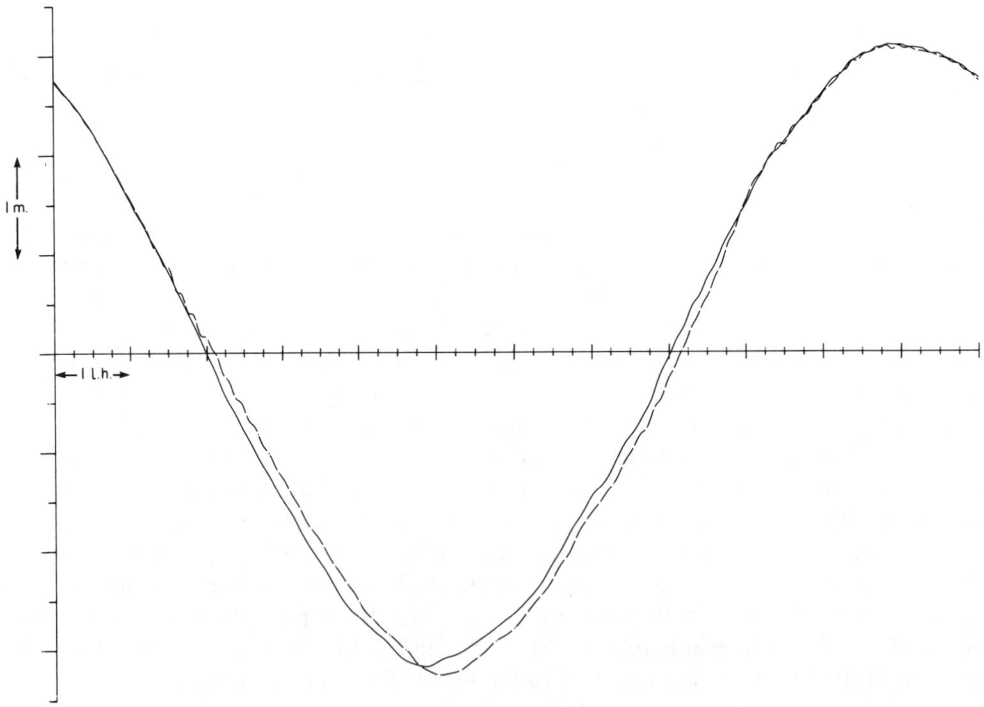


FIGURE 7. Comparison of computed surface elevations at point (I,J) = (16,18) from runs C41 (—) and C2 (- - - -).

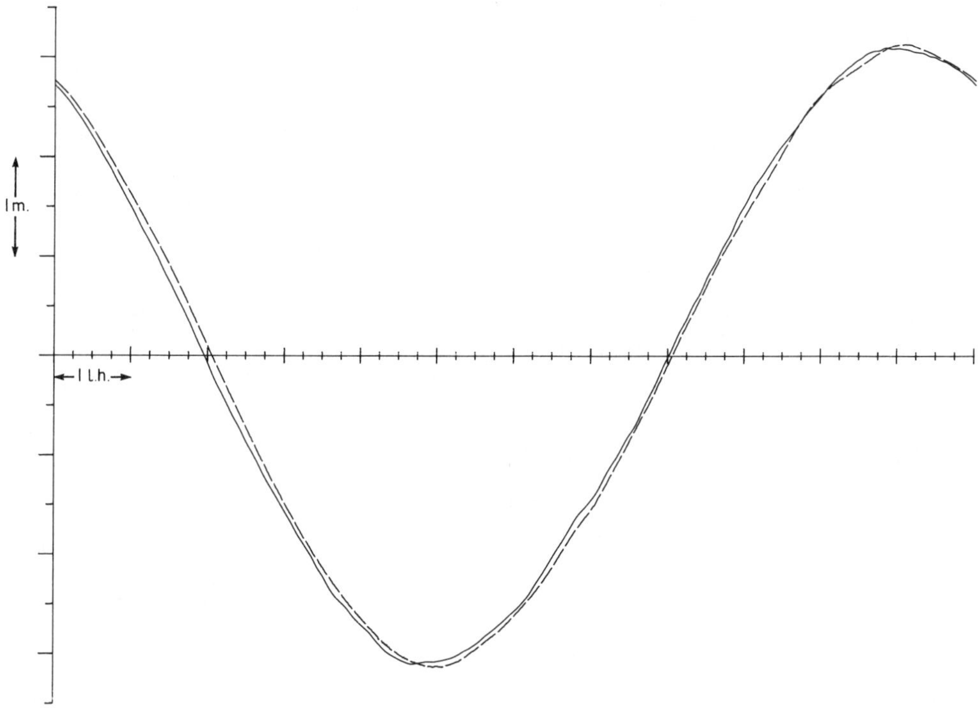


FIGURE 8. Comparison of computed surface elevations at point (18,19) from run C41 (—) and at point (45,78) from run F2 (- - - -).

$M_2$ phases at various locations shown in Table 1. The predictions of the phase at Halfway Shoals, which can be considered to be at the entrance to the deep-water channel, just outside the study region, are very similar and in good agreement with the observations. The phase advance of run C41 compared to the other runs increases along the channel, reaching 3° at Ramsden Dock, where the predicted $M_2$ phase is in good agreement with the observations. Pedersen[11] shows that propagation of a wave along a saw-tooth-shaped channel suffers a phase shift as a result of reflections from the lateral boundaries of the channel. The results of this work seem to suggest that Scheme 4 may avoid this problem, allowing the tidal wave to propagate diagonally, along the channel, more satisfactorily than the other schemes.

One interesting feature of run C1 is shown in the u current-velocity time series for grid point (17,19), Figure 9. Grid boxes (17,19) and (18,19) are "ponded" between approximately 3 and 8 lunar hours. The sudden drying of the surrounding grid boxes initiates a small amplitude seiche in the ponded grid boxes, which is slowly damped by bottom friction.

The current-velocity time series from runs C31 and C32 generally exhibit more "noise" than those from runs C2 and C41. The time series from run C32 are slightly smoother than those from run C31, especially on the flooding tide, when the lower value of $\epsilon$ used in run C32 leads to smaller amplitude oscillations as grid points become wet again, as discussed earlier for elevations.

Figure 10 shows u velocity time series for runs C2, C32, and C41 at u-point (17,19). The velocity components are much more sensitive to changes in bathymetry and the treatment of drying than the surface elevation, with the result that the time series of current velocities within the study region are much more variable than those of elevation. The velocity time series for runs C2, C31, C32, and C41 generally exhibit similar characteristics. At some locations, the results of runs C31 and C32 in particular show differences either side of high water; for example in Figure 10, the computed current in run C32 changes direction from

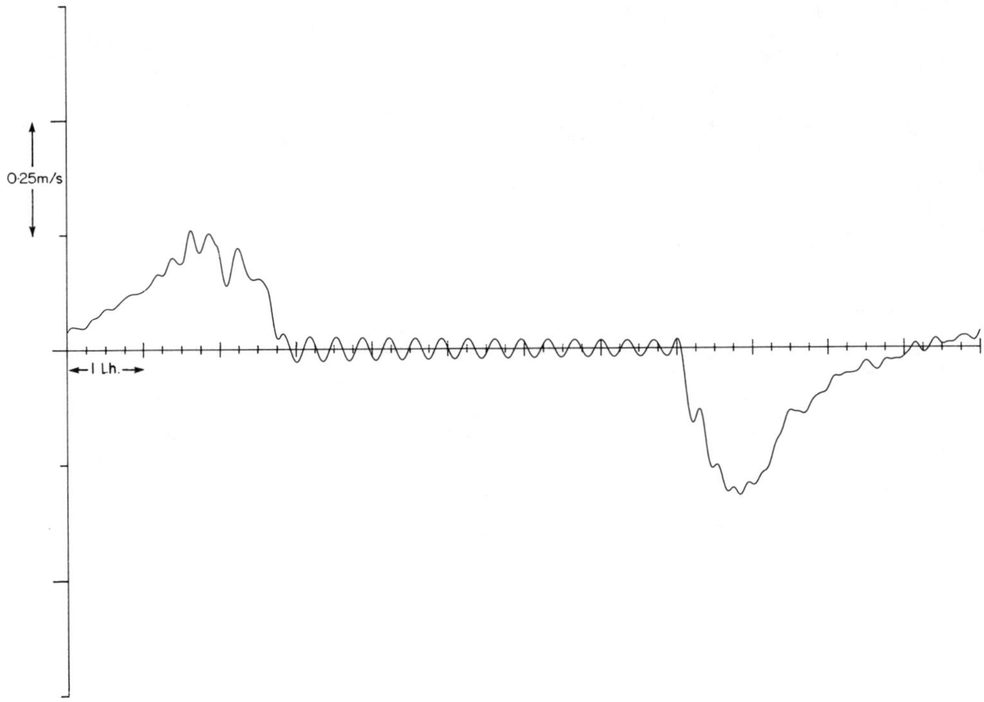


FIGURE 9. Computed u-component of current at point (17,19) from run C1.

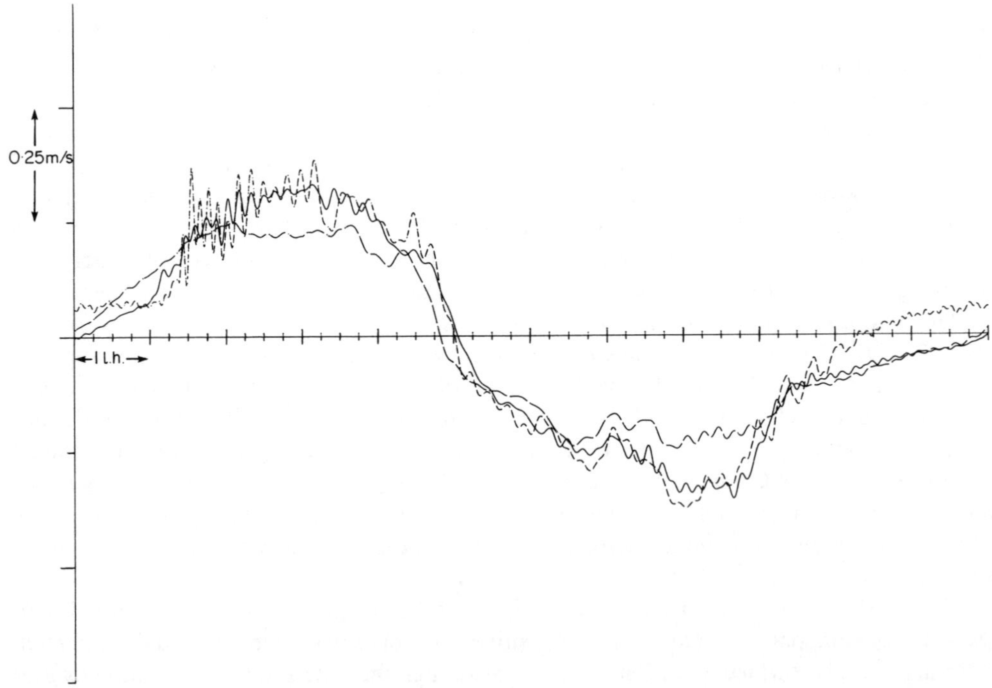


FIGURE 10. Comparison of computed u-components of current at point (17,19) from runs C2 (—), C32 (- - - -), and C41 (— — —)

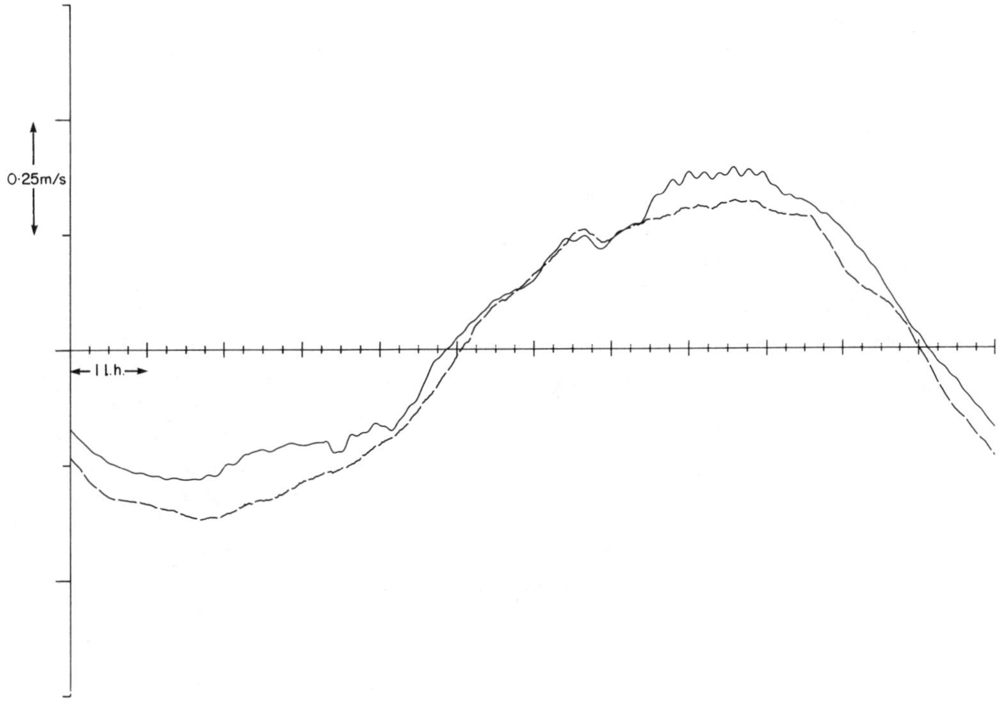


FIGURE 11. Comparison of computed v-components of current at point (20,20) from run C41 (—) and at point (62,88) from run F2 (— — —).

E-W to W-E about 1.5 h earlier than that in runs C2 and C41. The largest differences in magnitude appear to be between runs C2 and C41. In Figure 10, for example, the E-directed current-velocity of run C41 is generally weaker than that of run C2 throughout the whole tidal cycle, although that is not everywhere the case. The current-velocities of runs C2, C31, and C32 are similar in magnitude.

The strongest currents in the study region are in the area close to the southern tip of Walney Island and just east of Hawes Point, of the order of 1.00 to 1.25 $ms^{-1}$. This area is at the edge of the study region and so the differing effects of the various schemes are not particularly noticeable here.

Comparison of currents from the fine-grid model (run F2) with those from the coarse-grid model is not straightforward, since the computed time series represent essentially different things: in one case flow across the side of a grid box approximately 120 m wide, in the other an average over up to 1.2 km (depending on the scheme used). Where comparisons can reasonably be made between currents computed in runs C41 and F2, there is good qualitative agreement between the two; for example Figure 11 shows the v-component of velocity at point (20,20) of the coarse-grid model run C41 with the v velocity at grid point (62,88) of the fine-grid model run F2.

Current vector plots for a limited area of the fine-grid model at selected times are shown in Figure 12. These illustrate the substantial flow variations during the tidal cycle. Particularly noticeable is the flow along the deep water channel in NW and SE directions. The plot at 6 h shows the extensive areas of drying banks exposed at low water. As the tide floods the Bay, water flows up channels and spreads out onto the drying banks. A number of transient eddies, associated with bathymetric features and islands and usually persisting only for 1 or 2 h, can be seen in the current distributions. (The understanding and interpretation of the flow was aided by the use of a graphics workstation, permitting images such as those in

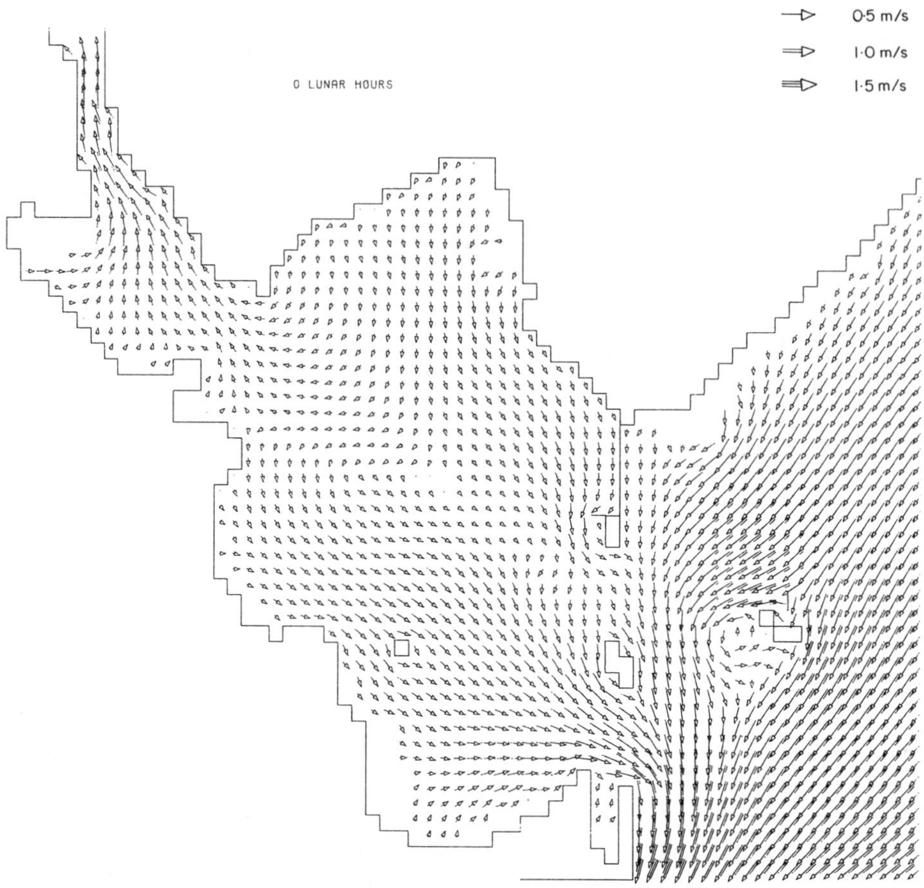


A

FIGURE 12. Distributions of computed current at selected times from the high-resolution model, run F2. (A) 0 Lunar hours; (B) 2 lunar hours; (C) 3 lunar hours; (D) 4 lunar hours; (E) 6 lunar hours; (F) 8 lunar hours; (G) 10 lunar hours; (H) 11 lunar hours.

Figure 12 to be displayed rapidly in sequence, thereby giving the effect of movement.) Regarding the coarse-grid model, these eddies are subgrid-scale motions so that the coarse-grid model can, at best, only approximate the flow along the channel and the covering and uncovering of the drying banks.

Figures 13 and 14 show current distributions within the study region computed in runs C1 and C41, respectively, for comparison with the fine-grid solution (Figure 12). Comparing Figures 13 and 12, it is clear that Scheme 1 is quite unable to predict the flow in the study area. In contrast, Scheme 4 (Figure 14), performs remarkably well considering the relatively very coarse resolution, reproducing not only the qualitative features of the flow but giving current speeds which are representative of the fine-grid velocities. It even suggests the development of eddies at 11 lunar hours, which correspond reasonably with those in the fine-grid solution.

## VI. CONCLUSIONS

After briefly reviewing available methods for representing drying processes in shallow-

FIGURE 12B.

water models, some modifications to the method of Flather and Heaps[1] have been proposed and tested.

Numerical experiments were performed to simulate the $M_2$ tide in Morecambe Bay focusing in particular on the region between Walney Island and Barrow-in-Furness, which contains a deep-water channel and large areas of tidal flats. A high-resolution model revealed details of the local variability of the motion and the covering and uncovering of the intertidal areas. The variability is clearly such that it could not be represented, other than in some average sense, with much coarser resolution.

The detailed bathymetry of the high-resolution model was used to examine four different schemes for approximating the bathymetry and representing drying processes in a lower-resolution model.

The first scheme employed averaged bathymetry in the study region and a drying scheme that essentially calculated a velocity at a velocity grid point if the total water depth at that velocity grid point was greater than zero. The predictions in the study region were not very realistic due to the smoothing of the bathymetric features.

The second scheme incorporated the same drying scheme but with the bathymetry adjusted to give a deep-water channel. The results with this scheme compared quite well with observations, indicating that some account must be taken of the essential bathymetric features, in particular deep conveying channels, in order to achieve accurate predictions.

The third scheme used the same bathymetry as the second scheme but included a narrow-channel approximation, incorporating (fixed) area and breadth factors, allowing the width

FIGURE 12C.

of the deep channel to be defined more realistically. The drying scheme was the same as before. This scheme yielded no apparent improvements over the second scheme in the present application and indeed the time series of elevations and current velocities were considerably noisier.

The fourth scheme attempted to take account of the subgrid-scale features by allowing the area and breadth factors to vary with the elevation. The drying scheme was also modified, vanishing cross-sectional area of flow being the criterion determining when and where a velocity component was set to zero. The wetted surface area of a grid box and cross-sectional area of flow vary continuously, so that the shocks and discontinuities associated with wetting and drying in the other schemes are avoided. Consequently, this scheme provided the smoothest time series.

Of the methods investigated, Scheme 4 gave the most satisfactory results, avoiding, or at least substantially reducing, the introduction of noise and discontinuities by representing the drying and flooding processes in a physically more realistic manner. Further investigation of this and some other possible approaches would be of interest, and in particular it would be useful to apply the various methods to some idealized test cases, such as those used by Sielecki and Wurtele.[9] Another practical question is that of how to apply an approach such as Scheme 4 in practice. Clearly, more detail of bathymetry than can be represented on a coarse grid in the standard manner (i.e., by one depth value per grid point) is required, yet it may not be practicable to digitize the bathymetry over the whole region at, say, ten times the basic resolution. One possibility might be to employ surface fitting techniques with

FIGURE 12D.

relatively low (and variable) resolution depth data or soundings, deriving the required area and breadth factors from the resulting local functional representation of the bathymetry.

With regard to the Morecambe Bay problem, observations of current would be useful for verification purposes, the computed currents being much more sensitive to the assumptions used in the model than are the elevations.

## ACKNOWLEDGMENTS

The authors are grateful to Mrs. Jackie Huxley and Mrs. Linda Parry, who typed the manuscript, and Mr. Robert Smith for preparing the figures.

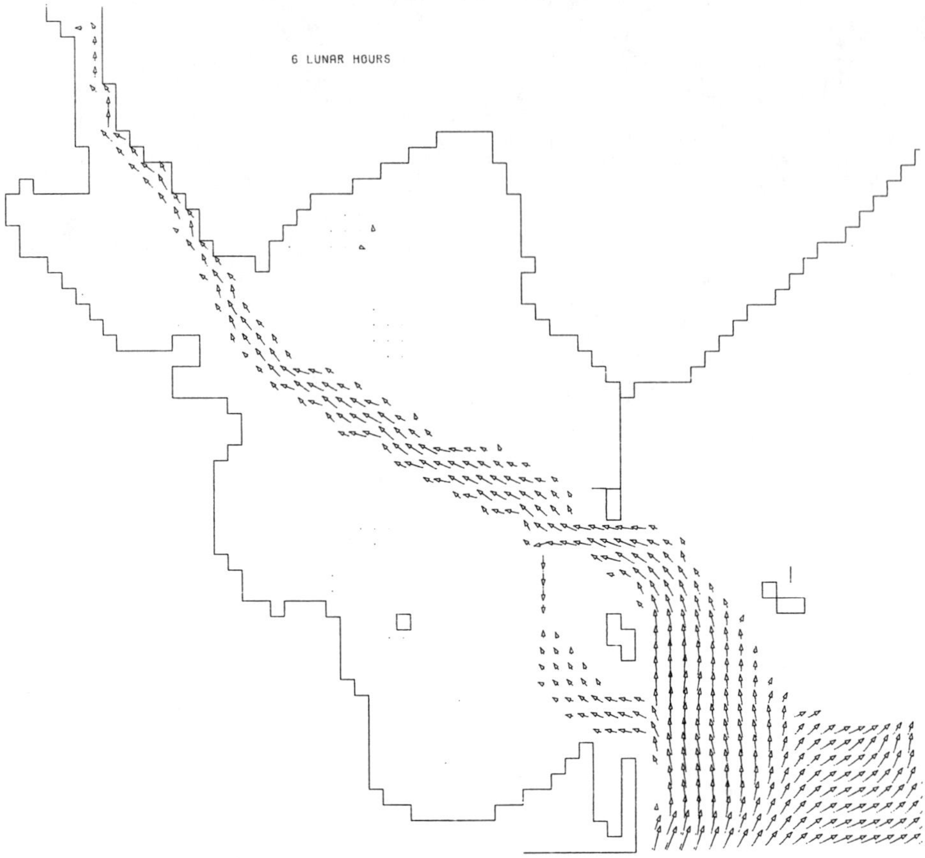


FIGURE 12E.

FIGURE 12F.

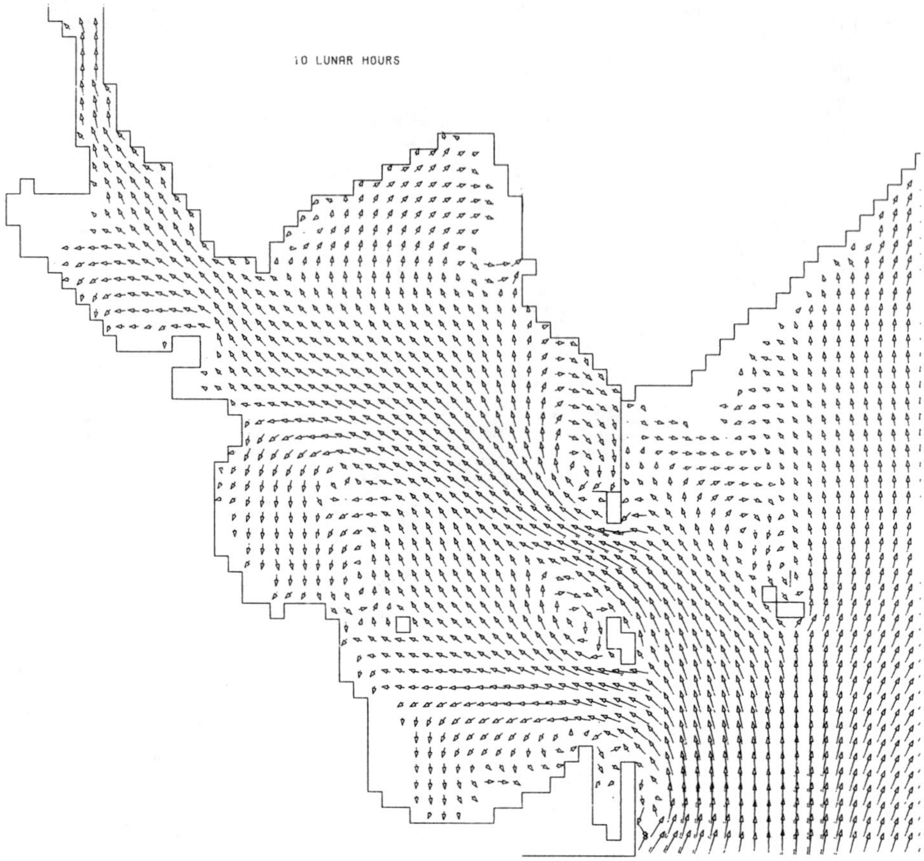


FIGURE 12G.

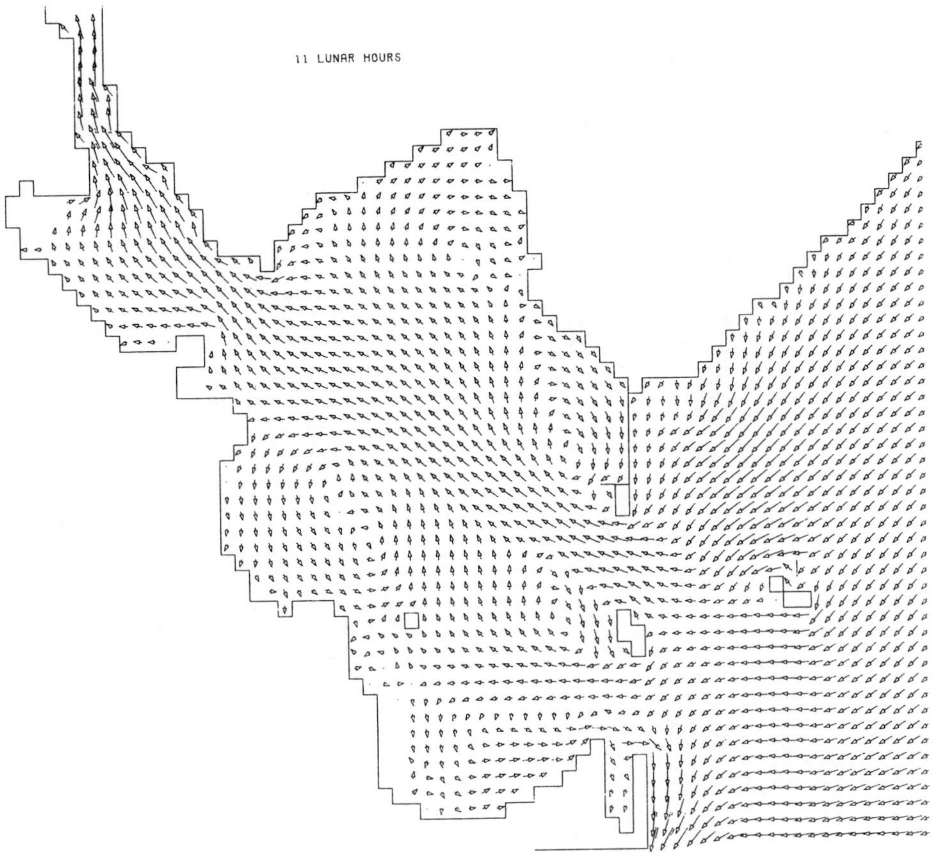


FIGURE 12H.

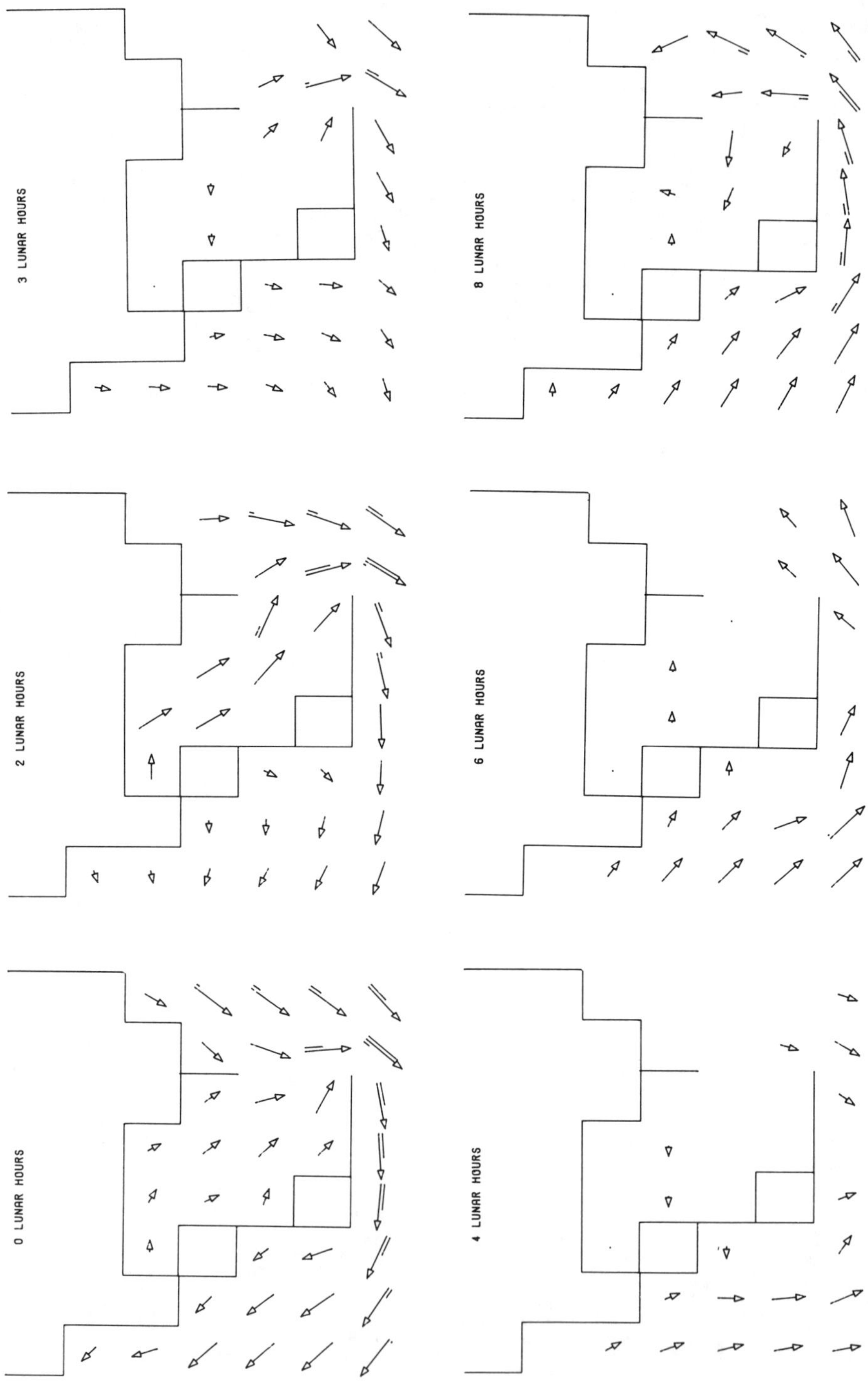
0 LUNAR HOURS
2 LUNAR HOURS
3 LUNAR HOURS
4 LUNAR HOURS
6 LUNAR HOURS
8 LUNAR HOURS

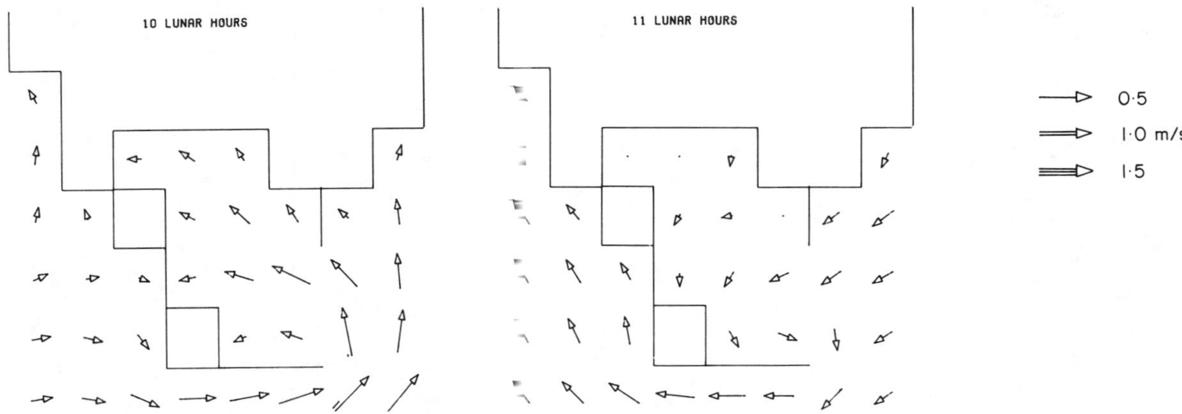


FIGURE 13. Distributions of computed current within the study area from the coarse-grid model, run C1.

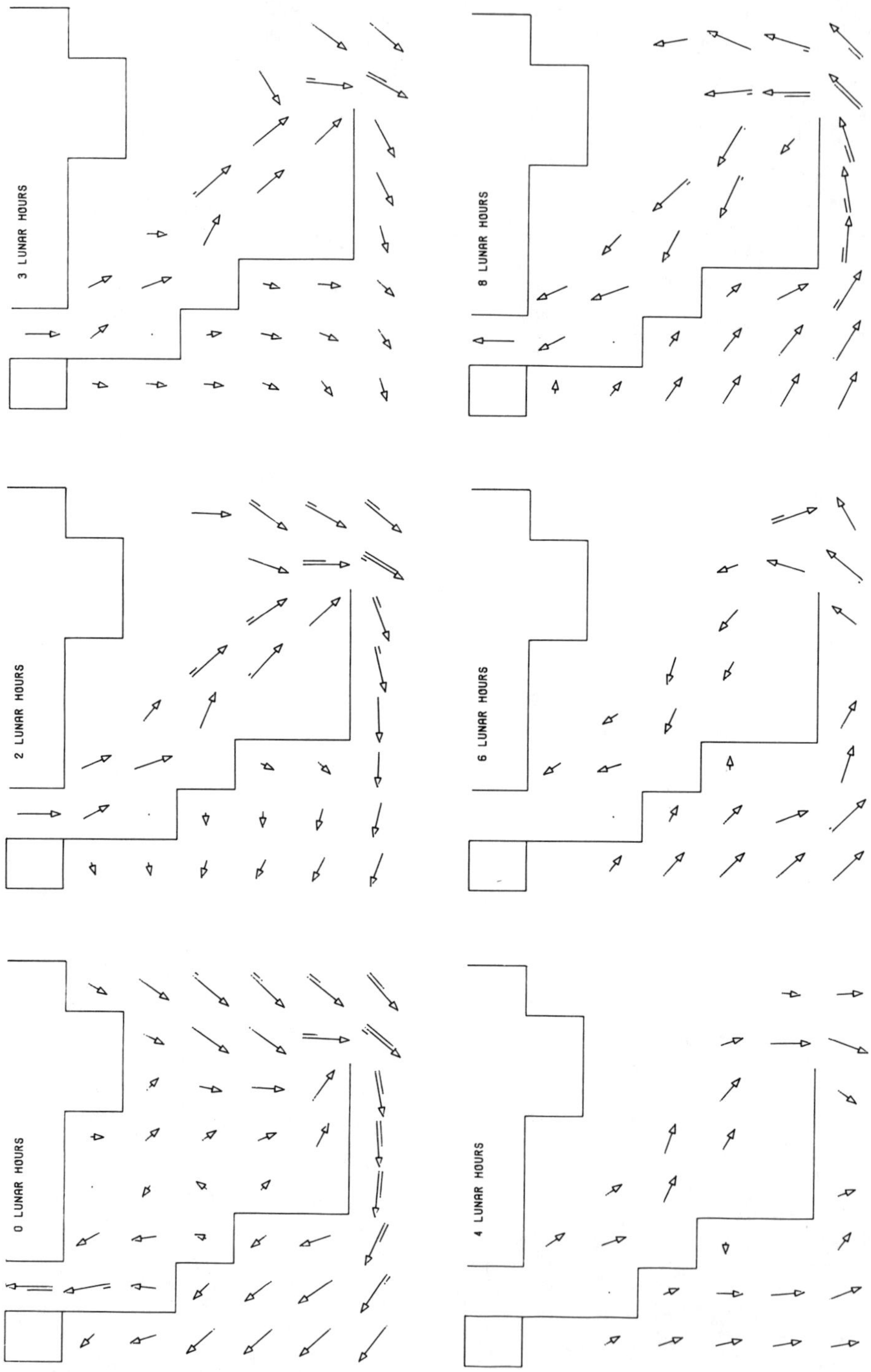
3 LUNAR HOURS
8 LUNAR HOURS
2 LUNAR HOURS
6 LUNAR HOURS
0 LUNAR HOURS
4 LUNAR HOURS

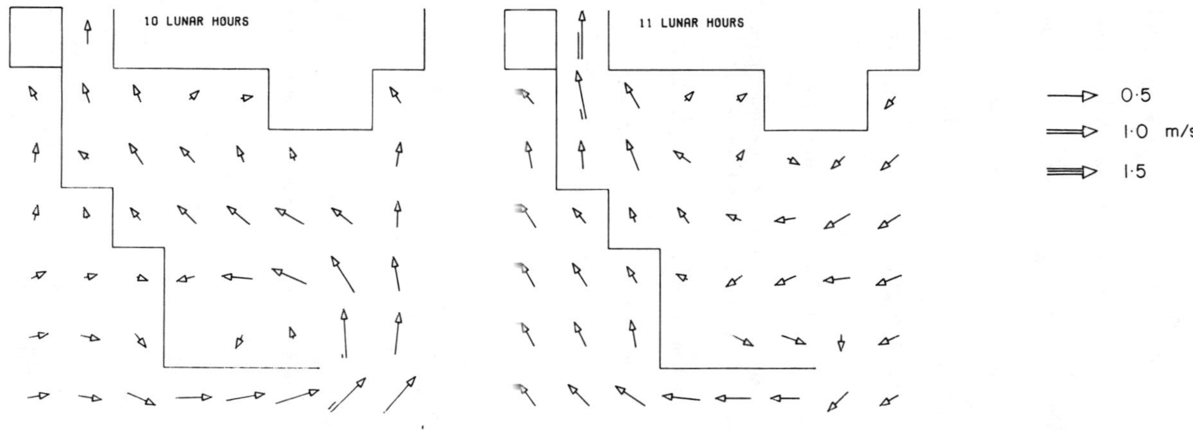


FIGURE 14. Distributions of computed current within the study area from the coarse-grid model, run C41.

# REFERENCES


1. **Flather, R. A. and Heaps, N. S.,** Tidal computations for Morecambe Bay, *Geophys. J. R. Astronom. Soc.,* 42, 489, 1975.
2. **Leendertse, J. J. and Gritton, E. C.** A water quality simulation model for well mixed estuaries and coastal seas, Vol. 11, Computation procedures, The Rand Corporation, New York, R-708-NYC, 1971.
3. **Stelling, G. S., Wiersma, A. K., and Willemse, J. B. T. M.,** Practical aspects of accurate tidal computations, *J. Hydraul. Eng.,* 112(9), 802, 1986.
4. **Thacker, W. C.,** Irregular-grid finite-difference techniques for storm surge calculations for curving coastlines, in *Marine Forecasting,* Proc. 10th Liège Colloq. Ocean Hydrodynamics, Nihoul, J. C. J., Ed., Elsevier, Amsterdam, 1979, 261.
5. **Lynch, D. R. and Gray, W. G.,** Finite element simulation of flow in deforming regions, *J. Comp. Phys.,* 36, 135, 1980.
6. **Falconer, R. A. and Owens, P. H.,** Numerical simulation of flooding and drying in a depth-averaged tidal flow model, in *Proc. Inst. Civil Eng.,* Part 2, 83, 161, 1987.
7. **Reid, R. O. and Bodine, B. R.,** Numerical model for storm surges in Galveston Bay, *J. Waterways Harbours Div., Proc. Am. Soc. Civil Eng.,* 94(WW1), 35, 1968.
8. **Yeh, G.-T. and Chou, F.-K.,** Moving boundary numerical surge model, *J. Waterway, Port, Coastal Ocean Div., Proc. Am. Soc. Civil Eng.,* 105(WW3), 247, 1979.
9. **Sielecki, A. and Wurtele, M. G.,** The numerical integration of the nonlinear shallow-water equations with sloping boundaries, *J. Comp. Phys.,* 6, 219, 1970.
10. **Lynch, D. R.,** Moving boundary numerical surge model, *J. Waterway, Port, Coastal Ocean Div., Proc. Am. Soc. Civil Eng.,* 106(WW3), 425, 1980.
11. **Johns, B., Dube, S. K., Sinha, P. C., Mohanty, U. C., and Rao, A. D.,** The simulation of a continuously deforming lateral boundary in problems involving the shallow water equations, *Comp. Fluids,* 10(2), 105, 1982.
12. **Stephens, C. V.,** Hydrodynamic Modelling Developments for the West Coast of the British Isles, Ph.D thesis, University of Liverpool, England, 1983.
13. **Davies, A. M. and Flather, R. A.,** Application of numerical models of the north west European continental shelf and the North Sea to the computation of the storm surges of November to December 1973. *Dtsch. Hydrogr. Z. Erg. Reihe A,* 14, 1978.
14. **Rossiter, J. R. and Lennon, G. W.,** Computation of tidal conditions in the Thames estuary by the initial value method, *Proc. Inst. Civil Eng.,* 31, 25, 1965.
15. **Pedersen, G.,** On the effects of irregular boundaries in finite difference models, *Int. J. Numerical Methods Fluids,* 6, 497, 1986.

Chapter 8

# A FINE-GRID THREE-DIMENSIONAL $M_2$ TIDAL MODEL OF THE EAST CHINA SEA

**Byung Ho Choi**

## TABLE OF CONTENTS

# I. INTRODUCTION

Fresh-water input from rivers, surface wind, solar radiation, surface cooling, and external water masses all play important roles in determining the patterns of water characteristics and currents in the Yellow Sea and the East China Sea continental shelf, but the region is most remarkable for the large tides along the west coast of Korea, Chinese coast, and for the complexity of tidal phenomena.

During the late 1920s and early 1930s extensive studies of this shelf sea area were reported and a considerable number of tidal measurements were made, from which co-tidal and co-range charts of diurnal and semidiurnal tides were prepared.[1] Russian scientists [2-4] reported the first application of two-dimensional mathematical models on the one hand to the Yellow Sea and on the other hand, to the Gulfs of Bohai and Liadong, employing boundary-value methods which call for much tidal data at external boundary points along the coast and open boundary.

During the past few years series of two-dimensional depth-averaged numerical tidal models based on initial value method have been reported and it was demonstrated by these models that the tides can be satisfactorily reproduced, thus providing general information on the tidal dynamics in this region.[5-9]

However, improvement on the accuracy of these models and verification of computed tidal currents were much limited due to the lack of long-term tidal observations. Also, these models are primarily based upon the vertically integrated equations; the vertical variation of tidal currents cannot be obtained.

Recently, Choi[10,11] has successfully used a three-dimensional tidal model with a grid resolution of 1/5° latitude by 1/4° longitude to study the vertical structure of four major constituents of tidal currents on the East China Sea continental shelf adopting the Galerkin spectral tidal modeling technique by Davies.[12]

With the advent of expanding computer power and increasing necessity of resolving the flow with enough details, the previous three-dimensional model of the East China Sea has been further improved to higher resolution of 1/15° latitude by 1/12° longitude covering the entire continental shelf. As a first step, the developed fine-grid model was utilized to reproduce the three-dimensional current structure of the $M_2$ tide representing the dominant tidal conditions of the shelf.

The computed results were then compared with the current-meter data taken as a part of joint mooring experiments in the Eastern Yellow Sea during the winter of 1986. Reasonable agreement was found between the observations and model results, thus supporting the computed vertical distribution of $M_2$ tidal currents in the region. Times and phases referred to in this study are referenced to 135°E longitude (time zone, −9h).

# II. HYDRODYNAMIC EQUATIONS

The equations of continuity and motion for homogeneous water neglecting nonlinear terms, shear in the horizontal and the direct influence of the tide-generating potentials, may be written in spherical polar coordinates

$$\frac{\partial \xi}{\partial t} + \frac{1}{R\cos\phi}\left[\frac{\partial}{\partial \chi}\int_0^h u dz + \frac{\partial}{\partial \phi}\int_0^h v\cos\phi dz\right] = 0 \tag{1}$$

$$\frac{\partial u}{\partial t} - \gamma v = \frac{-g}{R\cos\phi}\frac{\partial \xi}{\partial \chi} + \frac{\partial}{\partial z}\left(N\frac{\partial u}{\partial z}\right) \tag{2}$$

$$\frac{\partial v}{\partial t} + \gamma u = \frac{-g}{R}\frac{\partial \xi}{\partial \phi} + \frac{\partial}{\partial z}\left(N\frac{\partial v}{\partial z}\right) \tag{3}$$

where

| | |
|---|---|
| $\chi, \phi$ | east-longitude and north-latitude, respectively |
| z | depth below the undisturbed surface |
| t | time |
| $\xi$ | elevation of the sea surface above the undisturbed level |
| h | undisturbed depth of water |
| $\rho$ | the density of sea water |
| R | the radius of the Earth |
| $\gamma$ | Coriolis parameter ($\gamma = 2\,\omega \sin\phi$) |
| $\omega$ | angular speed of the Earth's rotation |
| g | the acceleration due to gravity |
| u,v | the east-going and north-going components of current at depth z |
| N | coefficient of vertical eddy viscosity |

In order to solve Equations 1, 2, and 3 for $\xi$, u, and v, the appropriate boundary conditions at the sea surface and sea bed have to be specified.

For tides, the sea surface condition is

$$-\rho\left(N\frac{\partial u}{\partial z}\right)_0 = -\rho\left(N\frac{\partial v}{\partial z}\right)_0 = 0 \tag{4}$$

with the subscript zero denoting elevation at z = 0. Assuming a slip boundary condition at the sea bed (z = h) and using a quadratic law of bottom friction yields

$$-\rho\left(N\frac{\partial u}{\partial z}\right)_h = K\rho u_h(u_h^2 + v_h^2)^{1/2} \tag{5a}$$

$$-\rho\left(N\frac{\partial v}{\partial z}\right)_h = K\rho v_h(u_h^2 + v_h^2)^{1/2} \tag{5b}$$

where K is the coefficient of quadratic bottom friction, taken as constant (0.0030). By expanding the two components of velocity u,v in terms of depth-dependent functions (basis functions) and horizontal space- and time-dependent coefficients, Equations 1, 2, and 3 can be solved using the Galerkin method in the vertical and finite difference grid in the horizontal.

For the choice of basis functions, Davies and Furnes[13] have shown that an expansion of only ten cosine functions is sufficient to accurately reproduce the depth variation of current. Choi[10,11] also has applied such an expansion to the computation of the $M_2$, $S_2$, $K_1$, and $O_1$ tides of the East China Sea continental shelf in the previous model. A full description of the Galerkin method is given in Davies[12,13] and will not be restated here.

The particular case in which the vertical eddy viscosity N is independent of depth coordinate z is considered. Along a closed boundary the normal component of current is set to zero, for all $t \geq 0$. Along the open boundaries of the model, $M_2$ tidal input interpolated from the two-dimension model[6] is specified and a radiation condition is employed to allow disturbances from the interior of the model to pass outward.

This condition involves a prescribed relation between the total normal component of depth-mean current q and total elevation $\xi$ given by:

$$q = q_T + \frac{c}{h}(\xi - \xi_T) \tag{6}$$

where $c = (gh)^{1/2}$. The change in sea-surface elevation $\xi_T$, arising from the $M_2$ tide is given by:

$$\xi_T = H_{M_2}\cos\{\sigma_{M_2}t - \kappa_{M_2}\} \tag{7}$$

and

$$q_T = Q_{M_2}\cos\{\sigma_{M_2}t - \gamma_{M_2}\} \tag{8}$$

In equations 7 and 8 $H_{M2}$ and $Q_{M2}$ denote the amplitude of $M_2$ tidal elevation and the amplitude of the normal component of depth-mean $M_2$ tidal current, respectively. Also $K_{M2}$ and $\gamma_{M2}$ denote the phase of $M_2$ tidal elevation and the phase of the normal component of depth-mean $M_2$ tidal current. These values were also interpolated from the previous results of two-dimensional model.[6]

As shown in Figure 1 the grid spacing of the present model is about 4 nautical miles resulting from one third refinement of the grid system in the previous course shelf model. The total number of grid points is over 35,000 and 60% of total grids are interior points. The time step for stable difference solutions according to the Courant-Friedrichs-Lewy criterion was chosen as 69.003 s and has the period of the $M_2$ constituent as an integer number of time steps (648). The coefficient of bottom friction used has a value of 0.003 uniformly over the region. The average depth for each grid was determined from navigational charts. Depth contours corrected with reference to mean sea level for model bathymetry is shown in Figure 2. Along with a perspective plot of depth fields illustrated in Figure 3, the region has very complex coastline and bottom topography.

## III. CURRENT OBSERVATION

The current meter data used in this study come from the part of a joint current monitoring experiment in the Eastern Yellow Sea by Florida State University and Sung Kyun Kwan University, Korea. The six moorings with one or two Burst Sampling Current Meters (BSCM) from Florida State University were set in January 1986 during the cruise of R/V Thompson. Four complete and one partial mooring were recovered in April 1986 during the recovery leg by Korea Hydrographic office vessel Pusan 802. The sixth mooring with 2 BSCM was recovered in June 1986 by R/V Washington.

The location of the moorings is shown in Figure 2, while Figure 4 displays mooring design. The anchors on the mooring were made of lead to minimize the magnetic influence on the compass system of the nearest BSCM. Coordinates of the moorings, as given in the statistics and in Table 1, are the Loran C position of the ship taken when the anchor was thrown from the deck.

A detailed description of compilation of data from the moorings is given in report by Harkema and Hsueh.[14] The compiled data of sampling interval of 7.5 min were subjected to a low-pass filter (half length of 40) and then reconstructed to an equivalent 1 h sampling interval. The time of the first observation in local time is given for each record. The results of the standard least-square tidal analysis of the data are listed in the Table 2.

## IV. MODEL RESULTS

The general propagation of the $M_2$ tide which represents average tidal condition is illustrated as co-amplitude and co-phase lines in Figure 5 and Figure 6, respectively. The $M_2$ tides have more or less constant phase at the shelf break of the East China Sea, with phases and amplitudes increasing landward. Amplitudes are high (more than 120 cm) south

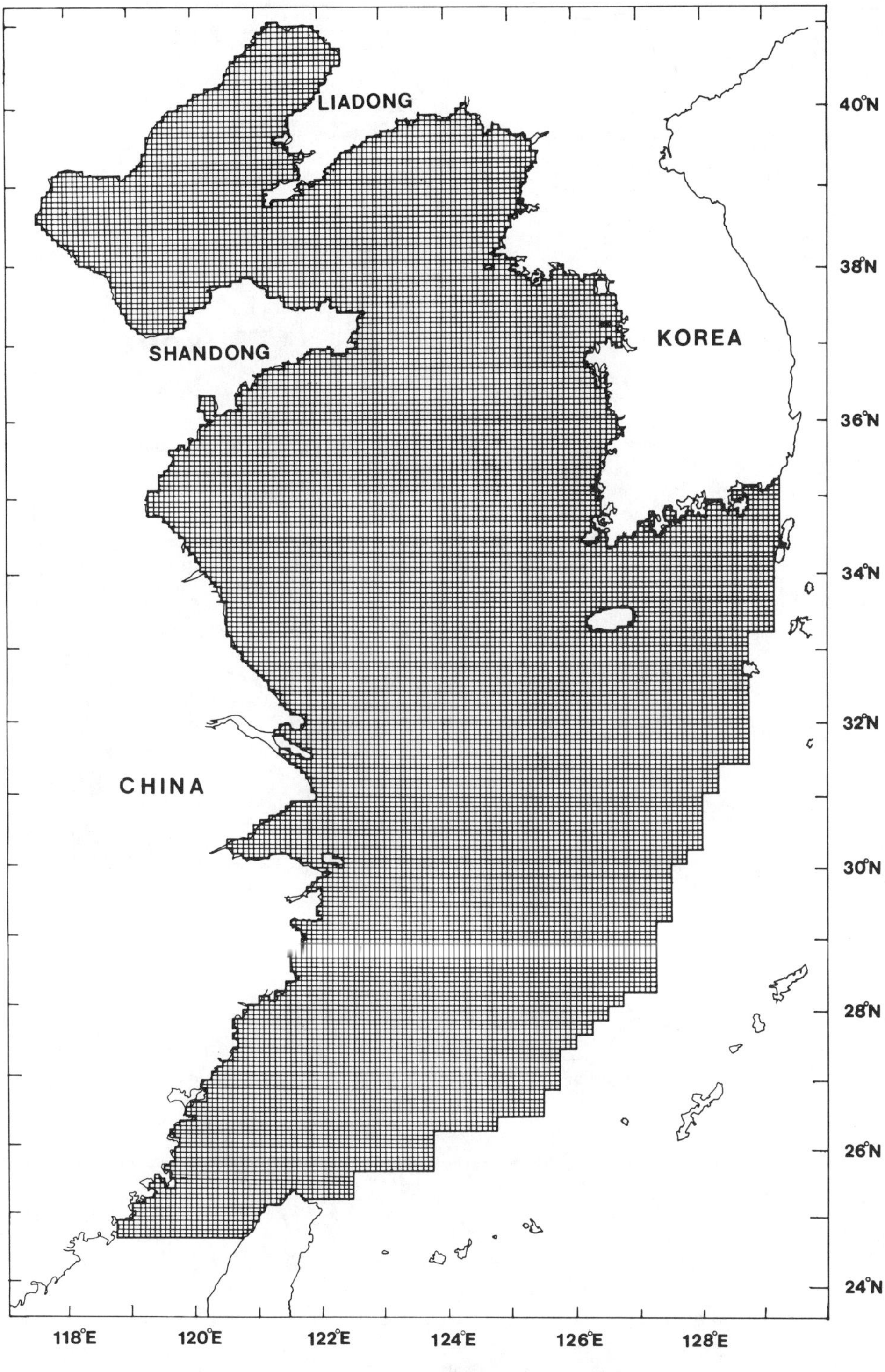


FIGURE 1. Finite difference grid of the three-dimensional East China Sea model.

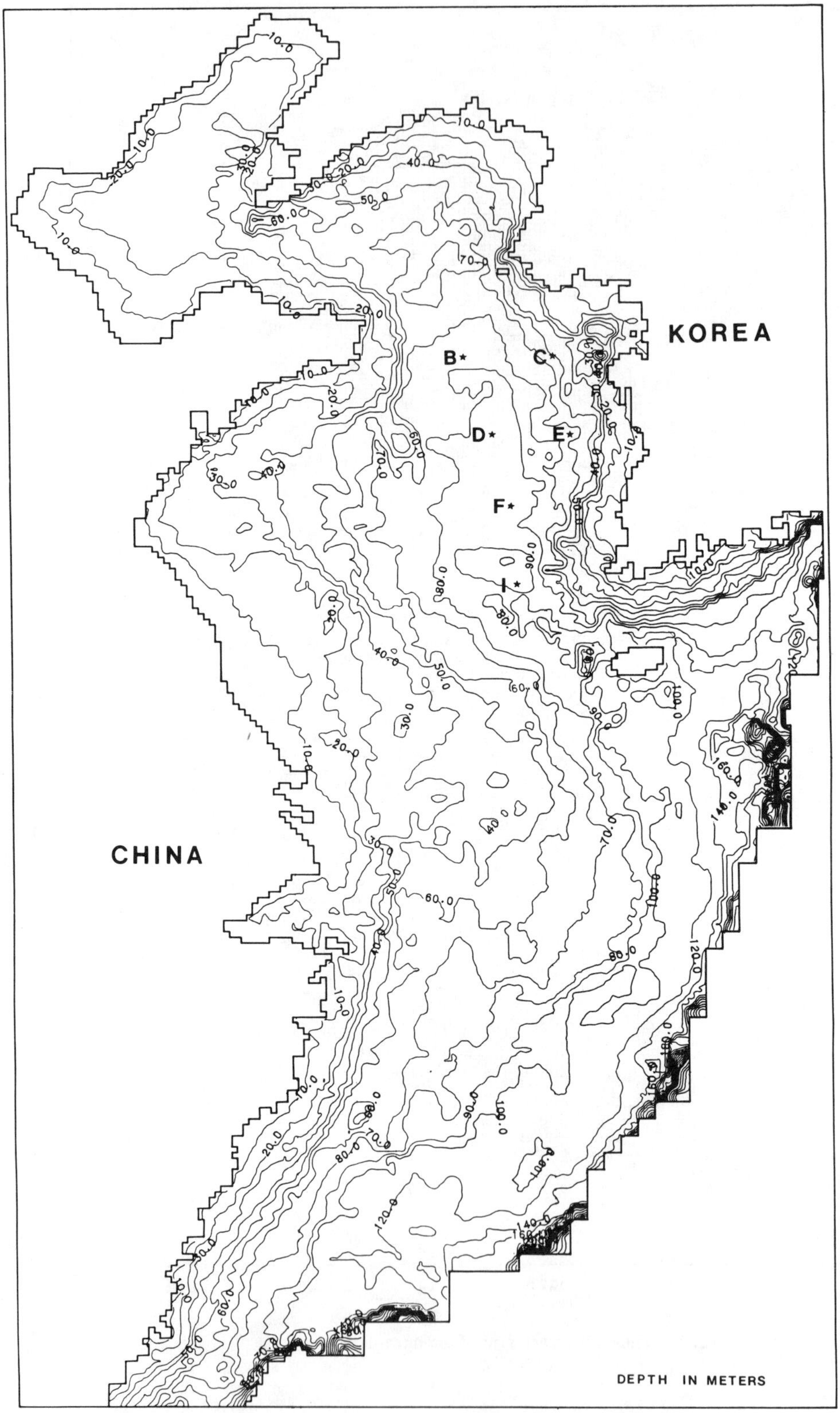


FIGURE 2. Bottom topographic map of the East China Sea showing the position of current-meter moorings in the Eastern Yellow Sea.

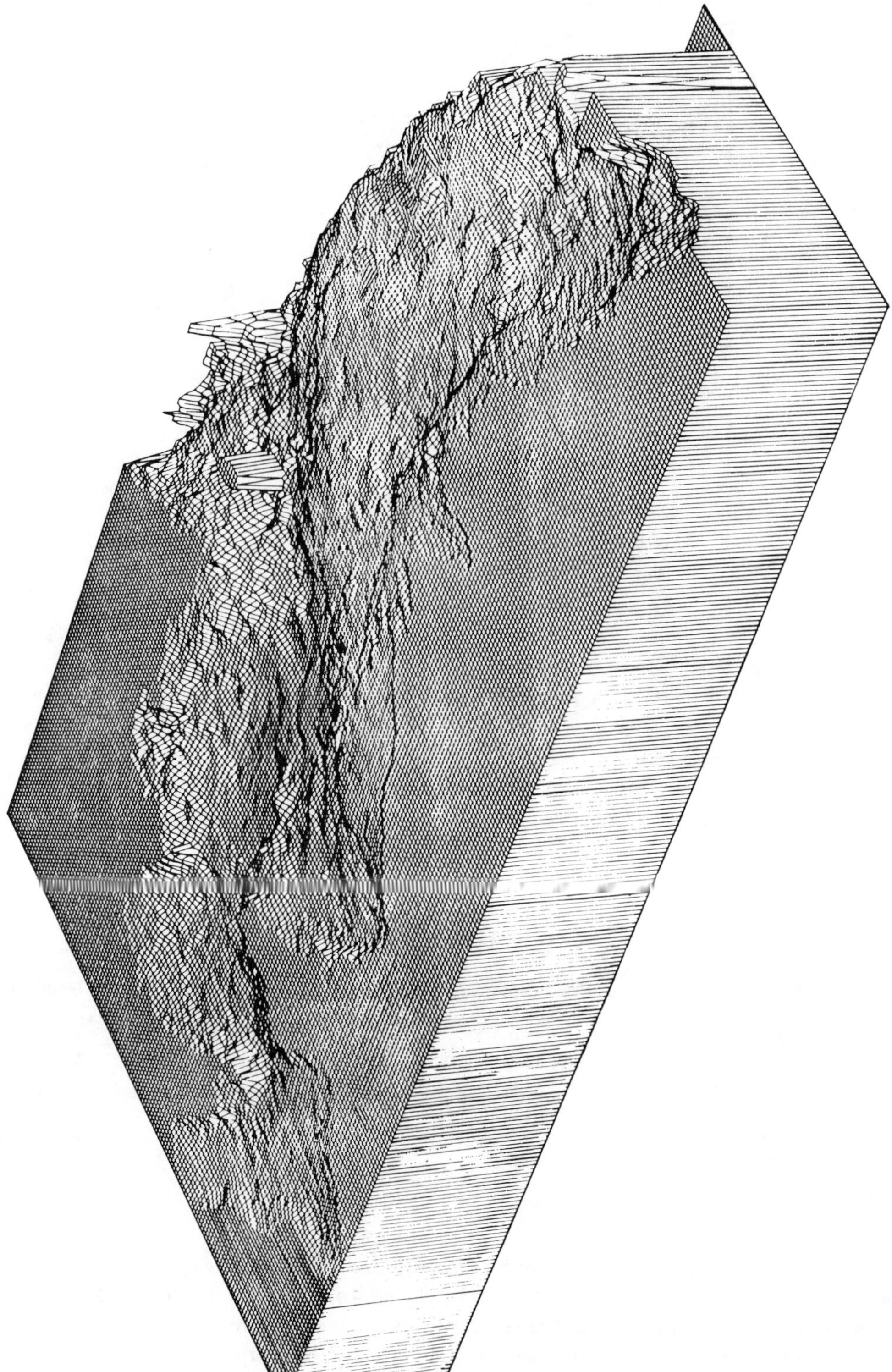

FIGURE 3 Perspective plot of modeled bottom topography.

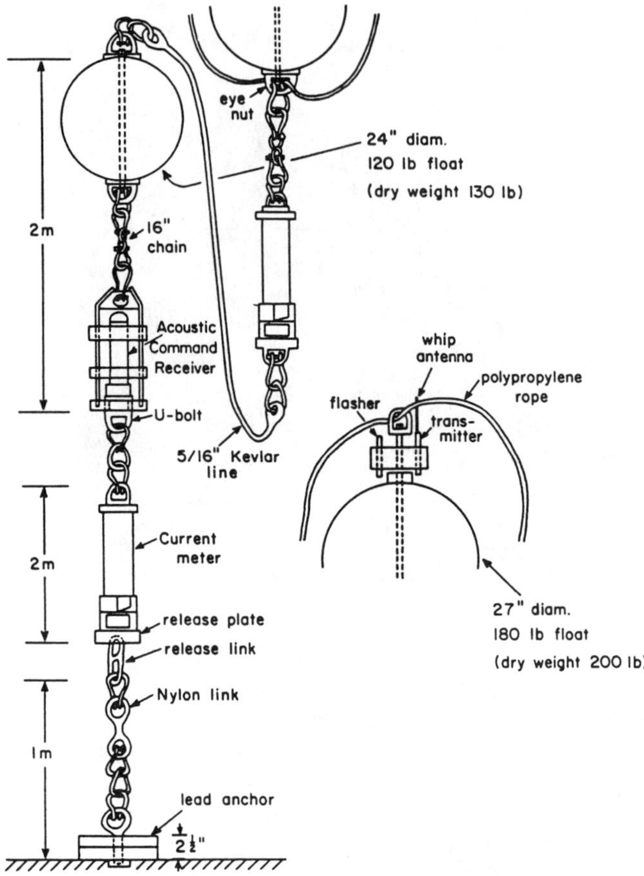


FIGURE 4. Mooring designs for current measurements in the Eastern Yellow Sea.

**TABLE 1**
**Locations of Moorings**

| Mooring | Meter depth (m) | Water depth (m) | Longitude | Latitude | Record length (d) |
|---|---|---|---|---|---|
| B | 38.0, 74.0[a] | 75.0 | 124° 05.0′ | 36° 56.9′ | 80 |
| C | 52.0 | 53.0 | 125° 24.8′ | 36° 57.0′ | 44 |
| D | 41.0, 86.0 | 87.0 | 124° 34.9′ | 36° 00.1′ | 81 |
| E | 63.0 | 64.0 | 125° 35.9′ | 36° 01.7′ | 79 |
| F | 70.0, 94.5 | 95.5 | 124° 44.3′ | 35° 13.8′ | 57 |
| I | 48.0, 92.8 | 93.8 | 124° 41.5′ | 34° 18.1′ | 21 |

[a] Lost current meter.

propagation of which the pattern was set by the influence of the Earth's rotation and sudden shallowness of continental shelf to outer sea by a factor of 30 or so. These results will not be discussed further since similar charts have been derived before [6,11] and have general agreement with empirically drawn charts based on coastal observations.[1] There are still discrepancies in the Gulfs of Bohai and Liadong underestimating the amplitudes. Position of amphidromic point in Bohai Gulf is improved due to the present fine-grid schematization.

The strength of the computed tidal currents at surface and sea bottom are illustrated by

## TABLE 2
## Harmonic Constants of Observed Tidal Currents in the Eastern Yellow Sea

Current meter rig: B
First record 00:00 day 13 1986, last record 23:00 day 72 1986

| | 38 m | | | | 74 m | | | |
|---|---|---|---|---|---|---|---|---|
| | u | | v | | u | | v | |
| **Tidal constituents** | **H** | **κ** | **H** | **κ** | **H** | **κ** | **H** | **κ** |
| $M_2$ | 18.6 | 4 | 32.8 | 146 | 11.8 | 12 | 22.4 | 136 |
| $S_2$ | 7.9 | 72 | 12.2 | 200 | 5.3 | 73 | 8.6 | 191 |
| $K_1$ | 2.4 | 144 | 5.9 | 272 | 1.8 | 147 | 3.7 | 272 |
| $O_1$ | 1.4 | 102 | 3.9 | 237 | 1.3 | 110 | 2.1 | 242 |
| $N_2$ | 2.8 | 341 | 5.4 | 120 | 1.9 | 349 | 3.6 | 113 |

Current meter rig: C
First record 00:00 day 12 1986, last record 23:00 day 65 1986

| | 52 m | | | |
|---|---|---|---|---|
| | u | | v | |
| **Tidal constituents** | **H** | **κ** | **H** | **κ** |
| $M_2$ | 29.5 | 4 | 26.0 | 88 |
| $S_2$ | 12.8 | 69 | 11.0 | 142 |
| $K_1$ | 2.7 | 153 | 4.8 | 259 |
| $O_1$ | 1.8 | 108 | 3.1 | 220 |
| $N_2$ | 4.2 | 346 | 4.3 | 65 |

Current meter rig: D
First record 00:00 day 12 1986, last record 23:00 day 71 1986

| | 11 m | | | | 06 m | | | |
|---|---|---|---|---|---|---|---|---|
| | u | | v | | u | | v | |
| **Tidal constituents** | **H** | **κ** | **H** | **κ** | **H** | **κ** | **H** | **κ** |
| $M_2$ | 15.2 | 11 | 26.1 | 63 | 15.5 | 10 | 17.0 | 60 |
| $S_2$ | 7.6 | 79 | 12.2 | 113 | 7.5 | 72 | 7.2 | 109 |
| $K_1$ | 1.5 | 220 | 8.8 | 257 | 2.0 | 169 | 5.1 | 262 |
| $O_1$ | 0.9 | 169 | 5.9 | 213 | 1.2 | 121 | 3.6 | 225 |
| $N_2$ | 2.1 | 358 | 5.1 | 46 | 2.1 | 3 | 3.0 | 51 |

**TABLE 2 (continued)**
**Harmonic Constants of Observed Tidal Currents in the Eastern Yellow Sea**

Current meter rig: E
First record 00:00 day 13 1986, last record 23:00 day 72 1986

| Tidal constituents | 63 m | | | |
|---|---|---|---|---|
| | u | | v | |
| | H | κ | H | κ |
| $M_2$ | 20.2 | 348 | 25.4 | 43 |
| $S_2$ | 8.9 | 49 | 11.9 | 95 |
| $K_1$ | 1.7 | 164 | 5.5 | 242 |
| $O_1$ | 1.4 | 118 | 3.4 | 203 |
| $N_2$ | 2.9 | 329 | 4.3 | 27 |

Current meter rig: F
First record 00:00 day 13 1986, last record 23:00 day 72 1986

| Tidal constituents | 70 m | | | | 94.5 m | | | |
|---|---|---|---|---|---|---|---|---|
| | u | | v | | u | | v | |
| | H | κ | H | κ | H | κ | H | κ |
| $M_2$ | 5.0 | 7 | 40.0 | 19 | 8.4 | 329 | 26.0 | 350 |
| $S_2$ | 2.7 | 90 | 19.6 | 71 | 4.3 | 57 | 12.1 | 66 |
| $K_1$ | 2.4 | 127 | 10.4 | 252 | 2.1 | 147 | 6.6 | 251 |
| $O_1$ | 2.4 | 54 | 7.3 | 211 | 1.6 | 51 | 4.8 | 190 |
| $N_2$ | 0.8 | 329 | 6.8 | 3 | 1.1 | 281 | 4.0 | 324 |

Current meter rig: I
First record 00:00 day 12 1986, last record 23:00 day 41 1986

| Tidal constituents | 48 m | | | |
|---|---|---|---|---|
| | u | | v | |
| | H | κ | H | κ |
| $M_2$ | 13.4 | 112 | 43.8 | 359 |
| $S_2$ | 7.1 | 164 | 18.4 | 54 |
| $K_1$ | 2.7 | 79 | 12.7 | 253 |
| $O_1$ | 2.7 | 42 | 8.4 | 207 |
| $N_2$ | 2.1 | 94 | 8.0 | 340 |

*Note:* Local phase lag in degrees referred to longitude of current meter moorings, amplitude in cm/s

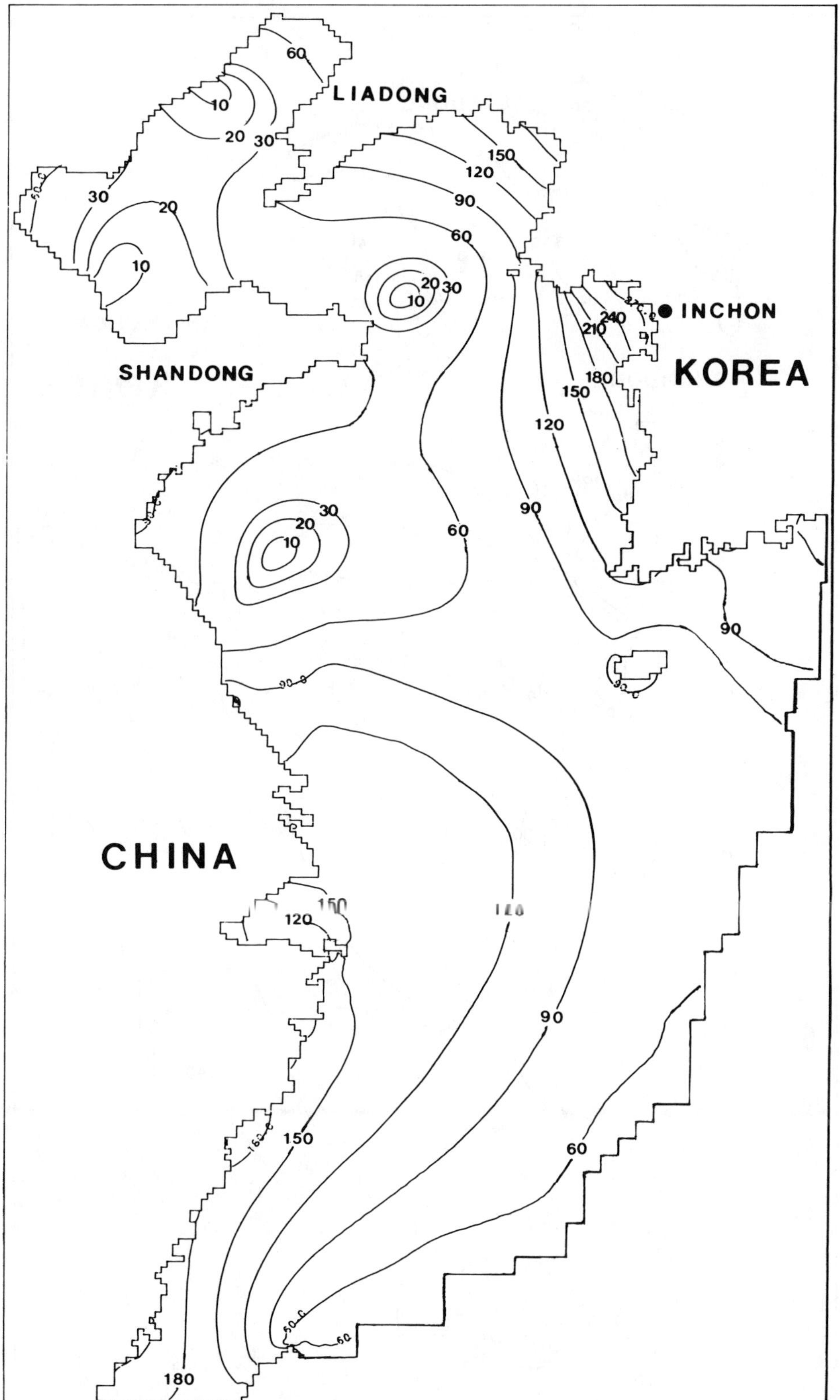


FIGURE 5. Computed co-amplitude chart for the $M_2$ tide. (Values are given in cm.)

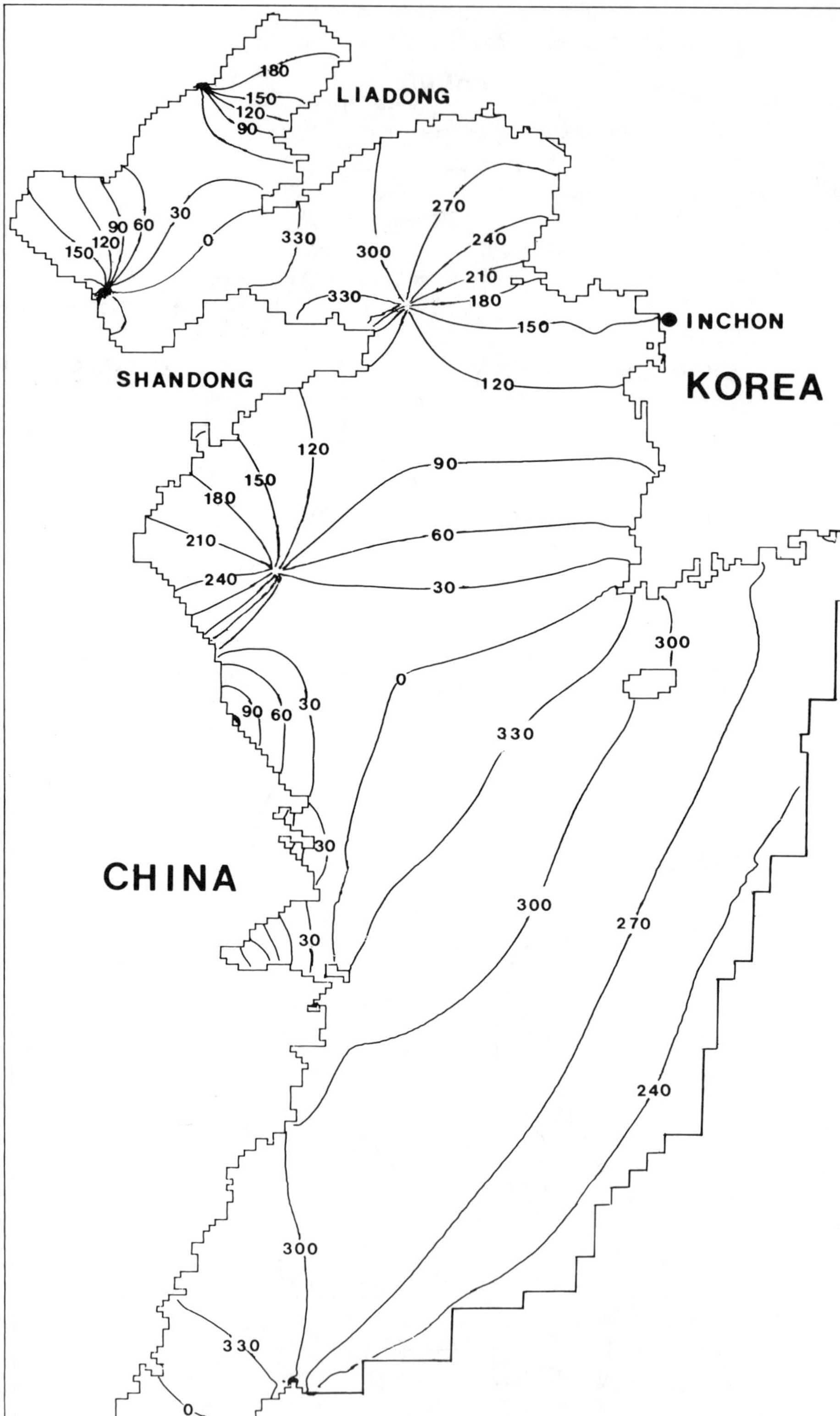


FIGURE 6. Computed co-phase chart for the $M_2$ tide. (Values are given in degrees relative to the moon's transit at 135° E.)

of 32°N and increase again into the Yellow Sea where there is an amphidromic circulation of semidiurnal tides. The large tides developed in Inchon Bay are due to the near-resonant response of the Bay system acting as a quarter wave length resonator for the entering tidal Figure 7 and Figure 8, obtained at 3 lunar hours. As seen in Figure 8, the current at sea bottom in Inchon Bay and Hangzhou Bay shows considerable magnitude suggesting the high transport potentials.

In order to make a comparison between computed and observed currents, amplitudes and phases of easterly and northerly components of currents were examined. In Table 3, computed and observed values of amplitude H (cm/s) and phase κ (degree referred to 135°E) for nine current meter rigs are given. It is seen from this table that good agreement between the observations and model results have been demonstrated. However, it is worth mentioning that a noticeable phase error in the easterly component of current in rig F may be due to the existence of a nodal point of easterly current in the nearby region.

These computed tidal currents can best be represented by current charts in the form of tidal ellipses. Spatial distribution of principal axes of ellipses at mid-depth and sea bottom are illustrated by Figure 9 and Figure 10, respectively. These ellipses give an overall impression of the magnitude and direction of the $M_2$ tide representing the maximum and minimum velocities as semimajor and semiminor axes, respectively. In these figures the rotatory currents are seen in the lower part of the Yellow Sea and over the East China Sea shelf and strong rectilinear currents were computed along the west coast of Korea and Hangzhou Bay both at mid-depth and at sea bottom. The sea floor swept by these strong currents forms a strong dynamic sedimentation regime on the shelf.[15]

## V. CONCLUDING REMARKS

A fine-grid three-dimensional model has been formultated to represent continuous vertical structure of the $M_2$ tidal currents in the East China Sea with sufficient details. It was demonstrated that computed currents are in good agreement with limited current observations made during the winter of 1986, thus supporting the computed tidal distribution from the model. With this fine-grid model the tides other than the $M_2$ component can also be modeled in a more detailed manner to provide better understanding of tidal physics in the shelf. However, shelf edge tidal observations and extensive current measurements are necessary to achieve better accuracy of the model.

## ACKNOWLEDGMENTS

The East China Sea modeling work was initiated during my visit to Bidston Observatory. The author is indebted to scientists in the dynamic modeling group, especially to the late Dr. N. S. Heaps. The joint current observation program was funded by Korea Science and Engineering Foundation and numerical computations by IBM 3083 was supported by Software Engineering Center at System Engineering Research Institute, KAIST.

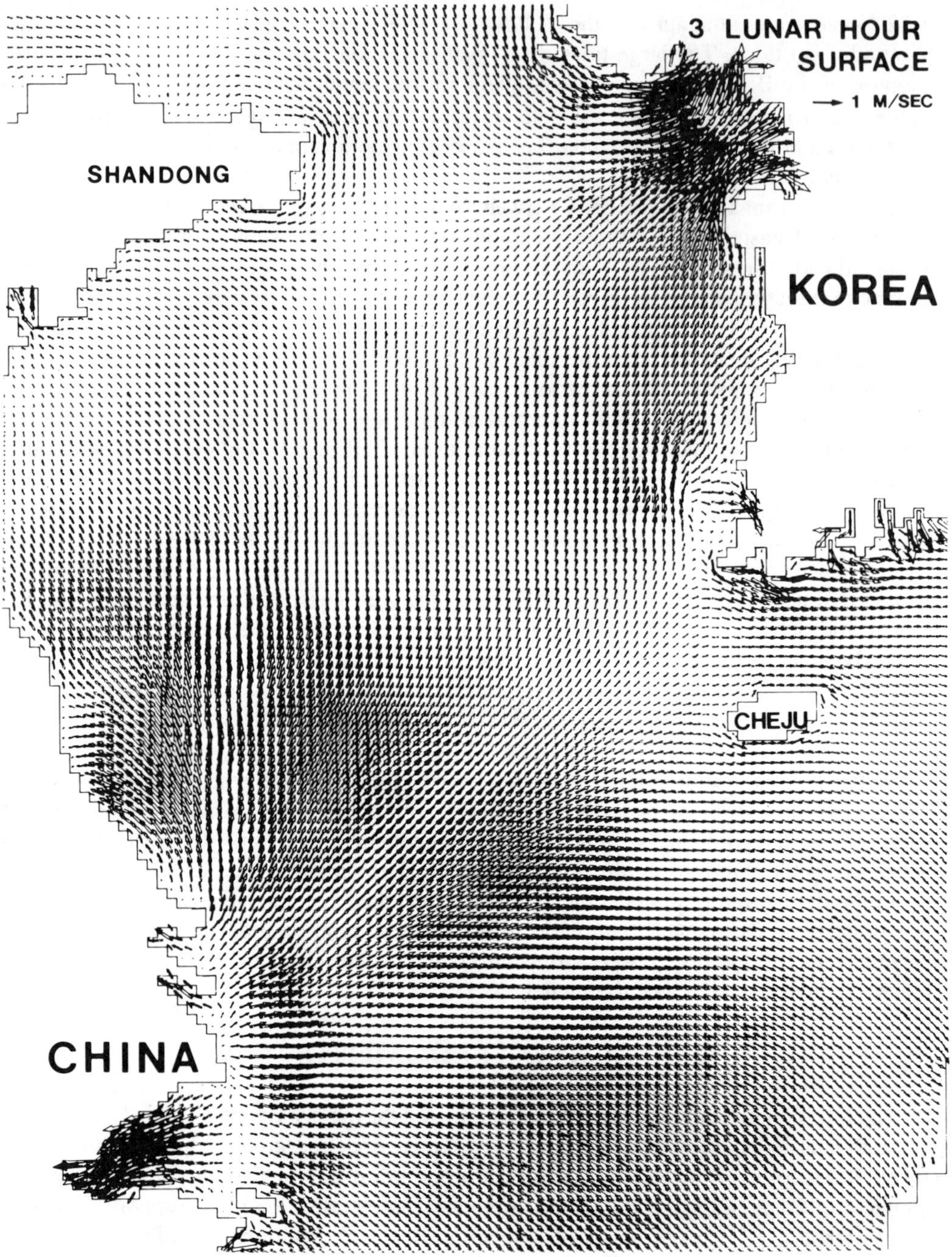


FIGURE 7. Current vectors at surface for the $M_2$ tide at 3 lunar hours.

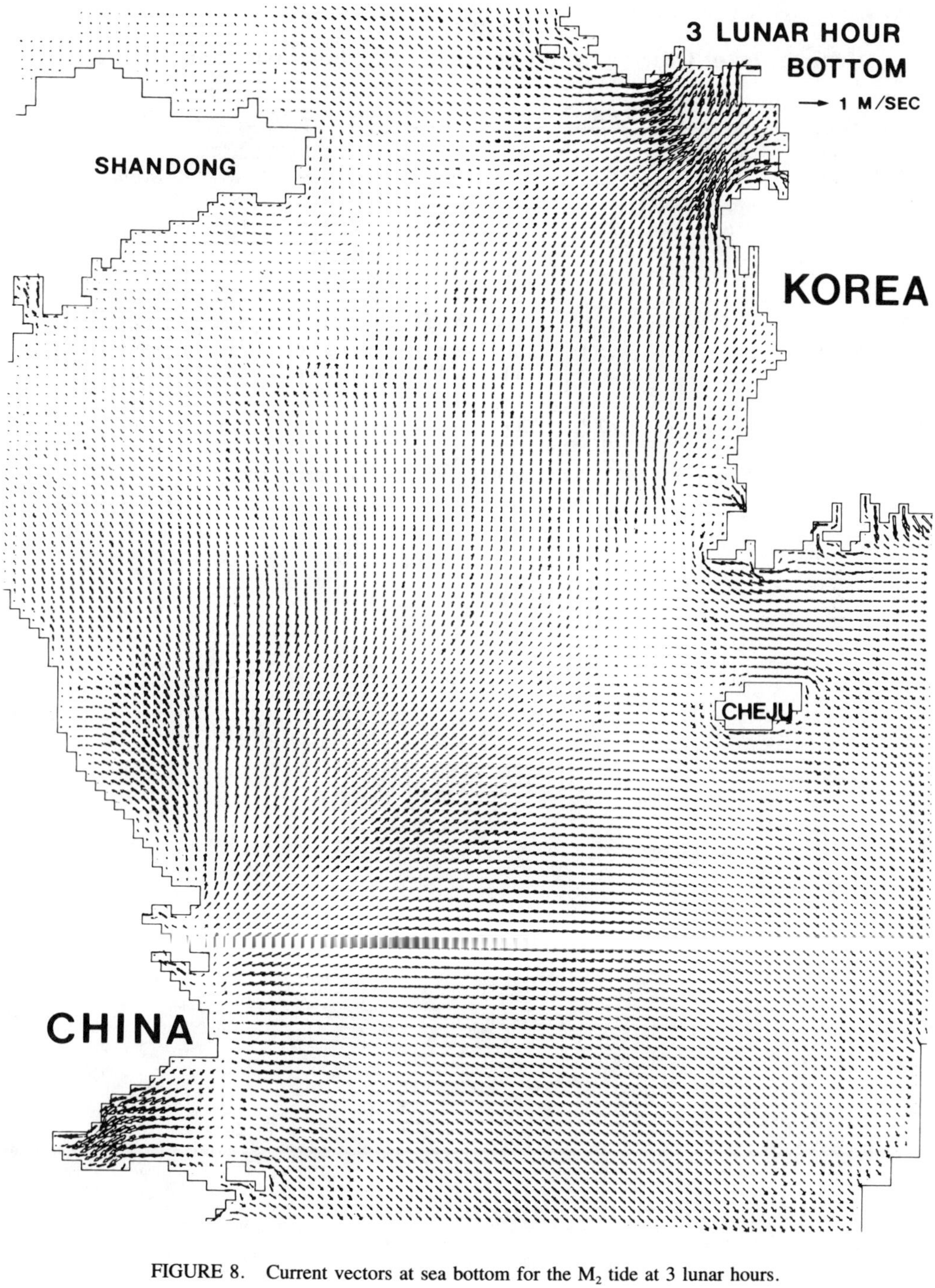


FIGURE 8. Current vectors at sea bottom for the $M_2$ tide at 3 lunar hours.

## TABLE 3
## Comparison of Observed and Computed Amplitudes (cm/s) and Phases

| Current meter rig | Observed | | | | Calculated | | | | Depth |
|---|---|---|---|---|---|---|---|---|---|
| | u | | v | | u | | v | | |
| | H | κ | H | κ | H | κ | H | κ | |
| B | 18.6 | 26 | 32.8 | 168 | 17.9 | 42 | 26.2 | 163 | Mid-Depth |
| B | 11.3 | 34 | 22.4 | 158 | 12.9 | 33 | 16.2 | 149 | Bottom |
| C | 29.5 | 23 | 26.0 | 107 | 31.5 | 36 | 23.6 | 93 | Bottom |
| D | 15.2 | 32 | 26.1 | 84 | 13.9 | 40 | 27.7 | 75 | Mid-Depth |
| D | 15.5 | 31 | 17.0 | 81 | 9.2 | 17 | 19.2 | 66 | Bottom |
| E | 20.2 | 7 | 25.4 | 61 | 18.3 | 10 | 28.0 | 56 | Bottom |
| F | 5.0 | 28 | 40.0 | 40 | 8.4 | 65 | 43.4 | 43 | Mid-Depth |
| F | 8.4 | 350 | 26.0 | 11 | 4.6 | 16 | 31.0 | 34 | Bottom |
| I | 13.4 | 132 | 43.8 | 19 | 16.9 | 144 | 49.8 | 23 | Mid-Depth |

*Note:* Degree referred to 135° E for u and v component of $M_2$ tidal current.

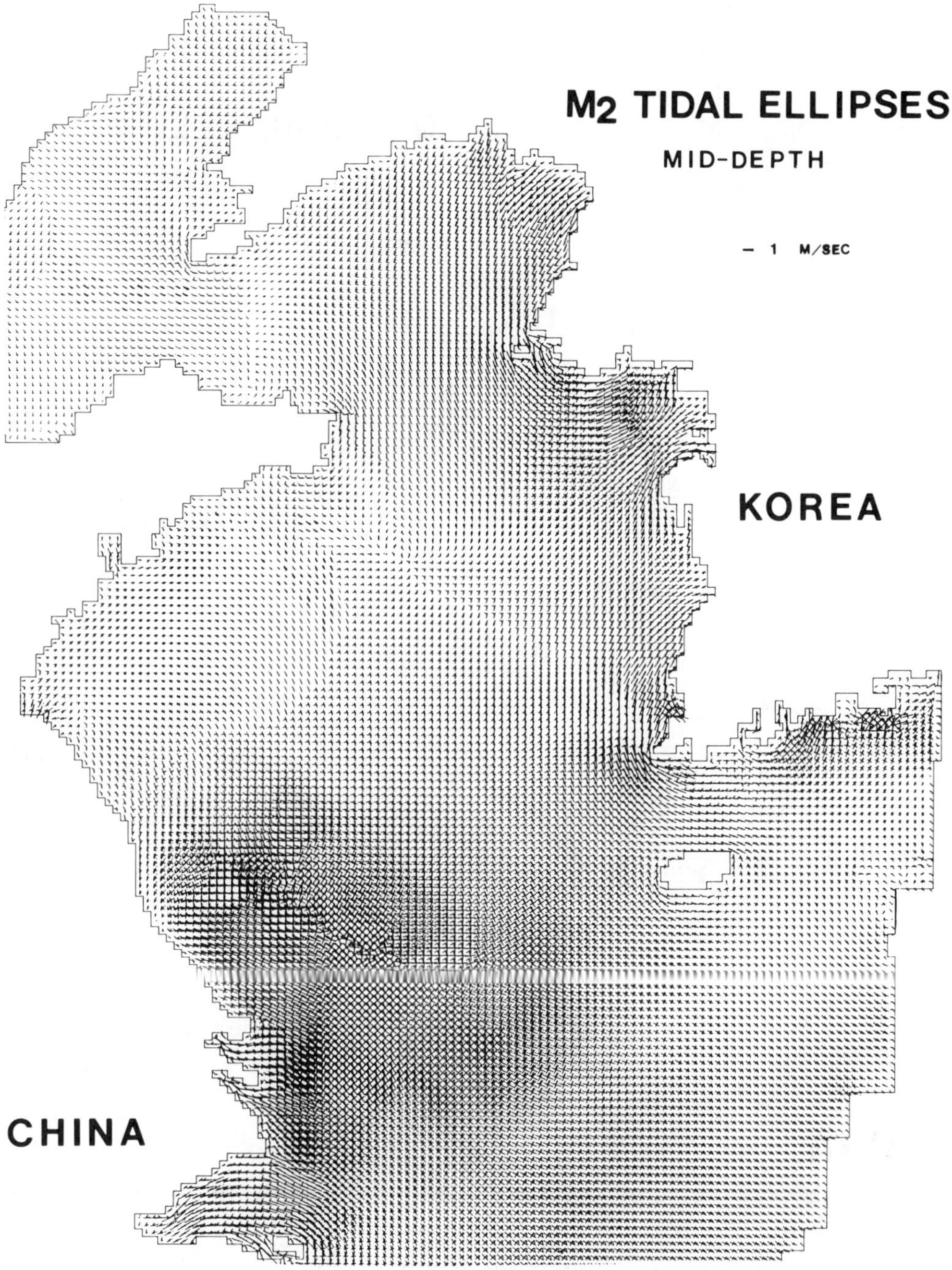


FIGURE 9. Tidal current ellipses at mid-depth for the $M_2$ tide.

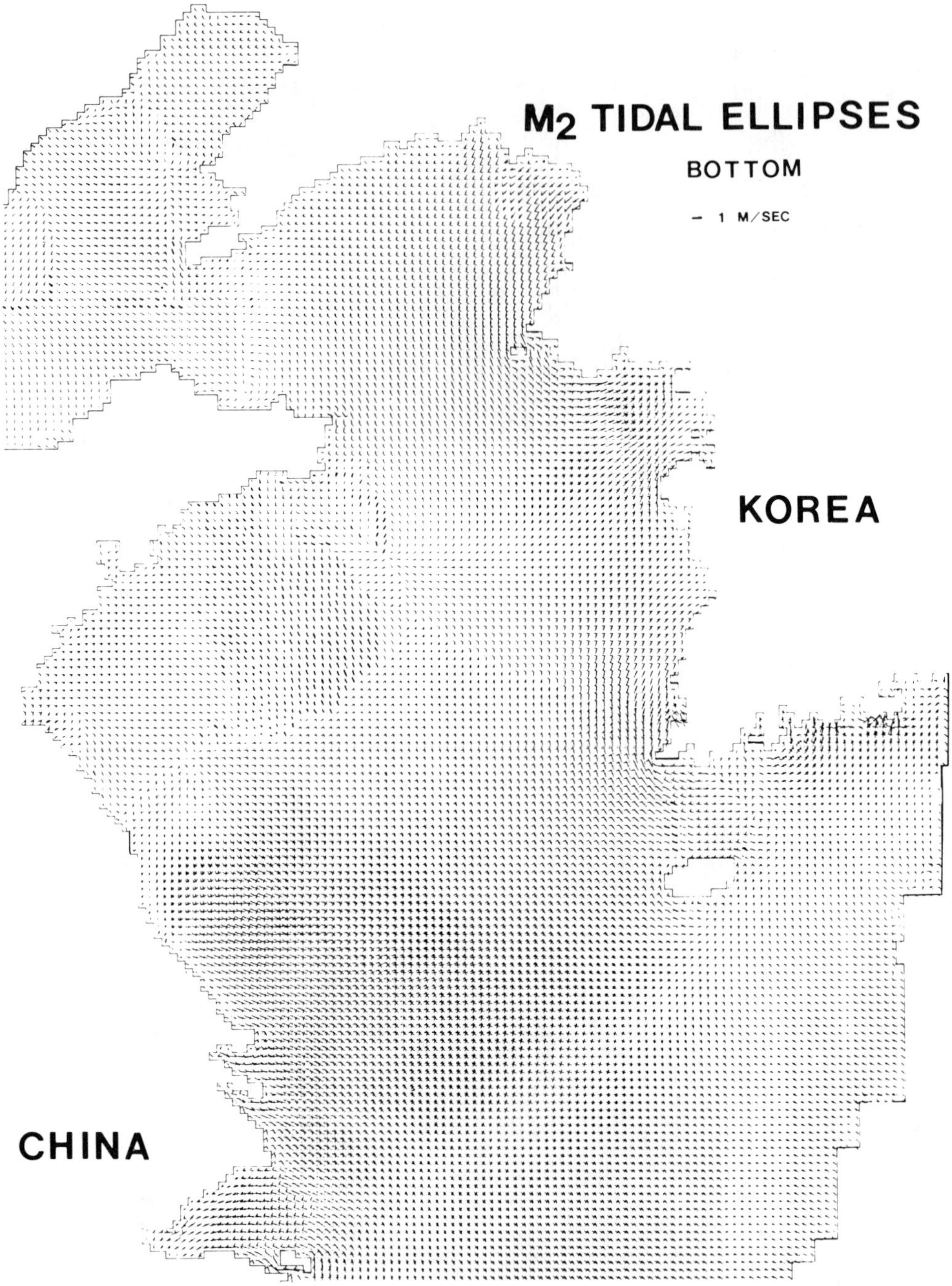


FIGURE 10. Tidal current ellipses at sea bottom for the $M_2$ tide.

# REFERENCES


1. **Ogura, S.** The tides in the seas adjacent to Japan, No. 7, Bulletin of Hydrographic Department, Imperial Japanese Navy, 1933, 189.
2. **Boris, L. I.,** Calculation of the tides and tidal currents in the Yellow Sea, *Tr. Leningr. Gidrometeorol. Inst.,* 1958.
3. **Tsiklauri, I. D. and Boris, L. I.,** An experiment to calculate tidal phenomena by means of "URAL-1" digital computer, *Tr. Leningr. Gidrometeorol. Inst.,* Rep. No. 10, 1961.
4. **Stepanov, S. I. and Boris, L. I.,** On the effect of errors in the original data on the accuracy of the solution of boundary value problems of tidal equations, *Tr. Leningr. Gidrometeorol. Inst.,* Rep. No. 17, 1964.
5. **An, H. S.,** A numerical experiment of the $M_2$ tide in the Yellow Sea, *J. Oceanogr. Soc. Jpn.,* 33, 103, 1977.
6. **Choi, B. H.,** A tidal model of the Yellow Sea and the Eastern China Sea, Rep. 80-02, Korea Ocean Research and Development Institute, 1980.
7. **Shen Yujiang,** Numerical computation of tides in the East China Sea, *J. Shandong Coll. Oceanol.,* 11(1), 23, 1980.
8. **Xia, Z. and Wang, Z.,** A numerical model of the $M_2$ constituent in the Huanghai Sea, *J. Oceanogr. Hunghai Bohai Seas,* 2(1), 1, 1984.
9. **Fang Guohong,** A finite difference-least square technique for solving tidal wave equations with specific application to the modelling of $M_2$ tide in the Huanghai Sea, *Sci. Sin. (B),* 28(10), 1110, 1985.
10. **Choi, B. H.,** A three-dimensional model of the East China Sea, in *Ocean Hydrodynamics of the Japan and East China Sea,* Ichiye, T., Ed., Elsevier Oceanography Series No. 39, 1984, 209.
11. **Choi, B. H.,** Observed and computed tidal currents in the East China Sea, *J. Oceanol. Soc. Korea,* 20(1), 56, 1985.
12. **Davies, A. M.,** Application of the Galerkin method to the formation of a three-dimensional non-linear hydrodynamic numerical sea model, *Appl. Math. Model.,* 4, 245, 1980.
13. **Davies, A. M., and Furnes, G., K.,** Observed and computed $M_2$ tidal currents in the North Sea, *J. Phys. Oceanogr.,* 10, 237, 1980.
14. **Harkema, R. and Hsueh, Y.** A compilation of moored current meter data in the Eastern Yellow Sea January — April 1986. Tech. Rep. CMF-87.01, Department of Oceanography, Florida State University, Tallahassee, 1987.
15. **Choi, B. H.,** Predictions of sand transport directions of the offshore tidal sand banks in the Yellow Sea, Proc. 5th Cong. APD, Int. Assoc. Hydraulic Res., 1986, 231.

Chapter 9

# MODEL SIMULATIONS OF TIDES AND SHELF WAVES ALONG THE SHELVES OF THE NORWEGIAN-GREENLAND-BARENTS SEA

**Bjorn Gjevik**

## TABLE OF CONTENTS

# I. ABSTRACT

Based on simulations with barotropic depth-integrated models, the structure of tidal- and storm-induced currents along the main shelf areas in the Norwegian-Greenland-Barents sea have been examined. Predictions have been made for the dominant diurnal and semi-diurnal constituents and maps showing current ellipse and current vector rotation with a grid resolution of about 50 km are presented for the shelf west of Norway, the Barents Sea escarpment, and the Fram Strait region. Comparison between model predictions and current measurements have been made for key stations within each area and the agreement is found to be good. The results of the model simulations indicate that topographic shelf wave modes are important components of the diurnal tide. The effect of atmospheric forcing has also been examined and it is found that eddy structures are generated along the Norwegian Shelf and the Barents Sea escarpment by cyclones travelling over the shelf. The period of the current oscillations associated with these eddies are in the near-diurnal band (20 to 36 h) and the eddies are attributed to topographic shelf wave modes. Finally, the article summarizes some main results obtained by various other authors and draws some lines for future model development for the Norwegian-Greenland-Barents sea region.

# II. INTRODUCTION

The ocean basin surrounded by Norway, Iceland, Greenland, Spitsbergen, and the northern coasts of Russia and Siberia consists of three interconnected seas; the Norwegian-Greenland-Barents sea (NGB sea). It is an interesting ocean system for several reasons. (1) The NGB sea provides a connection between the North Atlantic and the Arctic Ocean and therefore plays an important role in the heat exchange between the equatorial and the polar regions. (2) The NGB sea has strongly varying geometry with several pronounced topographical features which favor wave guide splitting and conversion between various long periodic topographical wave modes. (3) The shelves of the NGB sea are rich in fish resources and are important fishing grounds for several European nations. More recently these shelves also have proved to have large resources of oil and gas. Oil and gas production from platforms constructed at depths down to about 200 m are operating off the coast of Norway and others are planned for installation at deeper and more hostile waters further north. This activity clearly requires profound knowledge of the environmental conditions and several research programs have been devoted to observational and theoretical studies of the marine environment in these regions. It is therefore time to summarize what has been learned during the last decade and draw some lines for further work. The research activity has covered a broad range of problems, surface waves, tides, wind-driven currents, stratification effects, sediment transport, and ice processes and it would be impossible to review all this in a single article. This paper will focus on numerical modeling of barotropic currents and sea level variations associated with tidal and atmospheric forcing. Observational data will mainly be discussed in relation to model verification.

The bathymetry of the NGB basin is shown in Figure 1. There are three main entrances to the basin from the North Atlantic: the deep Shetland-Faroe trench, the shallow ridge between the Faroes and Iceland, and the Denmark Strait between Iceland and Greenland. The main exchange with the Arctic Ocean is through the deep Fram Strait between Greenland and Spitsbergen and to a lesser extent through the shallow straits between Spitsbergen, Franz Josef Land, and Severnaya Zemlya. Along the Norwegian coast the shelf width is between 50 and 150 km with depths ranging between 200 and 400 m. Along the southwestern coast of Norway runs the Norwegian Trench, an interesting topographical feature with depths to 350 m and a steep slope on the west side toward the shallow plateau east of Shetland and Orkney where the depth is less than 150 m. To the north there is a steep escarpment between

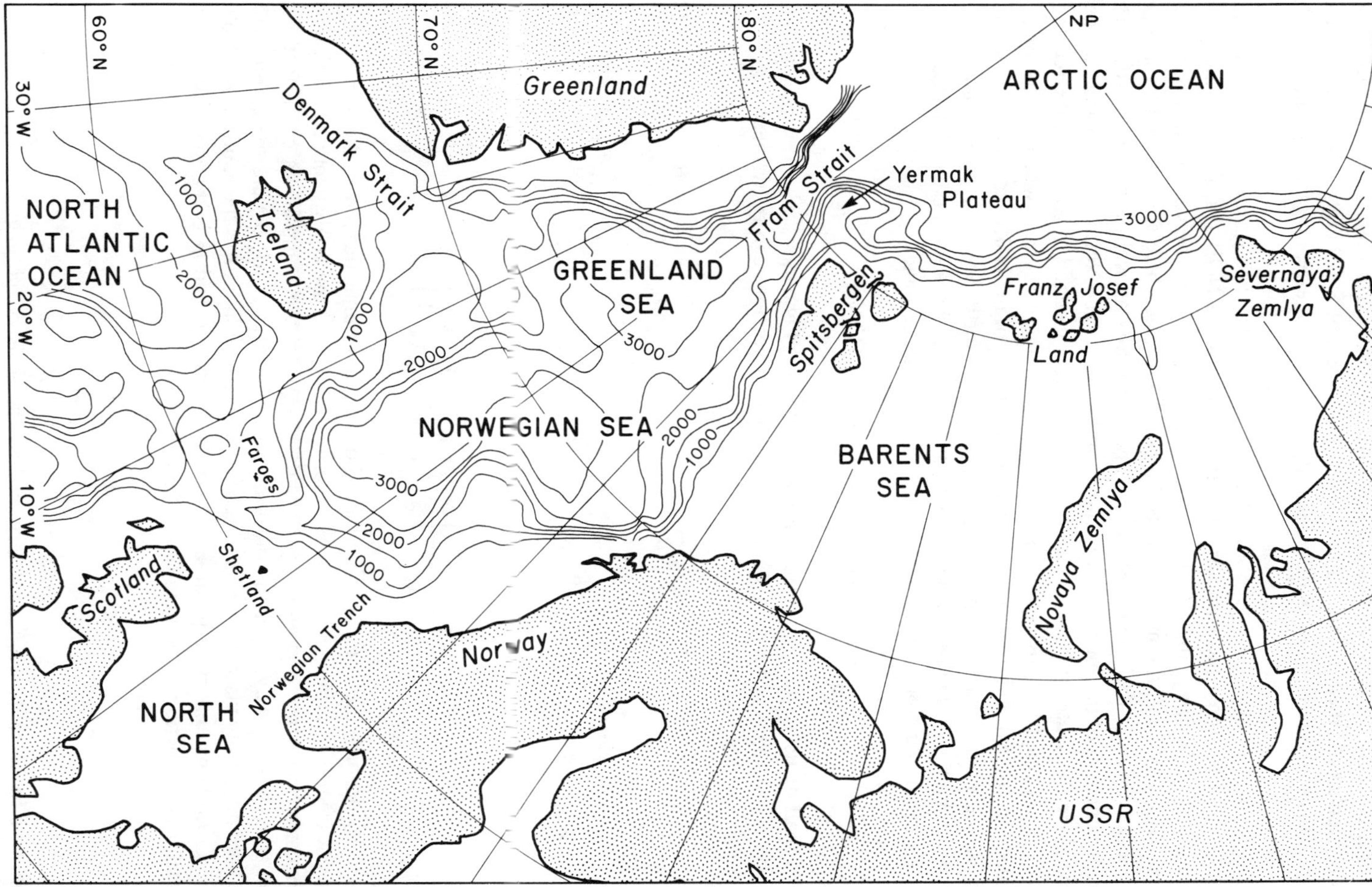


FIGURE 1. The m_in bathymetric features of the Norwegian-Greenland-Barents Sea.

Norway and Spitsbergen which forms the boundary between the shallow Barents Sea with depths ranging from 200 to 400 m and deep basin of the Norwegian Sea. As we shall see later, this escarpment gives rise to interesting flow patterns, especially for the diurnal tide and storm-induced currents. Along the east coast of Greenland there is a relatively broad shelf (particularly in the vicinity of the Fram Strait) north of Spitsbergen called the Yermak Plateau. This shelf area around Spitsbergen has an important effect on the flow through the Fram Strait.

This report will discuss in detail the structure of the tidal- and storm-induced barotropic currents within the three most important shelf regions of the NGB sea: the shelf along the western coast of Norway, the Barents Sea escarpment, and the shelves in the vicinity of the Fram Strait. Most of the model results which are presented here are obtained by a large-scale model covering the Arctic Ocean and the North Sea as well as the NGB sea.[1] The grid representation of the model area is shown in Figure 2. The North Sea area and the shelf regions along the southern boundary of NGB basin from Shetland to Iceland and Greenland will not be covered by this review and results from these regions will only be briefly mentioned in context with the discussion of the dynamics of the NGB sea proper. Tides and storm surges in the northeast Atlantic shelf and the North Sea area have been studied extensively, both with the use of model simulations and observational data.[2-5] These papers also give an introduction to other important studies of these topics.

## III. MODEL EQUATIONS

We shall be using the linearized depth integrated tidal equations which, expressed in a Cartesian coordinate system on a stereographic map projection, can be written[1]

$$\frac{\partial U}{\partial t} - fV = -mgh\frac{\partial}{\partial x}(\eta - \eta_e - \eta_s) + \frac{\tau_x^s}{\rho_w} - \frac{\tau_x^b}{\rho_w} + B_x \tag{1}$$

$$\frac{\partial V}{\partial t} + fU = -mgh\frac{\partial}{\partial y}(\eta - \eta_e - \eta_s) + \frac{\tau_y^s}{\rho_w} - \frac{\tau_y^b}{\rho_w} + B_y \tag{2}$$

with the continuity equation

$$\frac{\partial \eta}{\partial t} = -m^2\left[\frac{\partial}{\partial x}\left(\frac{U}{m}\right) + \frac{\partial}{\partial y}\left(\frac{V}{m}\right)\right] \tag{3}$$

Here (U,V) denotes the components of the horizontal volume flux, $\eta$ is the surface displacement from equilibrium, $\eta_e$ is the equilibrium tide, $\eta_s$ is the equivalent static height of the atmospheric surface pressure, $(\tau_x^s, \tau_y^s)$ and $(\tau_x^b, \tau_y^b)$, are the components of surface and bottom stress respectively. $\rho_w$ is the mean density of sea water $(B_x, B_y)$ are the components of the horizontal eddy viscosity terms which will be defined later, $h$ is the equilibrium depth, $f$ is the Coriolis parameter, $g$ is the acceleration of gravity, and $m$ is the map factor. The map factor is a function of latitude ($\phi$) defined by

$$m = \frac{1 + \sin\phi_s}{1 + \sin\phi}$$

with $m = 1$ for $\phi = \phi_s$ which is determined by the scale of the map.

The atmospheric forcing is provided by a cyclone model described by the pressure disturbance

$$p_s(x,y,t) = p_0(t)\exp\{-[(x - x_0(t))^2 + (y - y_0(t))^2]/R^2\} \tag{4}$$

where $p_0$ is the pressure perturbation at the center of the cyclone, $(x_0 y_0)$ is the instantanous position of the center, and $R$ defines the horizontal extension of the cyclone. Hence, the forcing terms due to the surface pressure field (Equations 1 and 2), are expressed by $\eta_s = p_s/\rho_w g$. The wind velocity components $(u_g, v_g)$ along the $x$ and $y$ direction, respectively, are evaluated from the geostrophic wind relations

$$u_g = -\frac{\sigma}{f\rho_a}\frac{\partial p_s}{\partial y} \tag{5}$$

$$v_g = \frac{\sigma}{f\rho_a}\frac{\partial p_s}{\partial x} \tag{6}$$

where $\rho_a$ is the density of air and $\sigma$ is a scaling factor. This factor can be specified to allow for wind reduction due to frictional effects and for simulating asymmetric wind fields. For the wind fields applied in the case studies (Section V) we have used

$$\sigma = 0.7\left(1 - \frac{2}{3}\sin\phi\right) \qquad 0 < \phi < \pi$$

$$\sigma = 0.7 \qquad 0 < \phi < 2\pi$$

where the angle $\phi$ defines anti-clockwise rotation from the direction where the cyclone is heading. For this cyclone model the maximum wind speed occurs at a distance $R/\sqrt{2}$ from the center. The wind stress is calculated from the empirical relations

$$\tau_x^s = \rho_a c_D (u_g^2 + v_g^2)^{1/2} u_g \tag{7}$$

$$\tau_y^s = \rho_a c_D (u_g^2 + v_g^2)^{1/2} v_g \tag{8}$$

where $c_D$ is the drag coefficient. The bottom stress has been related to the integrated volume flux by a quadratic friction law

$$\tau_x^b = -kU/h \tag{9}$$

$$\tau_y^b = -kV/h \tag{10}$$

where the friction coefficient $k$ is

$$k = \rho_w c_f (U^2 + V^2)^{1/2}/h \tag{11}$$

and $c_f$ is the drag coefficient. Enhanced bottom friction for the weak diurnal tidal current due to nonlinear interaction with the semidiurnal tide is obtained by substituting a characteristic velocity scale for the dominate semidiurnal current velocity in Equation 11. The horizontal eddy viscosity terms are expressed by

$$B_x = \nu m^2 \nabla^2 U \tag{12}$$

$$B_y = \nu m^2 \nabla^2 V \tag{13}$$

and the eddy viscosity is related to depth by

$$v = qh \tag{14}$$

where $q$ is a constant. Expressions for the components of the tide-generating force in the stereographic map projection $\left(\frac{\partial \eta_e}{\partial x}, \frac{\partial \eta_e}{\partial_y}\right)$ are given (Reference 1), which also gives further details on the formulation of the basic equations.

In a stereographic map projection the tidal energy flux vector ($\vec{F}$) can be computed by the formula

$$\vec{F} = \frac{\rho_w g \eta \vec{U}}{m} \tag{15}$$

where $\vec{U} = (U,V)$ and the mean value is obtained by averaging over one tidal period. Maps showing the flow of tidal energy in the NGB sea will be presented in Section IV.

The boundary of the model consists of the coastal boundary sections $\Gamma_c$ and the open boundary sections $\Gamma_0$ where the model basin is connected to the North Atlanic and adjacent seas through narrow passages. The boundary conditions along the coastal boundaries are zero volume flux normal to the boundary

$$U_n = 0 \qquad \text{along } \Gamma_c \tag{16}$$

Along the open boundary, $\Gamma_0$, one may either specify the surface elevation

$$\eta = \eta_0 \tag{17}$$

or the volume flux

$$U = U_0 \tag{18}$$

where $\eta_0$ and $U_0$ denote the prescribed boundary values which are obtained from observational data or previous model simulations. In this article we shall frequently be referring to model simulations of tides and storm-induced current for an area depicted in Figure 2. This grid system was adopted by Gjevik and Straume[1] for simulation of tides in the NGB sea and the Arctic Ocean and we shall here, for convenience, refer to it as the GS model. Boundary conditions along the open boundary for this model were obtained by a standard interpolation procedure from data provided by model simulations for the North Atlantic by Flather et al.[5] and Schwiderski's global tidal charts.[6,7]

The set of Equations 1 to 3 are solved by a finite-difference scheme with a space-staggered C-grid[8] where volume flux $(U,V)$ and surface displacement $(\eta)$ are evaluted at separate points. The grid size $(\Delta x = \Delta y)$ is defined as the distance between neighbor points for the same variable, and the distance between two neighbor points where elevation and flux are evaluated respectively is $\Delta x/2$. Further details about the numerical method, together with a discussion of the stability characteristics of the scheme are given in References 1 and 9. The numerical scheme which is being used for the GS model also enables specification of both $\eta$ and $U_n$ at the boundary with an adjustment procedure based on a radiation condition. This is found to have some smoothing effect on noise produced by inconsistency in the data. The effect which various types of boundary conditions have on the solutions for the semi-diurnal and diurnal tide has been discussed by Gjevik and Straume.[1]

The model parameters used for simulation of tides are

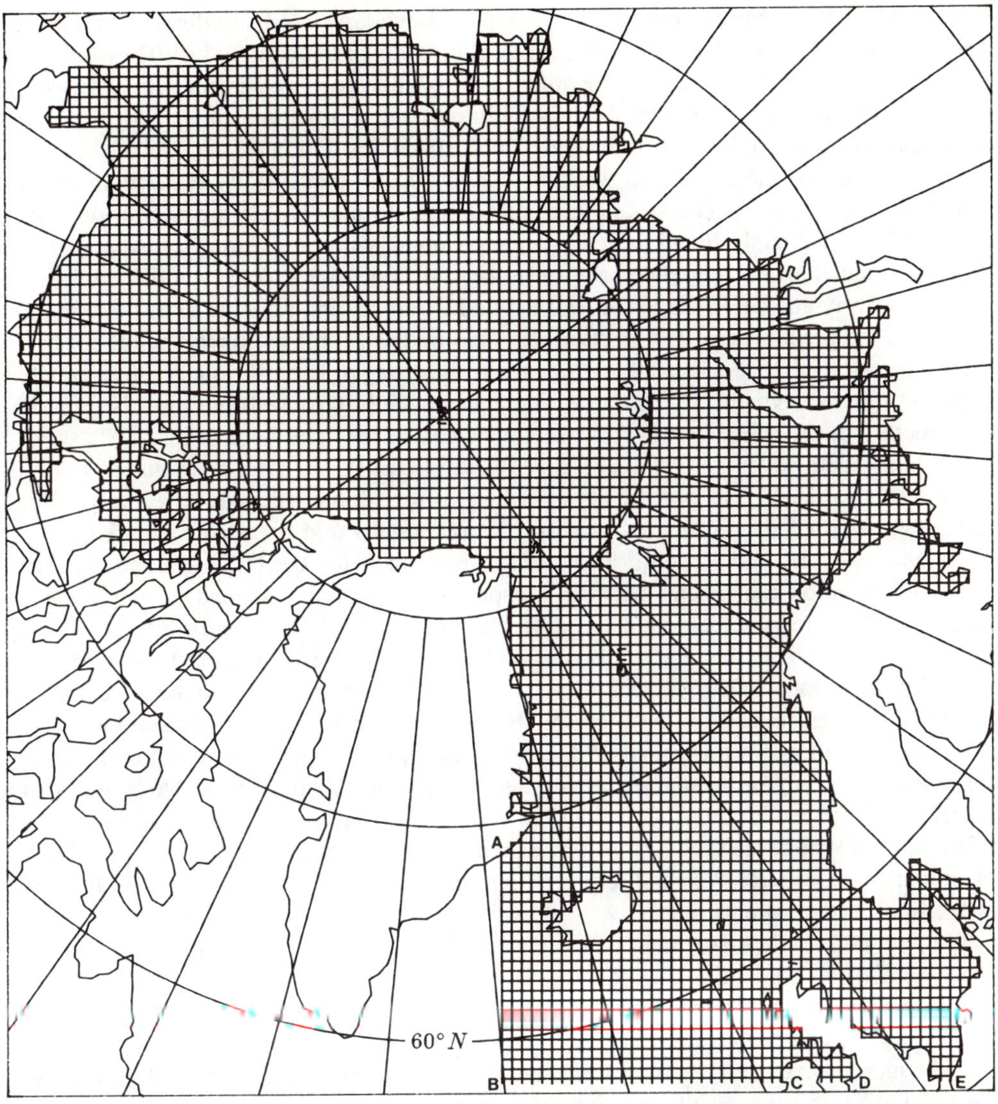


FIGURE 2. Grid representation of model area, grid size 50 km at 60°N, open boundaries at A-B-C, D, and E.

| | | | |
|---|---|---|---|
| Grid size (at 60°N) | $\Delta x = \Delta y$ | = | 50 km |
| Time step | $\Delta t$ | = | 168 s |
| Bottom friction | $c_f$ | = | 0.003 |
| Eddy viscosity factor | q | = | 100 $ms^{-1}$ |

The numerical integration was started from rest and steady tidal oscillations were normally obtained after a simulation time of 10 to 20 tidal periods.

## IV. TIDES

The NGB sea has previously been partly covered by tidal models developed primarily for adjacent oceans. The regions south of 72°N have been covered by the North Atlantic model by Flather et al.[5] and the area north of about 70°N has been treated as a part of an Arctic tidal model by Kowalik and Untersteiner.[10] Besides that, several high-resolution models for the continental shelf along the coast of Norway and the North Sea area have been developed.[11-13] Mathisen et al.[12] have applied boundary-fitted coordinates for the coast and shelf area all around Norway with grid resolution down to about 20 km. They computed tidal charts for the semidiurnal $M_2$ constituent and obtained good agreement with observations from coastal and pelagic stations. Main semidiurnal and diurnal constituent; $M_2$,$S_2$,$K_1$, and $O_1$ have been simulated by Martinsen et al.[13] with 10-km resolution for a model area covering the North Sea and the shelf area west of Norway up to about 68°N. Predictions of tidal amplitudes and phase for coastal and offshore stations located within the interior domain of the models are generally found to be in good agreement with observations. Comparison has also been made between computed and measured tidal currents at some stations on the shelf west of Norway and again the agreement is good. More testing remains to be done in order to explore the prediction power of these regional models. Since the open boundaries for all these models tend to be located in regions with strongly varying topography, which is important for the dynamics of the tide, it is questionable how well the models reproduce the structure of the tidal current in the regions close to the open boundaries.

A large-scale coverage of the entire NGB basin on a 1° × 1° grid resolution has been given by Schwiderski[14] as part of his modeling of the global tide. More recently, Gjevik and Straume[1] used the GS model descibed in Section III and calculated co-tidal and co-range charts for $M_2$ and $K_1$ with 50 × 50 km resolution for the NGB sea and the Arctic Ocean. Correlation coefficients between computed and measured $M_2$ and $K_1$ amplitudes for over 100 stations located all around the NGB sea were found to be 0.97 and 0.82, respectively, and the standard deviation for the corresponding linear regression analysis was 12 and 2 cm. With this model, also, current ellipse parameters, current vector rotation, and energy flux have been calculated for the entire NGB basin and for the Arctic Ocean. It has become clear that Schwiderski's tidal model[14] does not well reproduce the finer structure of the tide, particularly in a region near the Barents Sea escarpment. This is most likely due to the relatively coarse resolution of the model but since tidal measurements are available only for a few stations within the region, this may also have affected the accurancy of Schwiderski's interpolation procedure. We shall return to this point later in connection with the discussion of the tides in the Barents Sea.

In this article we shall mainly be concerned with dominant semidiurnal and diurnal constituents, i.e., $M_2$ and $K_1$, which are representative also for the other components and therefore demonstrate characteristic features of the tide in the NBG sea. The other important components $S_2$ and $N_2$ have recently been modeled by Straume.[15]

The mean energy flux vector for $M_2$ and $K_1$ has been computed for the NGB basin with the GS model and the results are shown in Figure 3. Gjevik and Straume examined the effect of different open boundary conditions and they found that the calculated tidal elevation and current to some extent were sensitive to the type of boundary conditions applied at the open boundaries of the model. The mean energy flux displayed in Figure 3 is obtained with the model version (V) which shows the best fit to observed tidal elevation data from the station along the Norwegian coast and the Barents Sea.[1] For this particular model version, surface elevation has been prescribed at the open boundary. The mean $M_2$ energy flux is relatively stable to small perturbations of the input data at the open boundary while the mean $K_1$ energy flux shows more small-scale variability and also is more sensitive to variations of the boundary input.

Both for $M_2$ and $K_1$ we see that the main source of tidal energy for the NGB basin is the influx from the North Atlantic, mainly through the channel between Shetland and the Faroes. The tide in NGB sea is therefore primarily a co-oscillating tide and the local effect of the tide-generating forces is small.[1] The tide-generating forces have, however, a relatively large effect on the elevation amplitude for $K_1$ in special regions, as for example, in the North Sea.[1,5]

From Figure 3 we see that there is a strong northeastward energy transport west of the

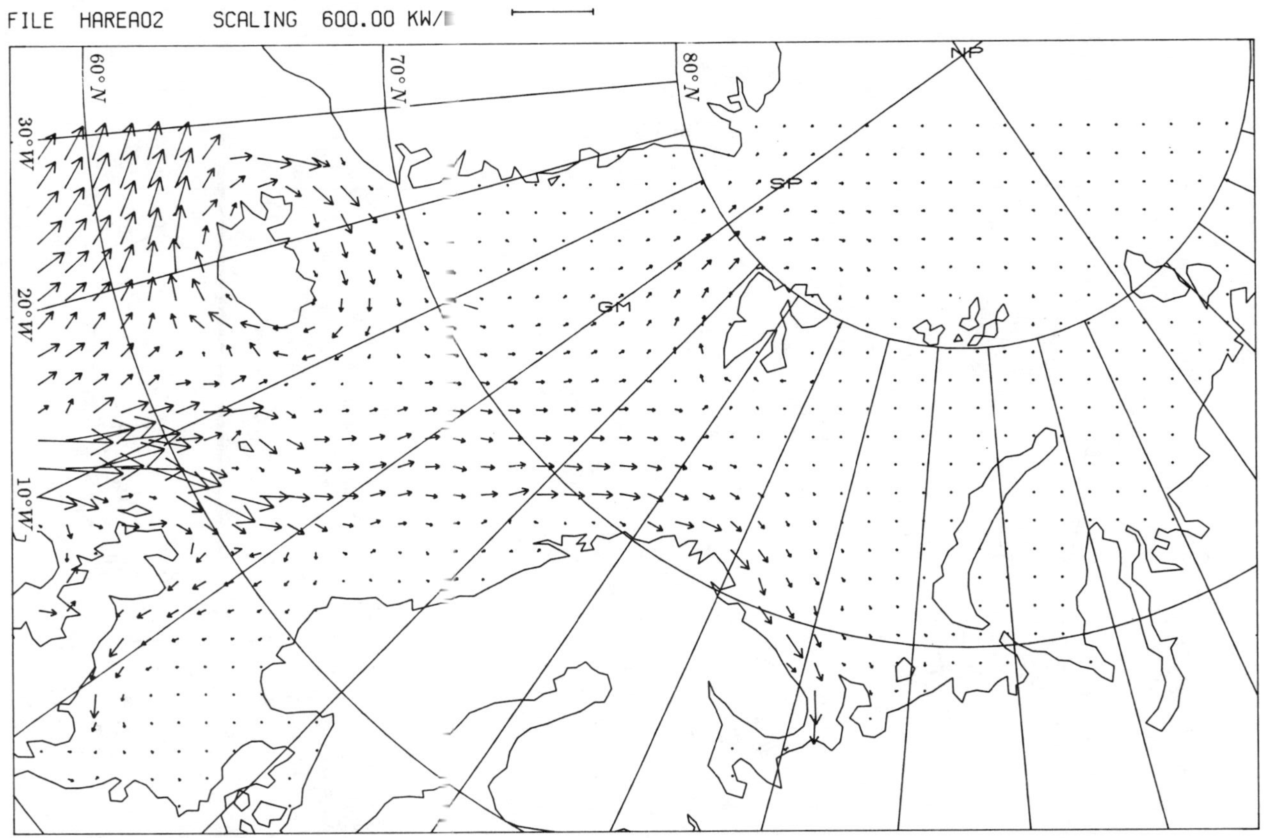


A

FIGURE 3. Mean energy flux vector for the (A) $M_2$ and (B) $K_1$ constituents.

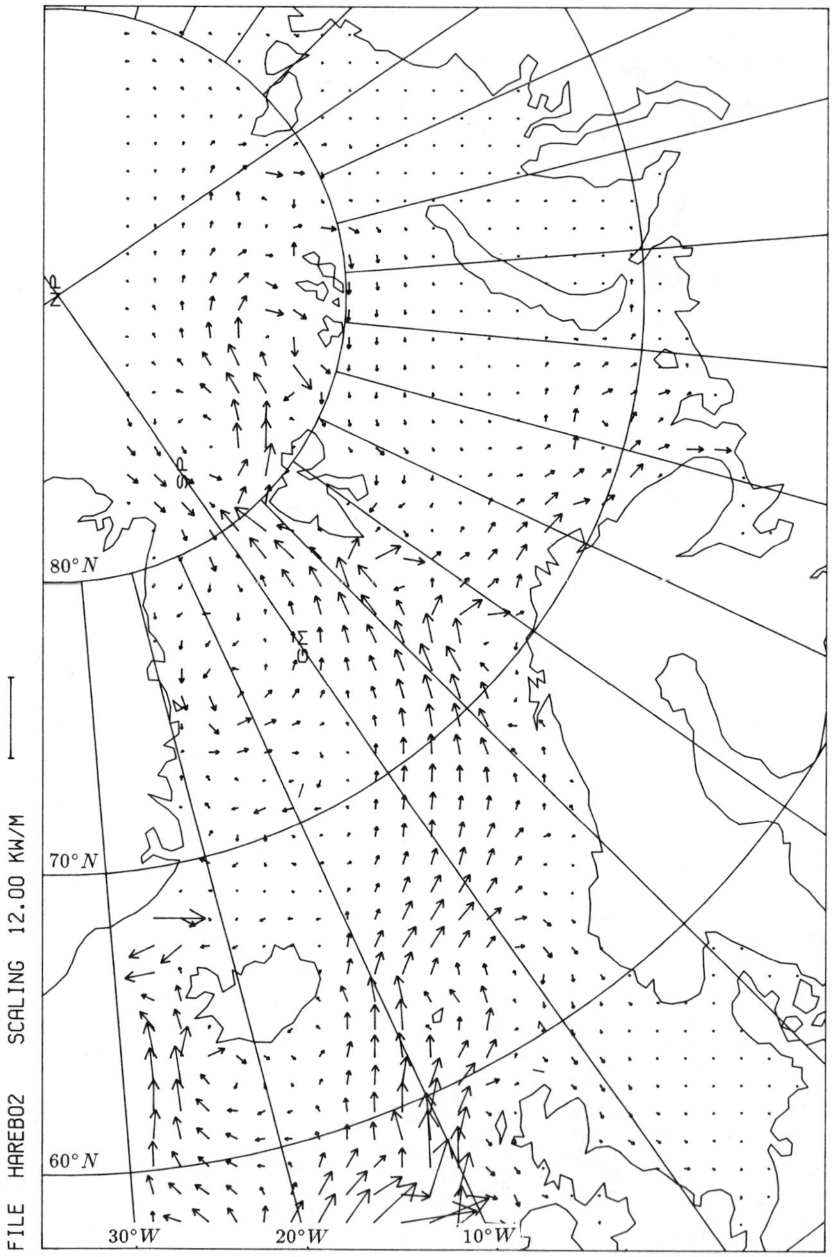


FIGURE 3B.

**TABLE 1**
**$M_2$ Energy Flux (Gigawatt)**

| Section | Energy flux (Gw) | Direction |
|---|---|---|
| Northern Norway-Spitsbergen | 38.4 | Eastward |
| Fram Strait | 25.4 | Northward |
| Spitsbergen-Franz Josef Land | 5.1 | Northward |
| Franz Josef Land-Sevemaya Zemlya | 4.3 | Northward |

Norwegian Shelf which bifurcates at the Barents Sea escarpment in an eastward-directed flow into the Barents Sea and a northward flow of energy toward Spitsbergen through the Fram Strait and into the Arctic Ocean. In addition to this general picture, which is strikingly similar for the semidiurnal and the diurnal components, we note the diffraction of energy north of Scotland and into the North Sea where it is being dissipated. We also note the anticyclonic transport of energy associated with the $M_2$ tide around Iceland. Since Iceland is relatively close to the open boundary, the results for this region may, however, be sensitive to the location of the boundary and some modifications are to be expected if the boundary is moved further west.

For the mean $K_1$ energy flux we see a pronounced influence of the Barents Sea escarpment and the flow develops two anticyclonic eddy features between Norway and Spitsbergen. There is also an interesting eddy pattern associated with the $K_1$ energy flow north of Spitsbergen and in the Fram Strait. Along the coast of Spitsbergen the mean energy transport is directed northward while the transport is southward along the east Greenland coast. This structure is clearly associated with a amphidromic point for $K_1$ which is found to be located in the Fram Strait.[1,7] The mean energy transport for the $M_2$ tide through some key sections of the model area is given Table 1. These results are based on averaged values obtained with the two model versions (model IV and V) which show the best overall fit to tidal measurements.[1] The model version IV utilizes flux data for open boundary input.

We shall now discuss some characteristic features of the semidiurnal tide within the three key shelf regions of the NGB sea: the shelf west of Norway, the Barents Sea escarpment, and the shelves along the Fram Strait. For most of the shelf stations where we shall be comparing model and measured current data, the depth is between 200 and 400 m and for the three stations in the Fram Strait (see below) the depth varies from 1100 to 2360 m. In all cases the depth is sufficiently large so that the bottom boundary layer is only a small fraction of the total depth. Depth-averaged mean velocities obtained with a two-dimensional model may therefore provide representative values for the free stream tidal current for the stations where we have been for comparing measured and computed current parameters, cf. Davies.[16]

The bathymetry of the shelf west of Norway is shown in Figure 4 and the mean current ellipses and mean current vector rotation for $M_2$ and $K_1$ are displayed in Figures 5 and 6. We see that the $M_2$ tidal ellipses become elongated and oriented in east-west direction, in the area around station H which also shows relatively large tidal energy transport from the deep ocean to the shelf. Within this area the current vector rotates in an anticyclonic (clockwise) fashion in contrast to broad surrounding areas with cyclonic (anticlockwise) rotation. This feature also appears in simulations with a high-resolution regional model for this shelf area[17] and the transition between anticyclonic and cyclonic rotation between stations Haltenbanken (H) and Traenabanken (T) is also confirmed by measurements (see Table 2). There is also a rather large difference in orientation of the ellipses between stations (H) and (T). With the GS model we find that the difference between the azimuth orientation of the major axis is about 70° which is somewhat less than Mathisen[18] obtained with a model utilizing boundary-fixed coordinates. A detailed comparison between model data and ob-

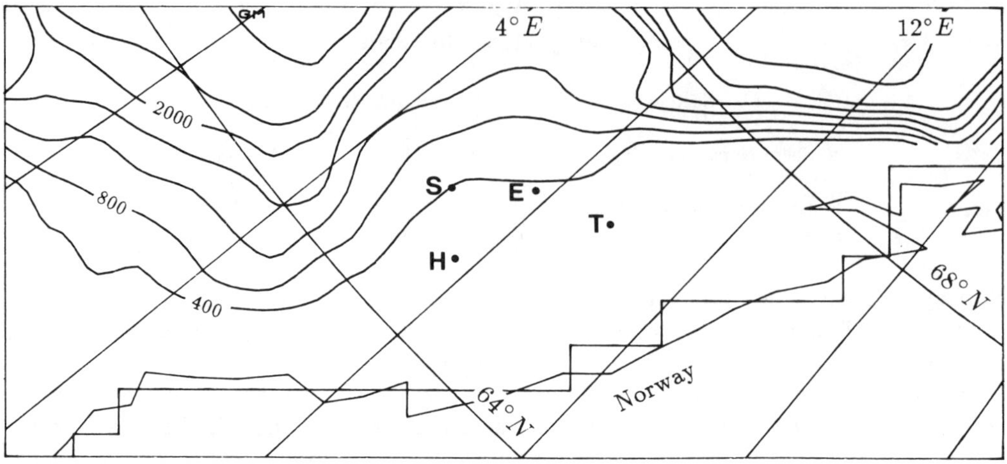


FIGURE 4. Bathymetry of shelf west of Norway with tidal measurement sites: Haltenbanken (H), Traenabanken (T), and a station at the shelf edge (E). Station S refers to the simulated time series for storm-induced currents (see Section V). Equidistance is 400 m.

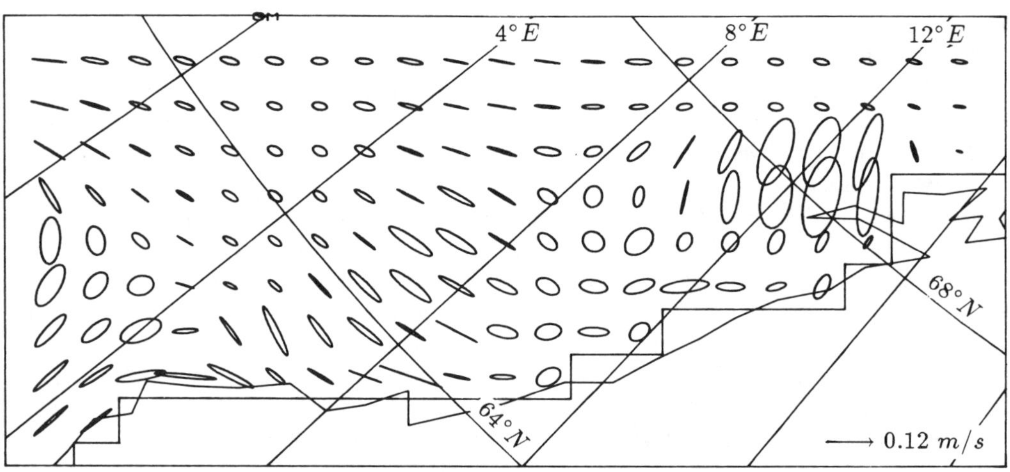


A

FIGURE 5. (A) $M_2$ tidal ellipses and (B) current vector rotation on the shelf west of Norway. Cyclonic rotation is indicated by ( − ), anticyclonic rotation is indicated by ( + ).

servation from three stations in the central region of the shelf is given in Table 2. The model data are obtained with the GS model,[1] model version IV (see above). The values for the measured current are obtained by averaging current measurements from different depth levels and data from different time periods. These data have been provided by the Oceanographic Center, Trondheim.[19] The agreement is very good for $M_2$ while the model underestimates the diurnal current. Two effects may be causing this discrepancy; measured diurnal tidal currents are contaminated by storm generated currents (see Section V) and the resolution of the model may be too coarse for resolving the currents on the shelf. The first effect is probably the most likely explanation but grid refinement may have some effect on the results.

The rotation pattern for the $K_1$ current vector (Figure 6) is closely correlated to the bottom topography with mainly anticyclonic rotation on the shelf and cyclonic rotation on

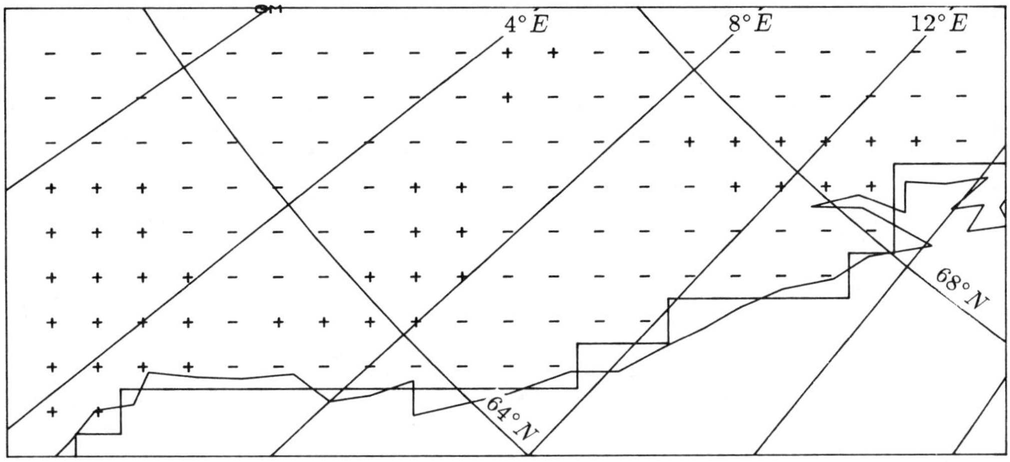


FIGURE 5B.

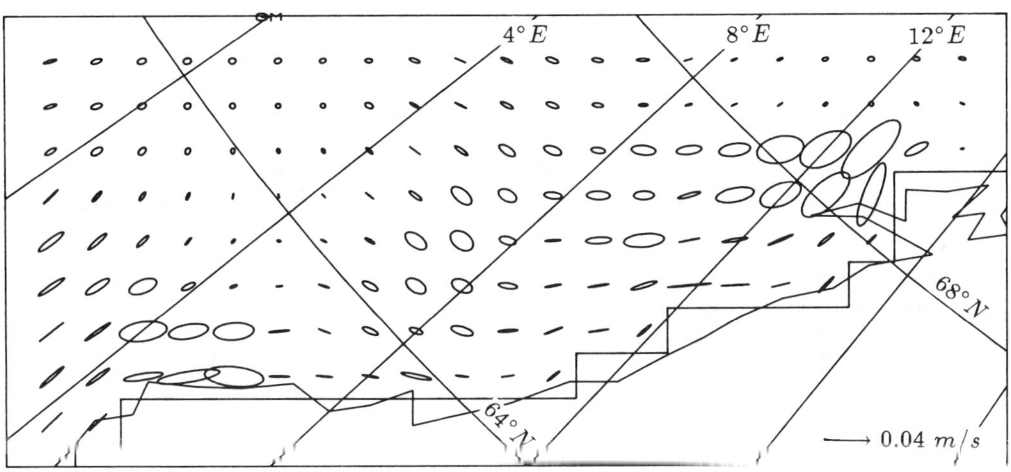


A

FIGURE 6. (A) $K_1$ tidal ellipses and (B) current vector rotation on the shelf west of Norway. Cyclonic rotation is indicated by (−), anticyclonic rotation is indicated by (+).

the slope toward the deep ocean. This indicates the existence of shelf wave components in the model solution for $K_1$ and that bathymetry of the shelf enhances the diurnal tidal current while the amplitude of the surface elevation is small. From Table 2 we see that, based on model data, the ratio between $K_1$ and $M_2$ peak currents is between 0.2 and 0.4, while the ratio between the corresponding surface elevation amplitude is only about 0.1. The model simulations show that the maximum diurnal current for the area occurs west of the Lofoten Islands at about 68°N, 13°E. The current speed is up to 4 cm/s but measurements are not available to confirm this prediction.

The bathymetry of the western part of the Barents Sea and the escarpment down to a depth of about 2200 m is shown in Figure 7. The strong effect which the escarpment has on the diurnal tide is demonstrated clearly by the plot of the $K_1$ energy flux vector which shows two anticyclonic eddy structures along the escarpment (Figure 8). East of the escarpment the model predicts amphidromes both for $M_2$ and $K_1$ (Figure 9). Recent pelagic

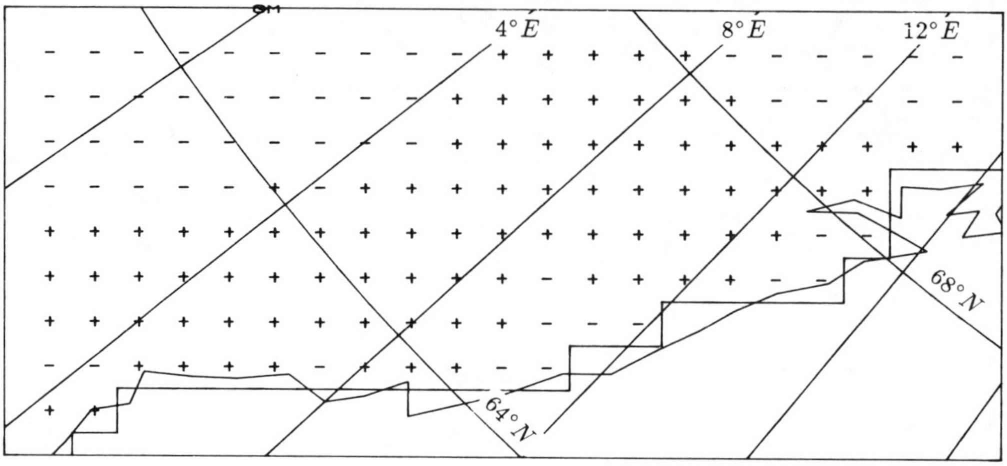


FIGURE 6B.

**TABLE 2**
**Tidal Ellipse Parameters, Shelf West of Norway**

| Station | Constituent | Major semiaxis (cm/s) Mod | Major semiaxis (cm/s) Obs | Minor semiaxis (cm/s) Mod | Minor semiaxis (cm/s) Obs | Azimuth degree Mod | Azimuth degree Obs | Rotation Mod | Rotation Obs |
|---|---|---|---|---|---|---|---|---|---|
| Haltenbanken (H) | $M_2$ | 5.6 | 5.6 | 0.7 | 0.7 | 74 | 74 | + | + |
| 65°03′N 7°35′E | $K_1$ | 1.0 | 2.4 | 0.6 | 1.6 | 70 | 107 | + | + |
| Traenabanken (T) | $M_2$ | 2.9 | 3.7 | 2.8 | 2.1 | 154 | 137 | − | − |
| 66°18′N 9°31′E | $K_1$ | 1.2 | 2.4 | 0.4 | 1.4 | 41 | 41 | + | + |
| Shelf edge (E) | $M_2$ | 3.3 | 3.5 | 1.3 | 1.7 | 75 | 108 | − | − |
| 66°02′N 7°32′E | $K_1$ | 0.8 | 1.6 | 0.4 | 0.9 | 54 | 58 | + | + |

*Note:* Azimuth: orientation of major axis degree from north positive eastward. Rotation: + anticyclonic; − cyclonic. Mod: model; Obs: observational data.

tidal measurements from Sentralbanken (S) support the model predictions. A comparison between measured and computed harmonic constant for this station and two key island stations in the same area is given in Table 3. In view of the good agreement between model and observations, we conclude that model provides a good description of the tides and that there are amphidromes both for $M_2$ and $K_1$ in this area. The existence of these amphidromes was first anticipated by Nekrasov[20] in his early tidal charts, but failed to be resolved by Schwiderski's model.[6,7,14]

The plot of the $M_2$ tidal ellipses in the Barents Sea (Figure 10) shows the strong gradients in the current over the escarpment and with strong local amplification on the shallow Svalbard Bank between Spitsbergen and Bear Island (Station B, Figure 7). The rotation of current vector is anticyclonic east of the escarpment and also in the deep-sea region southwest of Spitsbergen. The plot of the $K_1$ tidal ellipses (Figure 11) also shows a pronounced current amplification over the Svalbard Bank with maximum diurnal tidal current of about 16 cm/s. Since maximum semidiurnal current is about 38 cm/s in the same area, it implies that on the Svalbard Bank the ratio between the $K_1$ and $M_2$ peak current is about 0.4. This agrees well with observational data analyzed by Huthnance[21], which revealed strong diurnal currents near the Bear Island. The data analyzed by Huthnance refer to stations I and A (see Figure 7), which are located at the shelf slope south of Bear Island. For this particular location the

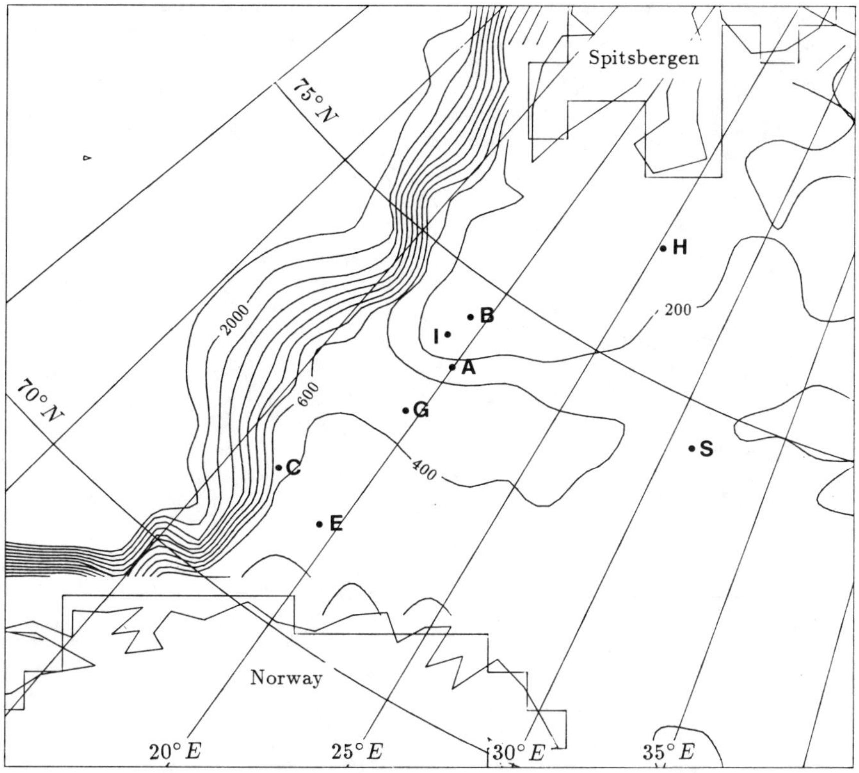


FIGURE 7. Bathymetry of the western part of the Barents Sea with position of sites for tidal elevation measurements: Hopen (H), Bear Island (B), and Sentralbanken (S). Stations I, A, G, and E indicate the position for tidal current measurements.[21] Station C refers to the simulated time series for storm-induced currents (see Section V). Equidistance is 200 m.

model predicts somewhat lower values for the current than observed. The reason for this discrepancy can probably be attributed to the coarse resolution of the model which does not resolve the steep slope south of Bear Island properly. For stations further south (G,E Figure 7) where the bottom slope is more gentle, the agreement between computed and measured current velocity is very good, as seen from Table 4. The rotation direction for the $K_1$ tide (Figure 11) shows a band with anticyclonic rotation on the shallow side of the escarpment with cyclonic rotation in the deep sea west of the escarpment. We also note the region with cyclonic rotation in the trench south of Bear Island and that station G is located near the transition zone between anticyclonic and cyclonic rotation. This may explain why the model simulation provides anticyclonic rotation for station G while observation indicates cyclonic rotation. Otherwise, the predicted direction of rotation is in close agreement with observations (Table 4). All the special features of $K_1$ diurnal tide along the Barents Sea escarpment which we have discussed above is a clear indication of the presence of a shelf wave mode in the simulated $K_1$ tide.

The bathymetry of the Fram Strait region is sketched on Figure 12 which reveals the shelf regions along the eastern coast of Greenland and north of Spitsbergen. A plot of the mean energy flux vector (Figure 13) shows pronounced eddy features related to the topography of Yermak Plateau north of Spitsbergen and the shelf north of Franz Josef Land. We also notice the strong shear in the energy flow across the Fram Strait and the deflection caused by the east Greenland shelf. The shear across the strait is clearly related to the amphidromic structure in the strait, but the topographic-induced deflections may be indicative

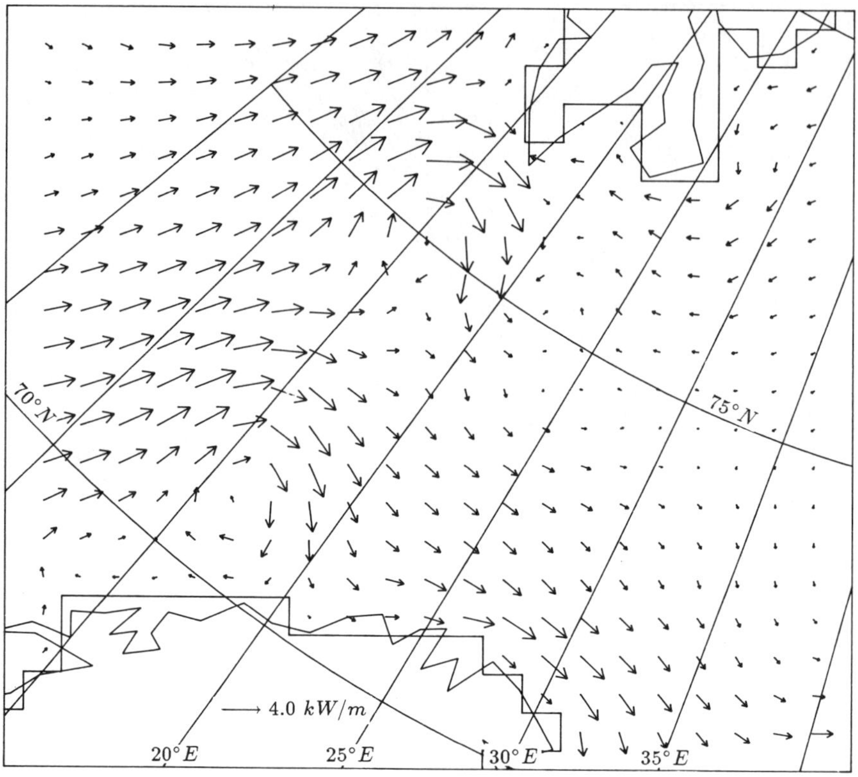


FIGURE 8. Mean $K_1$ energy flux vector along the Barents Sea escarpment (enlargement from Figure 3).

of a strong shelf-wave influence on the diurnal tide. Plots of $K_1$ current ellipses and current vector rotation (Figure 14) strengthens this suggestion.

The Arctic current measurement program conducted by Aagaard et al.[22] has provided data from the Fram Strait. Tidal ellipse parameters are available for three stations F1, F2, and F3 along a transect from the east Greenland shelf toward the center of the strait. The position of station F3 is plotted on the map in Figure 12 and F1 and F2 are located in more shallow water 15 to 25 km further west. Tidal constants are available for three depth levels: about 100 m, and 400 m, and near the bottom. For F1 and F2 baroclinic effects are noticeable, particularly near the surface, and the $M_2$ constants show some variation with depth from the 100 to 400 m level. The $K_1$ tide is apparently less influenced by stratification and the ellipse parameters vary little with depth. At station F3 which is located in deep water (2359 m) east of the Greenland shelf, the tidal current at the time of the measurement seems to be mainly barotropic. Computed and measured ellipse parameters and current rotation are presented in Table 5. In view of the comments above, the measured values corresponding to the near 400-m level recordings are assumed to be representative mean values and the agreement with model data is excellent. There is a surprisingly strong diurnal current with a ratio between $K_1$ and $M_2$ peak current of about 0.5, while the corresponding ratio for elevation amplitudes is only about 0.06. This also supports our suggestion (see text above) that shelf wave modes are an important component of the $K_1$ tide in the Fram Strait, although the low $K_1$ amplitude is a direct consequence of the amphidromic structure for the $K_1$ tide in the Fram Strait.[1]

Hunkins[23] has recently reported large diurnal currents with current reaching 30 cm/s over the Yermak Plateau north of Spitsbergen. He interpreted these oscillations as topographic

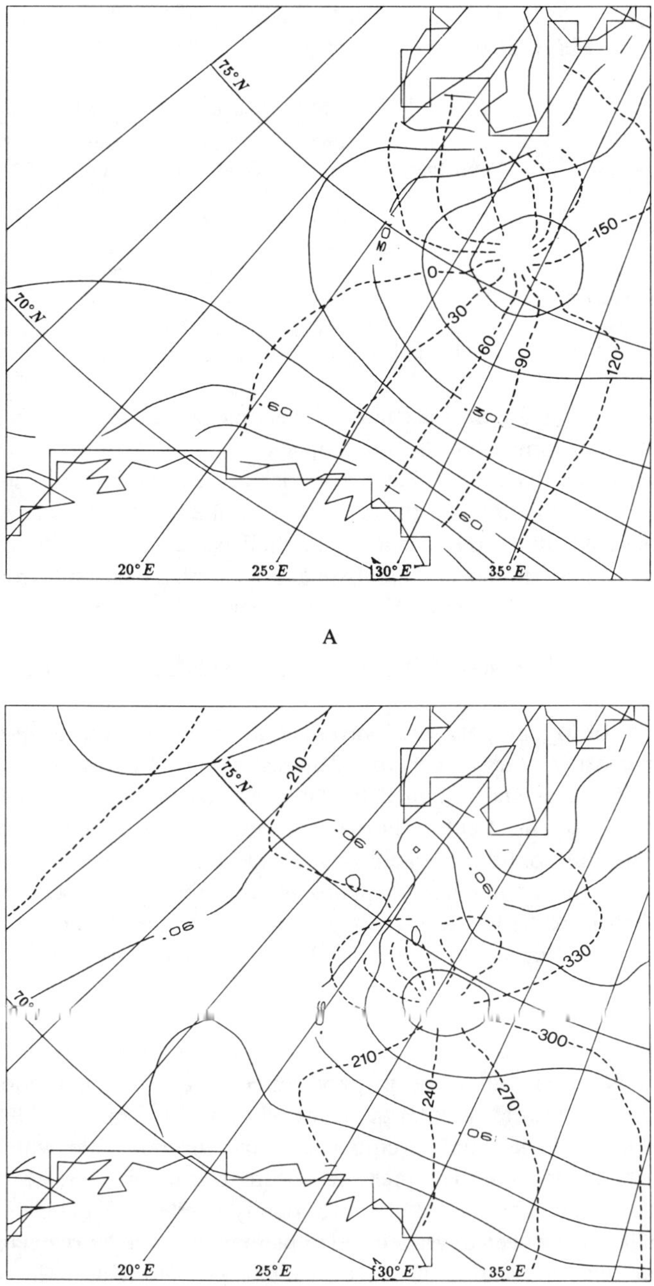


FIGURE 9. Co-range and co-tidal charts for (A) $M_2$ and (B) $K_1$ in the western part of the Barents Sea. Equidistance for co-range lines is 0.1 m in (A) and 0.02 m in (B).

waves resonantly forced by the deep-sea diurnal tide. Although the model shows an enhancement of the diurnal tide over the plateau, it does not reproduce the high current velocity observed by Hunkins. The reason may be that this is a very local phenomena caused by rather small-scale topographic features which are not resolved by the coarse grid of the model. On the other hand, the data presented by Hunkins[23] (his Figure 5) indicate that the

**TABLE 3**
**Tidal Elevation and Phase in the Barents Sea**

| | | Measurements | | Model predictions | |
|---|---|---|---|---|---|
| Station | Constituent | Amp (cm) | Phase GMT deg | Amp (cm) | Phase GMT deg |
| Bjørnøya | $M_2$ | 34.2 | 13 | 34.5 | 352 |
| 74°29′N 19°12′E | $K_1$ | 5.4 | 211 | 6.3 | 173 |
| Hopen | $M_2$ | 25.3 | 215 | 20.5 | 264 |
| 76°30′N 25°04′E | $K_1$ | 11.8 | 346 | 6.6 | 352 |
| Sentralbanken | $M_2$ | 15.8 | 92 | 13.2 | 86 |
| 74°32′N 30°58′E | $K_1$ | 4.0 | 322 | 3.2 | 296 |

strong diurnal oscillations are an intermittent phenomena and the stratification effects for this reason may be more important than anticipated.

It is clear from this discussion that the model for the diurnal tide shows small-scale variability and eddy formation along the shelves of all the three areas of primary interest in this study. We have attributed these features to shelf wave components in the diurnal tide and it may be added that similar effects also have been observed both from field data and model simulation of the shelf west of Vancouver Island, Canada.[24]

## V. STORM-GENERATED SHELF WAVES

The importance of topographic shelf waves for storm surges and storm-induced currents along the western coast of Norway was first demonstrated by Martinsen et al.[9] They applied a simple step shelf model and obtained tractable analytical expressions for the transient response due to moving wind fields. These expressions demonstrated clearly the important role of the fundamental continental-shelf wave mode for storm surges along the coast of western Norway. To the first order of appproximation the phase speed of the wave was found to be $fL$ where $L$ is the width of the shelf and the surface elevation decreases linearly from a maximum at the coast to near zero at the shelf break. Martinsen et al.[9] also solved the depth-integrated equation (Equations 1 and 2) numerically with $m$ = constant, a simple model representation of the dominant bathymetric features of the shelf, and with atmospheric forcing similar to the cyclone model (Equation 4). Prediction of sea-level changes during severe storm events was found to be in good agreement with observations from coastal stations. Maximum surface elevation at the coast was up to 1.5 m due to the storm, and the pressure disturbance and the wind forcing was found to contribute with about the same amount to the surface elevation. The pressure disturbance favors, however, a Kelvin wave response while the forcing by the wind stress mainly excites the lowest-order shelf wave mode. The model had since been upgraded and implemented at the Norwegian Meteorological Institute for operational predication of water-level changes and wind-induced currents along the Norwegian coast. Further documentation of the prediction capability of the model has been given by Martinsen et al.[13] and by Martinsen.[25] The former work examines the coupling between tidal- and the storm-induced current and provides an extensive comparison between predicted surface elevation, current speed, and direction for coastal and offshore stations in the region between western Norway, Shetland, and Scotland. Figure 15, which shows a comparison between computed and measured water-level changes during a storm event in November 1984, serves to demonstrate the very good performance of these depth-integrated models.

Water-level changes and currents induced by storms traveling across the Norwegian shelf have also been predicted with a two-dimensional version of the model described by Mathisen et al.[12] The model utilizes boundary-fitted coordinates for the coast and shelf areas

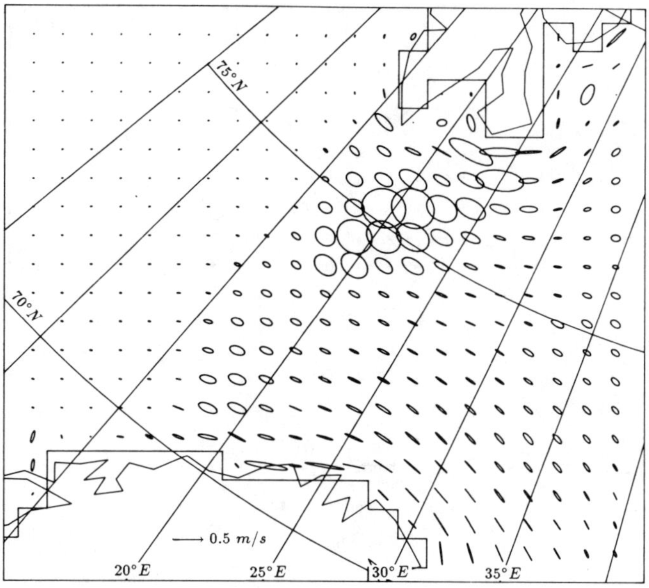


A

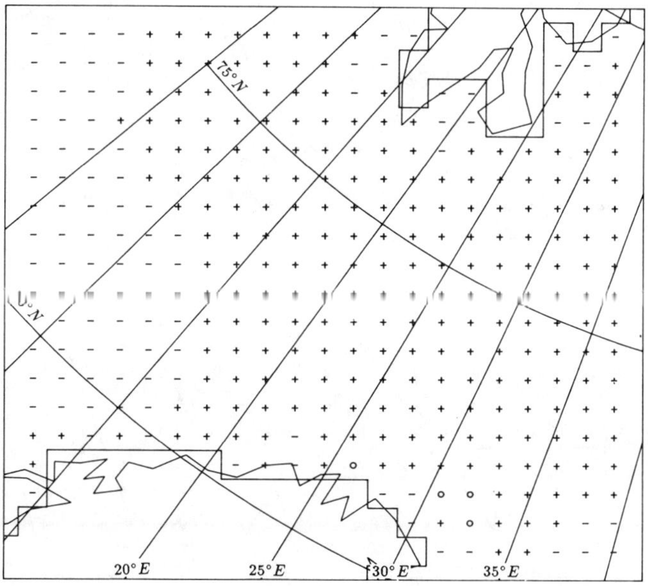


B

FIGURE 10. (A) $M_2$ tidal ellipse and (B) current vector rotation in the Barents Sea. Cyclonic rotation is indicated by (−); anticyclonic rotation is indicated by (+).

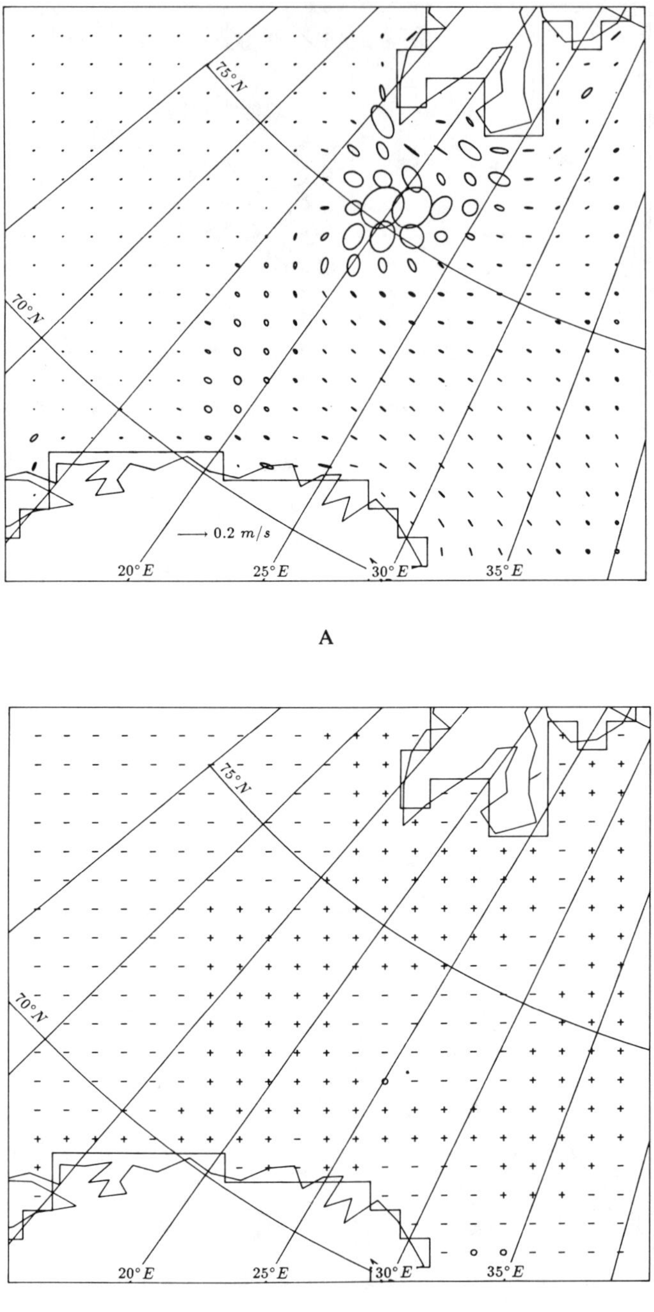


A

B

FIGURE 11. (A) $K_1$ tidal ellipse and (B) current vector rotation in the Barents Sea. Cyclonic rotation is indicated by ( − ); anticyclonic rotation is indicated by ( + ).

**TABLE 4**
**Tidal Currents in the Barents Sea**

| | | Max current (cm/s) | | Rotation | |
|---|---|---|---|---|---|
| **Station** | **Constituent** | **Mod** | **Obs** | **Mod** | **Obs** |
| I | $M_2$ | 23.0 | 42.0 | + | + |
| | $K_1$ | 5.9 | 19.0 | + | + |
| A | $M_2$ | 14.6 | 21.0 | + | + |
| | $K_1$ | 3.7 | 15.0 | + | + |
| G | $M_2$ | 9.7 | 10.0 | + | + |
| | $K_1$ | 2.0 | 1.8 | + | − |
| E | $M_2$ | 16.8 | 19.0 | + | + |
| | $K_1$ | 2.8 | 1.3 | + | + |

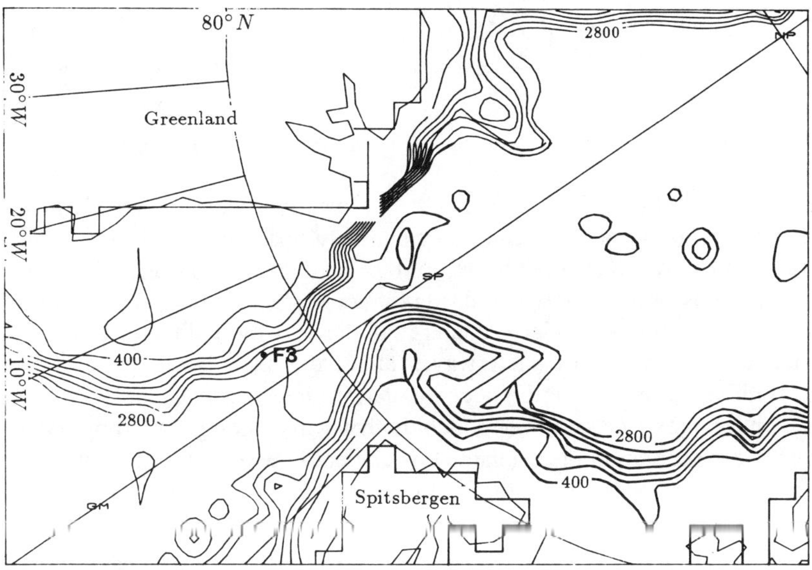


FIGURE 12. The bathymetry of the Fram Strait region. Equidistance is 400 m.

around Norway. Comparison with observations are only available for a few cases but the performance of the model seems to be good. The work on verification of the regional models developed for the Norwegian shelf has to a large extent been funded by oil companies, which in some cases have put restrictions on the use of the results. A full documentation of these models together with an assessment of their performance under various conditions are therefore not always available in the open literature.

Martinsen[25] noticed that topographic waves trapped to the west bank of the Norwegian Trench were generated by passing storms. Since the wavelength was much larger than the length of the trench only a single eddy spanning about 500 km and corresponding to one half wavelength could be identified from the numerical solution. The contribution to surface elevation from this eddy is small but the strong vorticity provided a considerable contribution to the current. Martinsen adopted the term "trench wave" and calculated the dispersion diagram for the lowest-order mode based on a simple step model of the bathymetric profile across the trench. The waves were found to propagate southward along the western bank of the trench and from his results we estimate the phase speed to be about 8.6 m/s for a wave

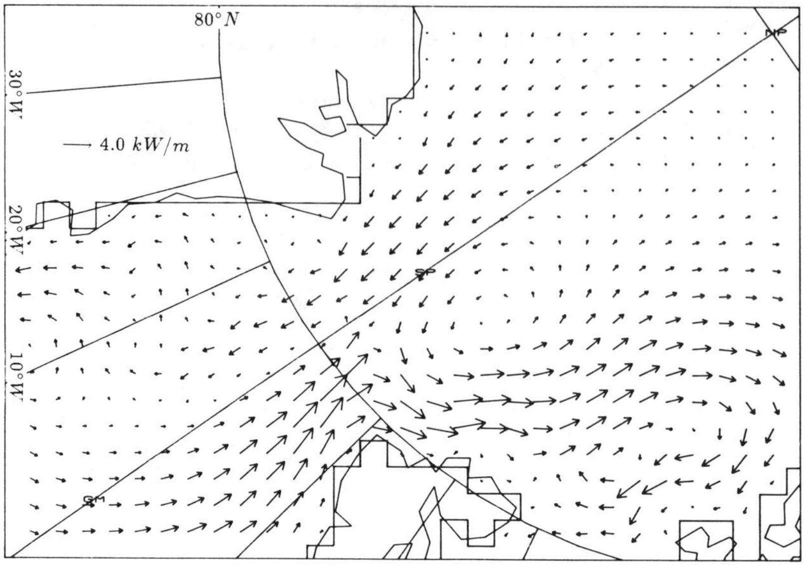


FIGURE 13. Mean energy flux for the $K_1$ tide in the Fram Strait region.

with length λ = 1000 km. This corresponds to a wave period of about 32 h, which agrees well with the typical period of oscillation frequently observed in current records from the western slope of the trench at about 60°N. Further north along the trench the observed period of oscillation seems to be shorter and current records from an oil field located close to the mouth of the trench show considerable energy in the near-diurnal band. Proctor et al.[26] have investigated the occurrence of shelf waves along the shelf break west of Shetland and Scotland. Both measurements and model simulations indicate that shelf waves with a period of about 24 h are preferred and they conclude that wind forcing by strong localized storms is the most likely generation mechanism. It is unclear how the current oscillations occurring along the shelf break west of Shetland are related to the oscillations oberved along the western slope of the Norwegian Shelf. The complex bathymetry in the area, with a sharp bend in the 100 to 400-m depth contours at the mouth of the trench, is likely to have pronounced effects on the wave propagation but a study of these aspects will require high-resolution numerical models. The deep-sea stratification in the trench may also have a considerable effect on the shelf wave period and propagation speed and stratification should be included in models developed for further studies of shelf waves in this area.

Most of the numerical studies of shelf wave generation in the North Sea area and along the shelf west of Norway have been performed with models which have open boundaries relatively close to the area of interest. These models may give a satisfactory description of the transient motion during the passage of the storm on a time scale of 1 to 2 d. It is, however, questionable if the subsequent adjustment on a time scale of 2 to 4 d is being modeled properly. We shall therefore present some results of simulation of storm-induced motion obtained with the GS model which covers a much larger area (see Section III) and therefore may be used to model long-term adjustment.

In the first example we shall examine the effect of the passage of a cyclone over the shelf west of mid-Norway. Instantaneous wind field and the track of the cyclone is shown in Figure 16. The wind field is computed from Equations 4 to 6 with $p_0$ = 40 mbar, $R$ = 500 km and $\sigma$ given by the formula in Section III. The total duration of the storm is 30 h with linear spin-up and decay over 6 h, respectively. Maximum wind speed is 28 m/s and the speed of the center of the cyclone is 15 m/s. This model serves to simulate characteristic features of a typical autumn-winter cyclone in the area and it corresponds closely to the

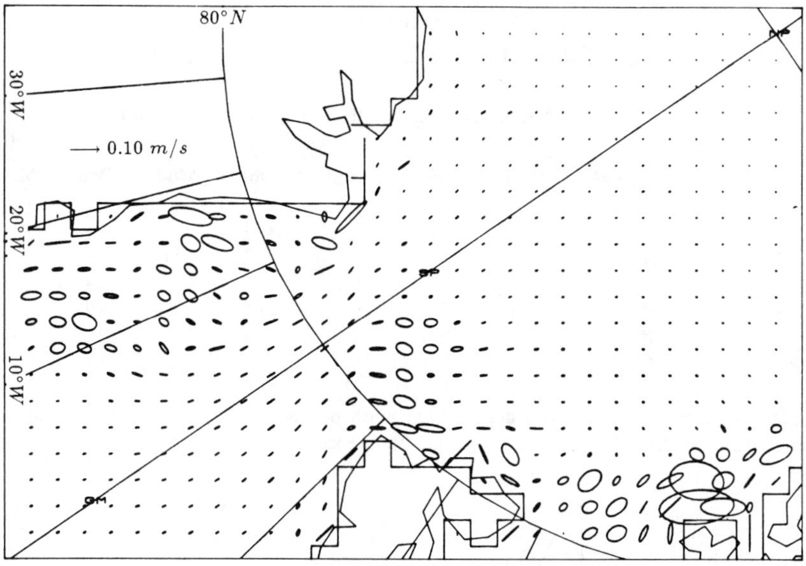


A

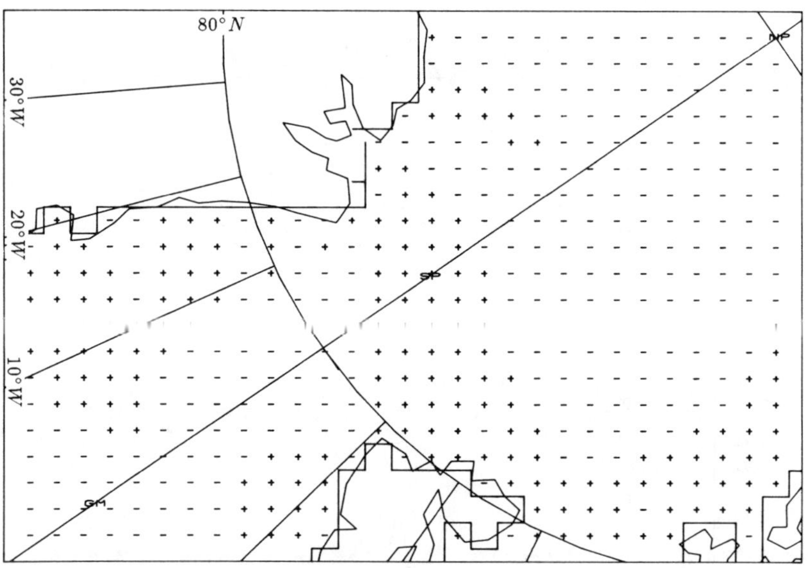


B

FIGURE 14. (A) $K_1$ tidal ellipse and (B) current vector rotation in the Fram Strait region. Cyclonic rotation is indicated by (−), anticyclonic rotation is indicated by (+).

cyclone model previously used for simulating extreme storm surges.[9] The barotropic response of the ocean due to the forcing by the cyclone is calculated with the GS model but without horizontal eddy viscosity terms and tidal forcing ($B_x = B_y = \eta_e = 0$). The surface drag coefficient is $c_D = 0.003$. Otherwise, the same values for the model parameters as used for simulation of tides has been applied (see Section IV). In Figure 17 instantaneous depth-averaged current field is shown for time 18 and 48 h after the origin of the storm. At 18 h

## TABLE 5
## Tidal Ellipse Parameters in the Fram Strait

| Station | Constituent | Major semiaxis cm/s | | Minor semiaxis cm/s | | Azimuth degree | | Rotation | |
|---|---|---|---|---|---|---|---|---|---|
| | | Mod | Obs | Mod | Obs | Mod | Obs | Mod | Obs |
| F1 | $M_2$ | 3.3 | 2.5 | 0.2 | 0.4 | 13 | 18 | + | + |
| 78°59′N 5°15′E | $K_1$ | 1.5 | 1.1 | 0.4 | 0.2 | −11 | −37 | − | − |
| F2 | $M_2$ | 2.6 | 2.5 | 0.2 | 0.3 | 10 | 14 | + | + |
| 79°00′N 4°25′E | $K_1$ | 1.3 | 1.2 | 0.3 | 0.1 | 1 | 19 | − | − |
| F3 | $M_2$ | 2.5 | 2.5 | 0.2 | 0.4 | 8 | 16 | + | + |
| 78°55′N 3°18′E | $K_1$ | 1.2 | 1.4 | 0.1 | 0.1 | 2 | −6 | − | − |

*Note:* Azimuth: orientation of major axis degree from north positive eastward. Rotation: + anticyclonic; − cyclonic. Mod: model; Obs: observational data. (Reference 22.)

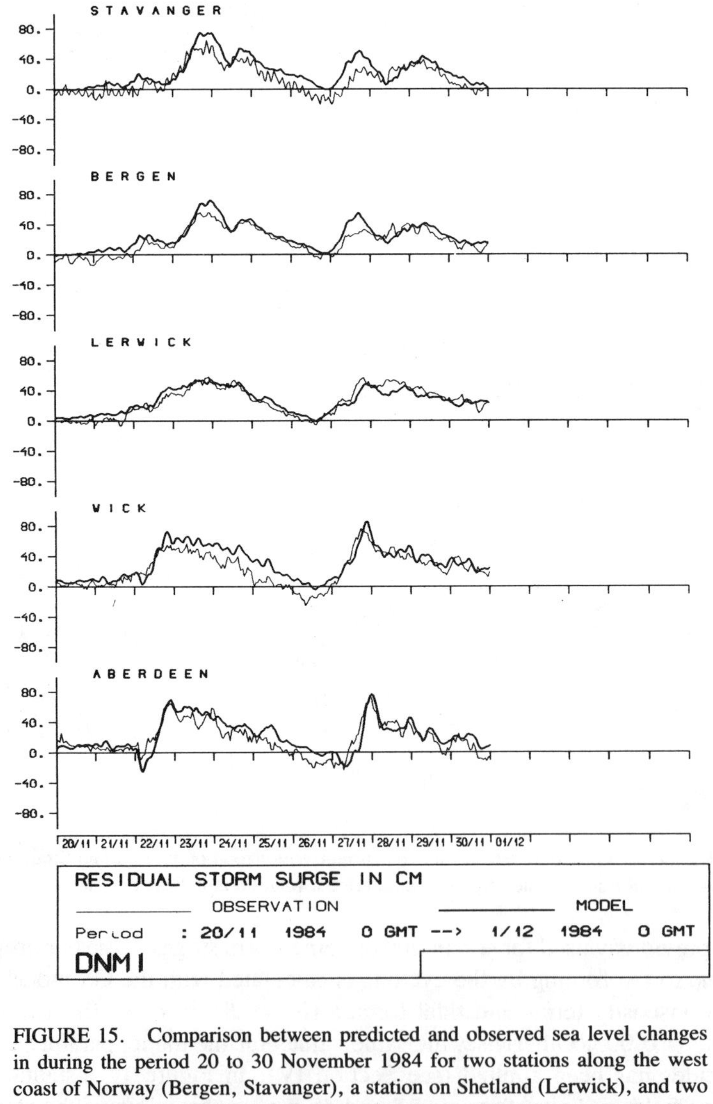


FIGURE 15. Comparison between predicted and observed sea level changes in during the period 20 to 30 November 1984 for two stations along the west coast of Norway (Bergen, Stavanger), a station on Shetland (Lerwick), and two stations in Scotland (Wick, Aberdeen). (Courtesy of E. Martinsen, The Norwegian Meteorological Institute, Oslo.)

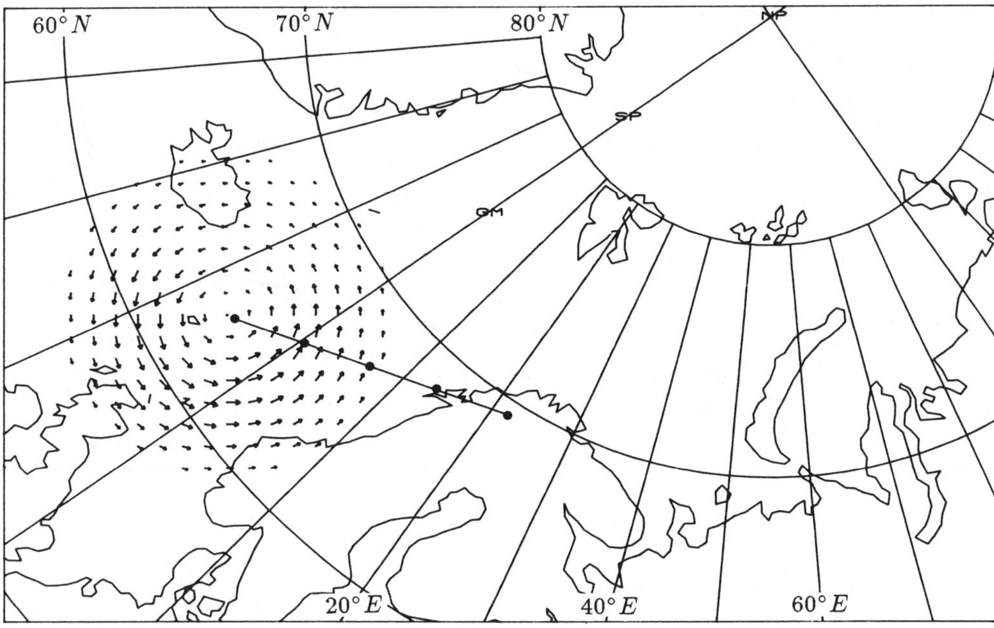


FIGURE 16. Wind field and track of cyclone used for simulation of shelf waves along the shelf wast of Norway. The position of the wind field 6 h after origin time is shown. The subsequent position of the center at intervals of 6 h are marked with dots along the track. Speed of cyclone is 15 m/s, total life time 30 h including 6-h spin-up and decay. Maximum wind speed 28 m/s.

the center of the cyclone is located on the shelf and there is a strong southwesterly wind field along the shelf west of Norway. This produces a longshore mean current up to 0.6 m/s and sea level elevation at the coast up to 1.5 m. The next panel displays the current field at 48 h when the storm has passed across the coastline and decayed. This shows that the longshore current eventually breaks up and develops an eddy structure with a pronounced eddy with diameter about 250 km located west of Haltenbanken at (64°30′N, 6°E). Time series of mean current speed and direction at station E (see Figure 1) are displayed in Figure 18. There is an anticyclonic rotation of the current vector with periods varying from 36 to 20 h and the oscillations are persistent for the total simulation time span of 5 d. Dispersion diagrams for the lowest-order shelf wave mode which can be trapped to the shelf west of Norway have been evaluated for a depth profile across the Haltenbanken area.[27] For a wave length of 500 km the corresponding wave period and phase velocity are found to be 29.7 h and 4.7 m/s, respectively, with a very small southward-directed group velocity. This is in good agreement with the results of the numerical simulations, although these indicate that a certain narrow band of frequencies are generated.

Current measurements from the Haltenbanken area[28] also show considerable energy in the near diurnal tidal band as illustrated by the rotary spectrum from Haltenbanken depicted in Figure 19. The rotation is generally anticyclonic on the shelf with a tendency for cyclonic rotation beyond the shelf break and the enhancement of the diurnal current is most likely associated with atmospheric forcing.

Finally we shall examine the current response at the Barents Sea escarpment due to a passing storm. We use the same cyclone model as in the previous case but choose a more northern track than before (Figure 20). The pressure at the center of the cyclone is $p_0$ = 45 mbar and with the same values for the radius $R$ and the parameter $\sigma$ as in the previous case, the maximum wind speed is up to 31 m/s. The total life time of the cyclone is 36 h with linear spin-up and decay over a time interval of 6 h. With these parameters the model

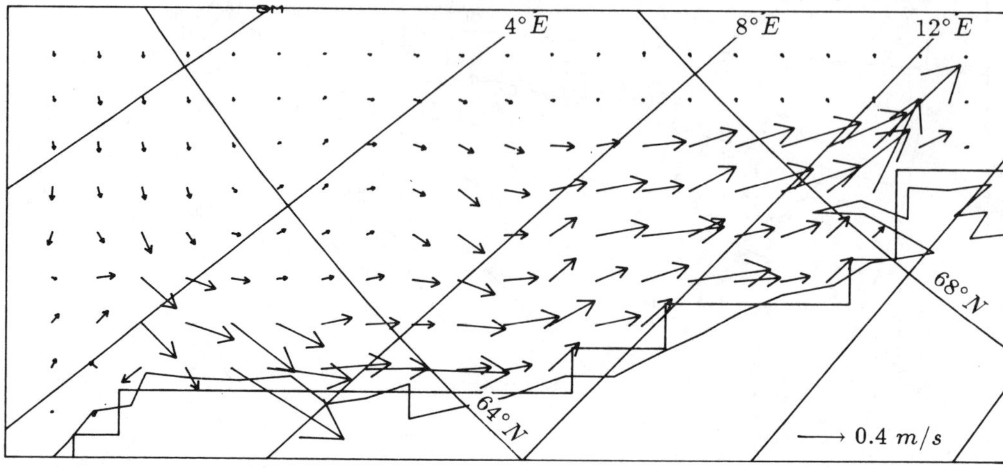


A

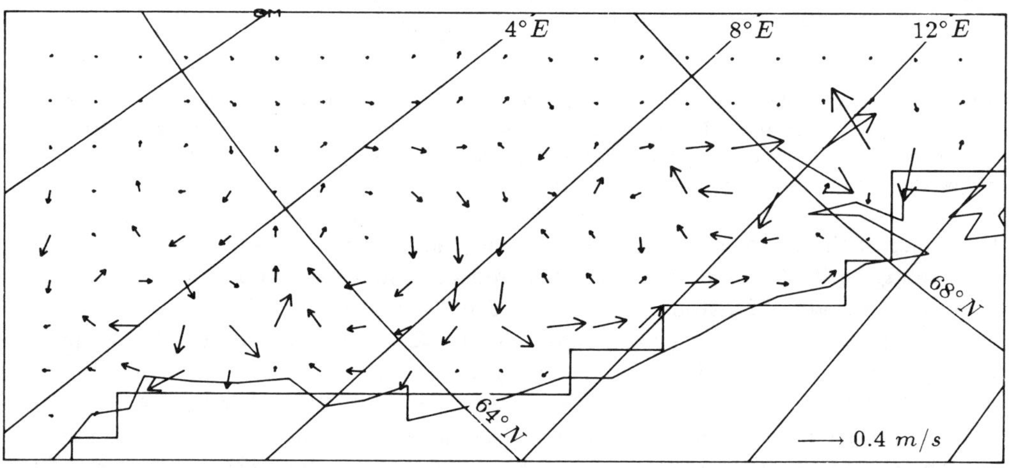


B

FIGURE 17. Model simulation of instantaneous current vector field on the shelf west of Norway generated by the cyclone depicted in Figure 16. Depth-averaged currents (A) 18 h and (B) 48 h after origin time of storm.

simulates characteristic features of a strong autumn-winter cyclone in the area. The barotropic response of the ocean is shown by the plots of the instantaneous depth-averaged current field depicted in Figure 21. The strong influence of the escarpment on the flow pattern is evident and at 36 h after origin time an eddy pair has formed with strong southernly currents along the escarpment.

The eddy with center on the escarpment carries anticyclonic rotation, while the eddy centered near the trench, which runs eastward from the escarpment, shows cyclonic rotation. In these plots also the current enhancement over the shallow Svalbard Bank is pronounced with mean current speed up to 0.8 m/s. Time series of the mean current speed and direction at station C on the escarpment (see Figure 7) are depicted in Figure 22. There is anticyclonic rotation of the current vector at this station with a period of oscillation varying from 24 to 32 h and with current speed up to 0.22 m/s. These results indicate that topographic vorticity waves with the structure of double Kelvin waves are being exited at the escarpment by the

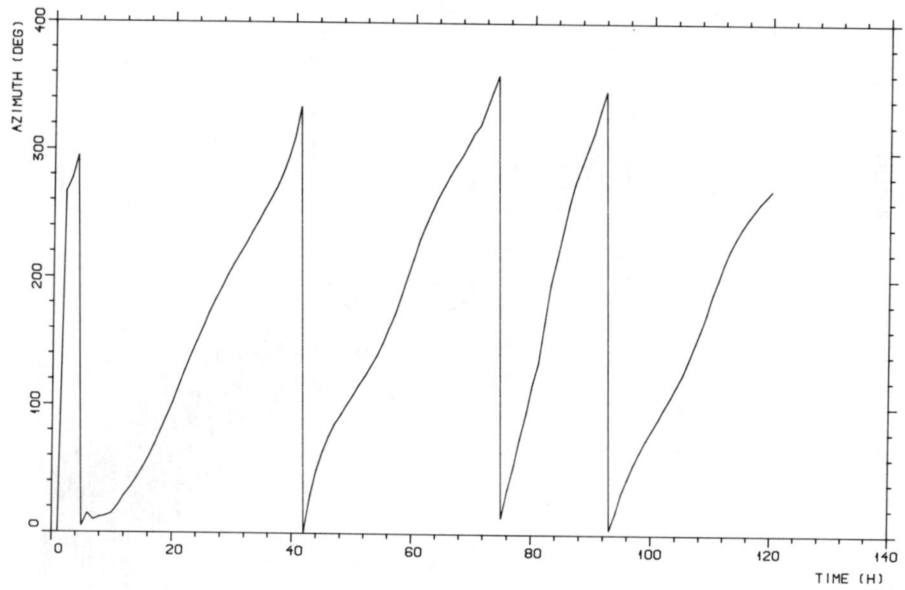


A

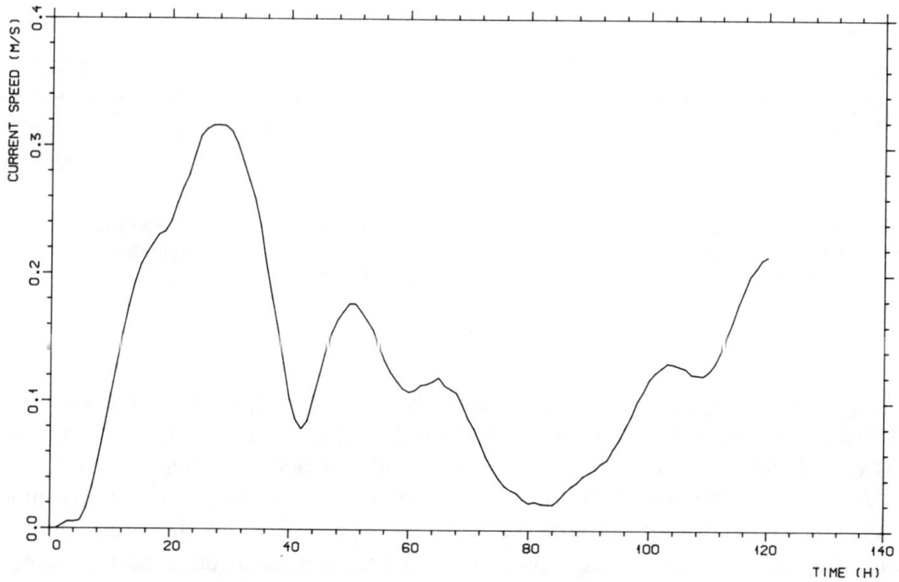


B

FIGURE 18. Time series of (A) current direction and (B) depth average current speed for station S on the shelf west of Haltenbanken (see Figure 4). Azimuth direction measured in degrees from north; 0° north, 90° east, etc.

travelling storm. At the same time the storm induces large-scale eddy features east of the escarpment and on the Svalbard Bank. Although the grid resolution used in the present model should be sufficient to resolve these large-scale features, it would be interesting to see the effect of grid refinements. It would also be interesting to analyze available current data from the area in view of numerical simulation of storm-induced current. A systematic experimental study will, however, require simultaneous current data collection along transects across the escarpment.

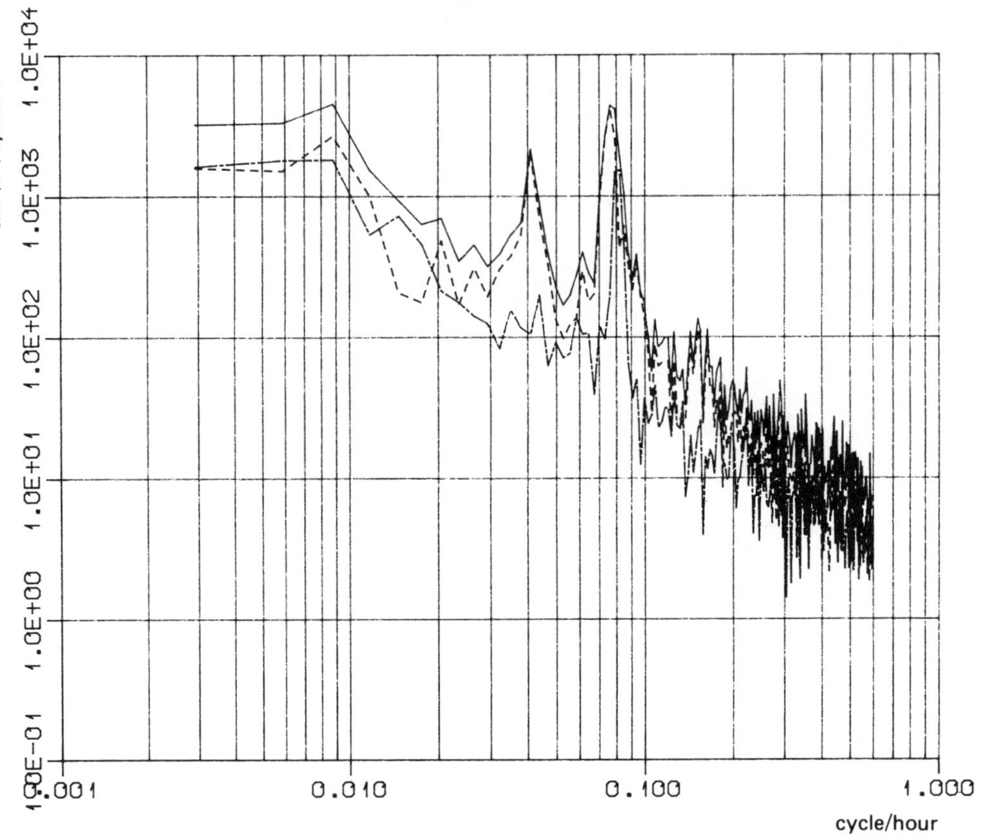


FIGURE 19. Rotary autospectrum for currents measured at 100-m depth on Haltenbanken (Station H, Figure 4) during intensive survey October—December 1985. (—) Total spectrum, (— — —) anticyclonic rotation, and (—·—·) cyclonic rotation. (With courtesy of J. P. Mathisen, Oceanographic Center, Trondheim.)

## VI. FUTURE WORK

In view of a large potential for further economic development on the shelves of the NGB sea, there will certainly be a demand for developing a next generation of predictive ocean models for this region within the next decade. Hence, statistics of current extremes and current profile parameters based on simulated time series may become a common and reliable procedure, as well as accurate short-term predictions of the ocean environment during storm conditions. In order to proceed toward these ultimate aims, model improvments are required and data from carefully planned experiments will be needed for model verification and for deepening our understanding of the fundamental dynamical and physical processes involved.

The work on improving existing barotropic depth-integrated models has already started and is evolving along the following lines: models which cover a large area like the GS model (see Section III) will be refined and run with a substantial finer-grid resolution and high-resolution regional models for a dedicated area will be nested into the large-scale models. This will certainly shed light on various interesting phenomena associated with topographical enhancement of tidal currents, for example, the strong diurnal currents observed at the shelf break south of the Svalbard Bank and the anomaly in the rotation of the $M_2$ current vector in the vicinity of Haltenbanken (see Section IV).

It is a well-established fact that depth-integrated models predict storm-induced changes in water elevation with a high degree of accuracy. Although the depth-integrated models

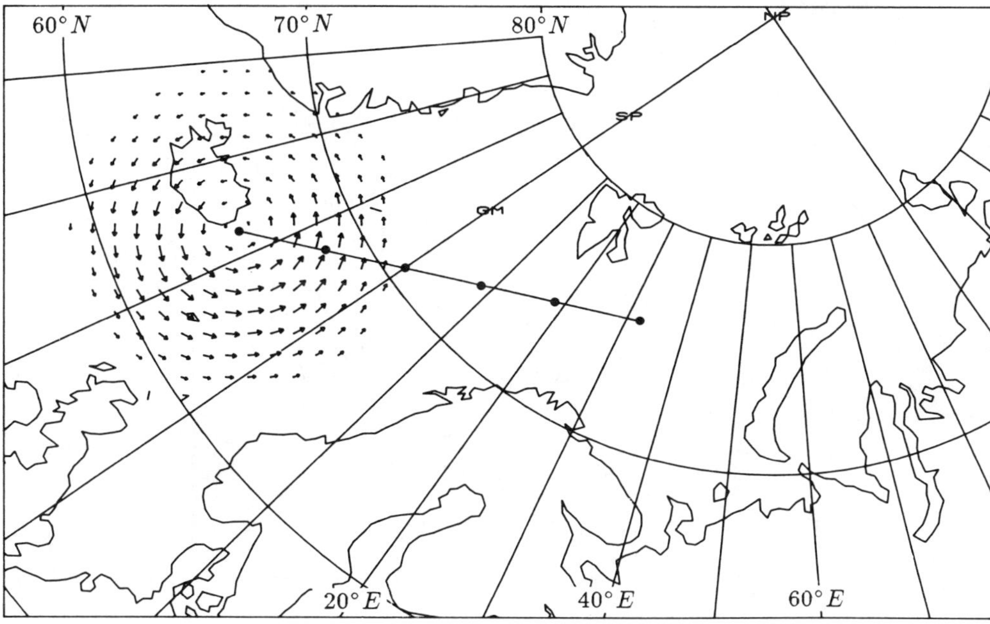


FIGURE 20. Wind field and track of cyclone used for simulation of shelf waves along the Barents Sea escarpment. The position of the wind field 6 h after original time is shown. The subsequent positions of the center at intervals of 6 h are marked with dots. Speed of cyclone is 15 m/s, total life time 36 h including 6-h spin-up and decay. Maximum wind speed 31 m/s.

also are known to predict storm-induced mean currents surprisingly well, the power of these models with respect to current prediction is not fully explored. Systematic verification studies based on simulation of long-time series of both elevation and mean currents are in progress as a joint effort between several European research institutions through the NESS project and as a part of the Norwegian ODAP program. The northern shelf areas, i.e., the Barents Sea escarpment and the Fram Strait region, have so far only received minor attention by modelers and verification studies for these areas would be particularly valuable. In the process of verifying the ocean models, the coupling to atmospheric models becomes an important issue and prediction of storm-induced currents is likely to benefit from new atmospheric models which also resolve mesoscale variability of the wind field. Updating the model simulation by data assimilation is another interesting aspect, particularly in view of the large amount of data which will be made available through various remote sensing techniques developed for earth-satellite platforms.

So far there have been few attempts to model the three-dimensional structure of tidal- and wind-induced currents in the NGB basin on a time scale of 1 to 5 d. The three-dimensional model implemented by Mathisen et al.,[12] with boundary-fitted coordinates for the Norwegian shelf, should have the potential for simulating current profiles but a systematic verification study of the prediction capacity of the model remains.

There is a wide range of well-documented numerical techniques available for three-dimensional ocean modeling and a review of the literature is given by Davies.[29] Verification of the performance of these models under the transient storm conditions typically met with on the shelves of the NBG sea is, however, largely an open question. Recently Davies[29] devised a computationally efficient method for deducing current profiles from two-dimensional models. Combined with more advanced turbulent closure models this may provide an interesting method which should be exploited further.

In connection with the development of new models there are several special physical

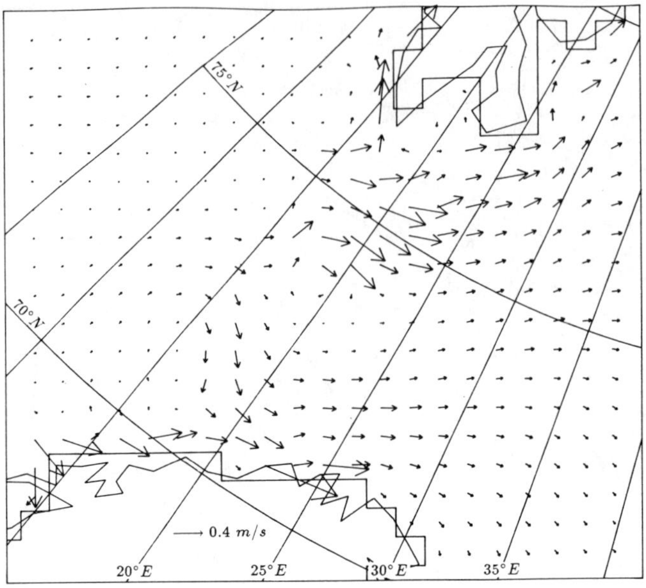


A

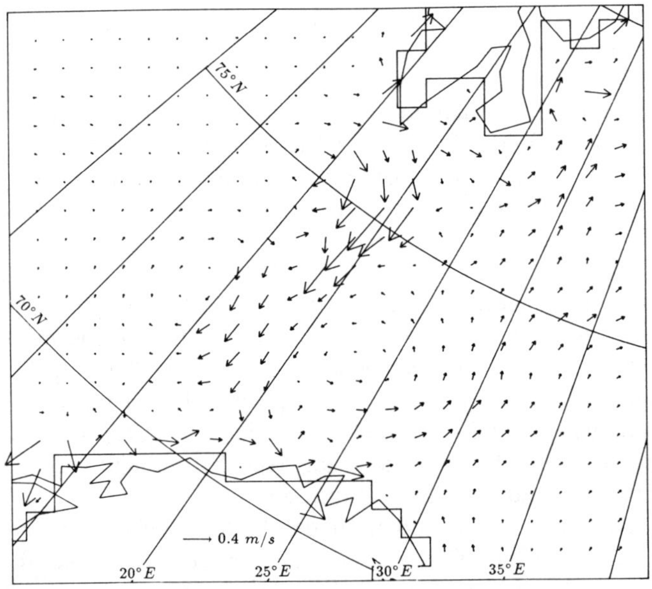


B

FIGURE 21. Model simulation of instantaneous current vector field along the Barents Sea escarpment due to the cyclone depicted in Figure 20. Depth-averaged currents (A) 30 h and (B) 36 h after origin time of storm.

and dynamical processes which need to be studied in more detail before these effects can be properly parameterized by numerical models. For modeling tides and storm-induced currents in the northern regions of the NGB sea the effect of sea ice needs to be included. The ice will primarily have a damping effect on the tide but this is believed to have only a minor effect on the tide in deep water. The ice cover is likely to have a more dominant

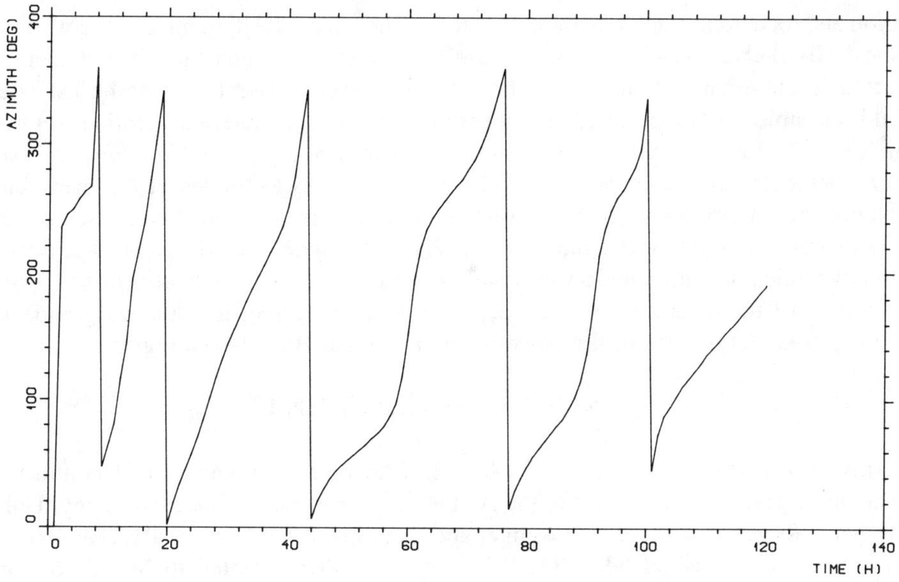


A

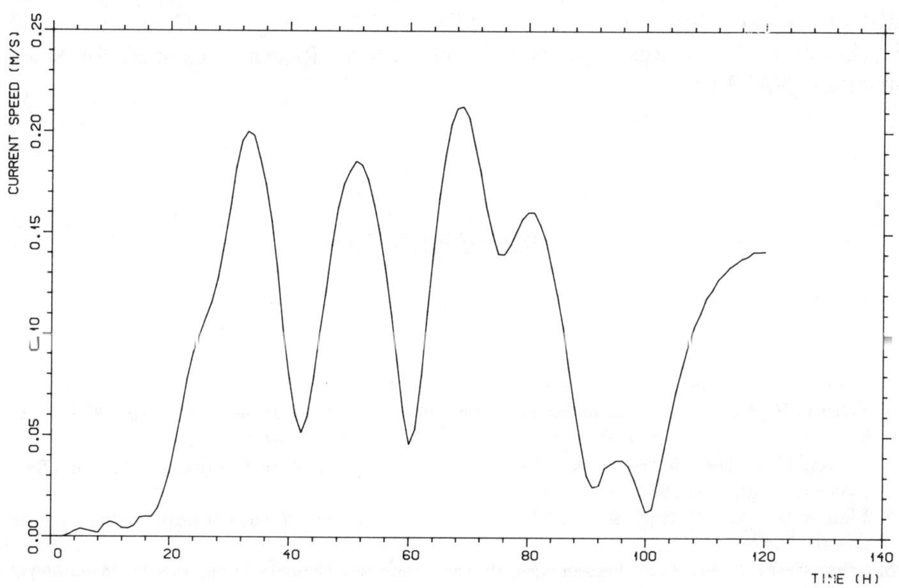


B

FIGURE 22. Time series of (A) current direction and (B) depth-averaged current speed for station C on the Barents Sea escarpment (see Figure 7). Azimuth direction measured in degrees from north; 0° north, 90° east, etc.

effect on the wind-induced current since the presence of the ice strongly will affect the rate of transfer of momentum from the atmosphere to the sea. This will in turn influence the drifting of the sea ice and the capability to predict the response of the sea ice cover to atmospheric forcing on a time scale of 1 to 5 d. This is an interesting aspect which needs to be clarified.

There are also a large number of problems connected to the combined effect of stratification and local topographic features and the influence on the dynamics of the water masses on the NGB shelves which need to be resolved. Despite the fact that shelf dynamics have attracted the attention of numerous mathematicians and oceanographers for the last 2 decades, and it is possible to draw on a large number of analytical and numerical solutions for idealized situations,[30,31] there are only a few results which may apply to shelf regions with strongly varying geometry as is the case for the NGB shelves. As examples of the many unsolved problems we may mention scattering of topographic waves at the Barents Sea escarpment and at the mouth of the Norwegian Trench. The effect which the deep-sea stratification has on the dynamics of topographic wave modes in these regions needs also to be clarified.

In view of these remarks it will be, in the years to come, a challenging endeavour to explore and model the sea in the Norwegian-Greenland-Barents sea region.

## ACKNOWLEDGMENT

This work has been supported by STATOIL, Stavanger, Norway through contract T4257. The author is grateful to members of the staff at the Department of Oceanography, University of Bergen, Norges Sjokartverk, Stavanger and Oceanographic Center, SINTEF, Trondheim, for providing unpublished tidal data. The author is also indebted to Mr. T. Straume for assistance with computer programming and graphics. The work was completed while the author was a visiting scholar at the Institute of Geophysics and Planetary Physics and Institute for Nonlinear Science, University of California, San Diego, with support in part by the DARPA Univ. Res. Init. under Appl. and Comp. Math. Program, Contract N00014-86-K-0758, Office of Naval Research, and the Norwegian Research Council for Science and Humanities (NAVF).

## REFERENCES

1. **Gjevik, B. and Straume, T.,** Model simulations of the $M_2$ and the $K_1$ tides in the Nordic seas and the Arctic Ocean, *Tellus,* 41A, 73, 1989.
2. **Cartwright, D. E., Edden, A. C., Spencer, R., and Vassie, J. T.,** The tides of the northeast Atlantic Ocean, *Philos. Trans. R. Soc. London,* 298(1436), 87, 1980.
3. **Flather, R. A.,** A tidal model of the northwest European continental shelf, *Mem. Soc. R. Sci. Liege,* 6(X), 141, 1976.
4. **Davies, A. M. and Furnes, G. K.,** Observation and computed $M_2$ tidal currents in the North Sea, *J. Phys. Oceanogr.,* 10, 237, 1980.
5. **Flather, R. A., Proctor, R., and Wolf, J.,** A tidal model of the northeast Atlantic Ocean, private communication.
6. **Schwiderski, E. W.,** Global ocean tides. II. The semi-diurnal principal lunar tide $M_2$, Atlas of Tidal Charts and Maps, Rep. 79-414, Naval Surface Weapons Center, Dahlgren, VA, 1979.
7. **Schwiderski, E. W.,** Global ocean tides. IV. The diurnal luni-solar declination tide ($K_1$), Atlas of Tidal Charts and Maps, Rep. 81-142, Naval Surface Weapons Center, Dahlgren, VA, 1981.
8. **Mesinger, F. and Arakawa, A.,** Numerical methods used in atmospheric models, GARP Publ. Ser., WMO, 17, 64, 1976.
9. **Martinsen, E. A., Gjevik, B., and Røed, L. P.,** A numerical model for long barotropic waves and storm surges along the western coast of Norway, *J. Phys. Oceanogr.,* 9(6), 1126, 1979.
10. **Kowalik, Z. and Untersteiner, N.,** A study of the $M_2$ tide in the Arctic Ocean, *Dtsch. Hydrogr. Z.,* 31, 216, 1978.
11. **Mathisen, J. P. and Johansen, Ø.,** A numerical tidal and storm surge model for the North Sea, *Mar. Geodesy,* 6(3-4), 267, 1983.
12. **Mathisen, J. P., Jenssen, Ø. O., Spalding, M., and Swanson, C.,** A three dimensional model for ocean currents where horizontal grid spacing is varied using boundary fitted coordinates, in Proc. SUT Conf. Modelling Offshore Environments, London, April, 1987.
13. **Martinsen, E., Engedahl, H., Kristensen, M., and Sandnes, H.,** CUMEX winter storm modelling. I. Numerical model and main results, Rep. Norwegian Meteorol. Inst. Blindern, Norway, 1987.

14. **Schwiderski, E. W.,** *The Nordic Seas,* Hurdle, Ed. Springer-Verlag, New York, 1986, 191.
15. **Straume, T.,** Model simulations of the $S_2$ and $N_2$, tidal components in the Nordic Seas and the Arctic Ocean, Rep. Department of Math, University of Oslo, Norway, 1987.
16. **Davies, A. M.,** On determining current profiles in oscillatory flows, *Appl. Math. Model.,* 9, 419, 1985.
17. **Nøst, E.,** Model simulation of the $M_2$ tide in the Haltenbanken area, Rep. Department of Math, University of Oslo, Norway, 1988 (in Norwegian).
18. **Mathisen, J. P.,** Three-dimensional model of ocean currents, ANODA, Summary Rep. no. 34, Oceanographic Center, SINTEF, Trondheim, Norway.
19. Oceanographic Center, SINTEF, Trondheim, Norway, Data report, 1987.
20. **Nekrasov, A. V.,** Preliminary charts of the main diurnal and semi-diurnal tidal waves in the Norwegian and Greenland Seas, *Trans. Leningr. Hydrometeorol. Inst.,* 10(1), 81, 1961 (in Russian).
21. **Huthnance, J. M.,** Large tidal currents near Bear Island and related tidal energy losses from the North Atlantic, *Deep-Sea Res.,* 28A, 51, 1981.
22. **Aagaard, K., Darnall, C., Foldvik, A., and Tørresen, T.,** *Fram Strait current measurements 1984—85,* Rep. No. 63, Department of Oceanography, University of Bergen, Norway, 1985.
23. **Hunkins, K.,** Anomalous diurnal tidal currents on the Yermak Plateau, *J. Mar. Res.,* 44, 51, 1986.
24. **Flather, R. A.,** A tidal model for the northeast Pacific, *Atmosphere-Ocean,* 25, 1, 22, 1987.
25. **Martinsen, E.,** The barotropic current due to storm surges along the coast of Norway, in *The Norwegian Coastal Current,* Saetre, R. and Mork, M., Eds., University of Bergen, Norway, 1981, 500.
26. **Proctor, F., Flather, R., and Gordon, R. B.,** Storm generated barotropic shelf waves on the Scottish Shelf, paper presented at the XIX General Assembly, IUGG, Vancouver, Canada, 1987.
27. **Huthnance, J. M.,** Personal communication, 1986.
28. **Mathisen, J. P., Bjerke, P. L., Olsen, R., and Lønseth, L.,** Current measurements Halten-Shelf Edge October—December 1985, Rep. 59.0002.02, Oceanographic Center, SINTEF/Oceanogr., Trondheim, Norway, 1986.
29. **Davies, A. M.,** Numerical modelling of marine systems, in *Numerical Modelling-Application to Marine Systems,* Noye, J., Ed., North-Holland, Amsterdam, 1986.
30. **Mysak, L. A.,** Recent advances in shelf wave dynamics, *Rev. Geophys. Space Phys.,* 18(1), 211, 1980.
31. **Huthnance, J. M., Mysak, L. A., and Wang, D. P.,** Coastal trapped waves. Baroclinic processes on continental shelves, *Coastal and Estuarine Science,* Vol. 3, American Geophysical Union, Washington, D. C., 1986.

Chapter 10

# MODELING SAND TRANSPORT ON CONTINENTAL SHELVES

**David A. Huntley and Anthony. J. Bowen**

## TABLE OF CONTENTS

## ABSTRACT

The prediction of sediment transport on continental shelves involves equations which attempt to relate net or gross sediment movement to readily measured parameters such as wave height and period, wind speed and direction, tidal currents and mean currents. However, the link between the moving sediment near the sea bed and these synoptic parameters is highly complex. To formulate satisfactory equations we must understand both the hydrodynamics of the boundary layer at the seabed and the sedimentary response to the boundary layer flow. In this paper, we review the current understanding of the relationships between synoptic parameters and sediment transport and discuss the possibilities for improving both our understanding of sediment transport and our predictive models.

## I. INTRODUCTION

Several different kinds of questions might be addressed by numerical models of sediment transport on the continental shelf. For example:

1. Does sediment of a given size move? Under what conditions does it move, and how often?
2. How does the sediment move? As suspended load or bedload? Are bedforms developed? To what depth does movement occur?
3. What is the rate of sediment movement? How is the total rate divided between bedload and suspended load? What is the transport rate of a particular size fraction? Under oscillatory flow, what is the rate and direction of the net transport?
4. What are the rates of erosion and deposition? What are the spatial distributions of regions of erosion and deposition?

These four groups of questions identify four increasing levels of complexity required in numerical models. For example, consideration of threshold conditions for sediment movement is central to the first group of questions, but must be addressed also by models in each of the other groups. As we shall see, even at this lowest level there are significant limitations on the accuracy of existing models, particularly where bedforms occur and where the sediment under consideration is cohesive. At the third and fourth levels quantitative estimates of sediment transport are required, involving a choice of suitable sediment-transport equations from among the wide variety currently available. The questions grouped last are the most difficult since erosion and deposition rates are dependent on spatial gradients of transport rates.

The temporal and spatial scales addressed by numerical models also vary over a wide range. Continental shelves are characteristically environments where the velocity field varies on a wide range of time scales (Figure 1).

For a storm-dominated shelf, interest in low frequencies is focused on the occurrence of the severe conditions associated with winter storms or summer hurricanes. Engineering design is related to storms of long return period, the 10, 50, or 100 year storms, for which direct observations are likely to be either absent or sparse. In these cases the modeling process will generally reproduce each storm separately. An important class of models is, therefore, one which runs for a relatively short time, a few days at most, with wind, tide, and wave input for the appropriate extreme conditions.

In contrast, a study of suspended-sand transport in strong tidal currents may totally ignore the effects of extreme events. They may be too infrequent to influence the overall transport patterns. Here the critical factors are the asymmetry of the tidal currents (the contrast between flood and ebb), the spring-neap cycle, and short-term changes in the wave field, all of which might be measured directly.[1]

A third type of model is needed if the primary interest is in the mean transport *paths* over a long period of time. Hill and Bowen,[2] for example, suggest a general pattern of movement off the shelf onto the continental slope. Investigation of very-long-term trends is of considerable geological interest but is probably not a priority for engineering studies, fortunately perhaps, as running a model for long periods, with realistic inputs, presents serious practical difficulties. To look at geological timescales, some type of stochastic modeling is necessary where the occurrence and intensity of various events is expressed in terms of return periods. Suitable combinations of events might then provide an estimate of the patterns of transport on geological time scales.

Relatively long runs of a deterministic model might be associated with experimental data, e.g., with a tracer experiment running for several months. In this case, the primary interest would be movement at one point, or at least within a small area. Nevertheless, prediction of the wind-driven bottom currents would require a large-scale modeling exercise and it would clearly be necessary to monitor the hydrodynamic conditions for the whole period of a tracer experiment.

Turning to the different spatial scales of models, the definition of the sedimentary environment being modeled involves horizontal length scales, the scale of any bedforms for example. The problem of spatial resolution is particularly acute in the context of group 4 questions, those concerned with erosion and deposition. To resolve the spatial derivatives of transport rates necessary to calculate erosion and deposition, at least four estimates of transport over the length scale of the topography of interest are needed. For example, to resolve the dynamics of a 400 m sand wave (characteristic of the southern North Sea) a model grid scale of 100 m would be required. The modeling of large areas on this scale is not really practical but a finer grid might be inserted within a larger coarse-grid model to resolve features of particular interest.[3]

With the differing aims for modeling and with such a wide range of possible time and space scales, a single universal model of sediment transport is not a practical goal. There is an infinite set of possible models, each addressing a different question and dealing with a different set of conditions. Nevertheless, all models have certain requirements in common. The first is an adequate representation of the hydrodynamic conditions on appropriate time and space scales. The second is an adequate description of the sedimentary environment, the grain-size distribution, availability of material, bottom roughness, bedforms, and general bathymetry. The final requirement is a suitable relationship between the hydrodynamic conditions and the sediment response, for example in the form of a predictive equation for the total sediment transport.

In the following sections we investigate each of these three model requirements. Each one presents problems and, taken together, they may lead to model predictions of sediment transport which can be widely different from each other as well as from the limited observations. It is becoming clear that any satisfactory model of sediment transport on continental shelves will have to contain descriptions of a wide range of complex processes involving the dynamics of both sediment and water. In the final section we review some recent numerical models which, though still limited, appear to point the way ahead.

## II. HYDRODYNAMIC INPUTS FOR MODELS

It is the *stress* at the seabed which is generally considered to be the agent responsible for moving sediment. However, most existing sediment transport equations model this stress in terms of water velocity at some standard height (usually 1 m) above the bed. The predicted sediment transport rate is then dependent on this velocity raised to some power. For example, Bagnold's[4] equations suggest transport rates proportional to $u^3$ or $u^4$, while the Einstein-Brown[5] equation uses $u^6$. With these high powers, it is evident that predictions are very sensitive to errors in input velocities. Accurate knowledge of the hydrodynamic conditions is therefore vital.

Unfortunately, the available input data is generally much less than ideal. Direct measurements of seabed stresses are rare,[6-9] and are subject to large experimental uncertainties. More commonly, water velocities at a standard height above the seabed are available or can be estimated. Frequently, for example, when modeling extreme storm conditions, only the general characterisitics of the particular meteorological situation are known. In this case, hydrodynamic parameters would have to be assumed or predicted.

In what follows, we discuss first the prediction of velocities at a standard height above the bed. It is then necessary to consider some of the complexities involved in converting these velocities to seabed stresses. The use of oversimplified models is undoubtedly responsible for some of the inaccuracies in predictions. Finally, to complicate matters further, we briefly review some results which suggest that seabed stresses, however accurately predicted, may in fact be inadequate for modeling sediment transport.

## A. THE FLOW FIELD ABOVE THE SEABED

### 1. Currents

Nearbed currents due to tides and meteorological forcing can be readily measured by moored current meters. Tidal currents can also be predicted from long-term current observations. However, measurements of currents sufficient to define the detailed pattern of the flow over an area of continental shelf are expensive to obtain.

In the absence of measurements, a variety of two- and three-dimensional numerical models exist to predict tidal- and wind-driven currents and their spatial patterns.[10-13] Tryggestad et al.[14] and Davis and Flather[15] have used numerical models to predict *extreme* bottom currents. However, such models are not without problems. For example, they require as input an appropriate functional form for the bottom stress. Predictions of low-frequency flows in shallow water are particularly sensitive to the form chosen. As we shall see later, the problem of choosing a correct functional form is greatly complicated by the presence of high-frequency wave currents or of appreciable bedforms, both common on continental shelves. Very few numerical models of currents on continental shelves take account of the influence of surface wave motion on the stress at the seabed, one exception being that of Clarke and Brink.[16] Hydrodynamic models are also subject to the same problems as sediment-transport models in regions of complex bottom topography. The problems are particularly acute in regions with bedforms too large to be considered simply as increased seabed roughness yet small compared to model grid scales. The sand wave field of the southern North Sea is an example of such a region.

In strong tidal flows, steady tidal residual currents, generated by nonlinear processes, can be significant for sediment transport, though the magnitude of the currents is generally small (typically a few centimeters per second). The patterns of these flows can be reasonably well described by numerical models, particular interest having been focused on flows associated with offshore sand banks.[17-19] However, some details of the flow near the seabed, particularly the cross-isobath flow, seem very sensitive to the model assumptions.

Long-term trends in sediment motion may be influenced by other oceanic processes (general shelf circulation, internal tides, etc.) which provide a background input, negligible during a storm event, but perhaps significant in the long term. Modeling all these separate oceanographic processes would be a truly formidable undertaking. Direct monitoring of the hydrodynamic conditions near the sea bed over a timescale of some months should provide further insight as to the relative importance of these processes. However, much longer deployments may be necessary to identify the longer term variability outlined in Figure 1.

### 2. Surface Waves

The high-frequency wave component of the flow field must also be considered. Direct

SHELF CURRENTS

| TIME SCALES OF VARIATION | WIND DRIVEN | WAVES | TIDES | DENSITY DRIVEN | DEEP OCEAN |
|---|---|---|---|---|---|
| Secs | | Orbital Velocity | | | |
| Hours | Surges | Longshore and Rip Currents | Tidal Currents | Internal Tides | |
| Days Weeks | Upwelling Shelf Waves | Drift Velocity | | | |
| Months | Circulation | | Spring-Neaps | Estuarine Circulation | Rings Eddies |
| Seasonal | Seasonal Patterns | Hurricanes Winter Storms | Low frequency variability | Seasonal Run-off. | Ocean Currents |
| Annual | Winter | Ice | | Insolation | |
| Long Term | Extreme Storm | Conditions | | | Oceanic Circulation |

FIGURE 1. Contributions to the velocity field on a continental shelf.

measurements of wave orbital velocity near the bed are obviously most desirable, but are expensive to obtain and consequently rare. Spectra of surface wave elevation, frequently measured using an accelerometer buoy, are more common and can be used to estimate seabed velocities. However, the directions of wave propagation, important for bottom stress estimation, are not usually measured.

Perhaps most commonly, wave information, when available at all, is only in the form of significant wave heights and periods, perhaps coupled in a matrix of joint occurrences. When used to estimate nearbed orbital velocities this information can be misleading. The decrease in wave influence with depth below the surface is strongly frequency dependent, with higher frequencies attenuated more rapidly. Consequently, for a spectrum of orbital velocities, lower frequency (longer period) waves will be more dominant at depth than at the surface. Hence, the mean wave period will be shifted towards a longer period. This effect will only be negligible if the wave spectrum is narrowly peaked about the significant wave period. For a broad spectrum the error involved in ignoring the shift in frequency can have a significant effect on the estimation of bottom stresses.

Numerical models for the prediction of wave conditions have been much studied in recent years. They range from simple parametric models such as the schemes of Darbyshire and Draper,[20] Bretschneider,[21] or JONSWAP,[22] to the complex numerical models used by the SWIM, SWAMP, and WAM groups.[23] None of these schemes is entirely satisfactory, and comparisons with observations suggest that considerable further work in needed, for example in parameterizing the energy losses as waves propagate into shallower water.[24] Nevertheless, as the importance of wave motion to sediment transport becomes recognized, the use of these models in prediction schemes will undoubtedly increase.

As yet, however, in many cases wave data are not available at all, as in the studies by Heathershaw[25] and Lees.[26] One reason for this lack of wave data has been an emphasis on measuring or modeling sediment transport in areas of strong tidal currents. A modeler lacking wave data has traditionally done one of the following:

1. Assumed that wave effects are small[25,27]
2. Computed theoretical influences for some conceptual waves using a simple model such as Bijker's[28]
3. Adjusted the threshold of movement, drag coefficient, or effective bottom roughness to try to compensate for the presumed wave activity[29]

Given the strong sensitivity of the bottom stress to wave motion none of these alternatives is likely to produce very accurate predictions.

### B. SEABED BOUNDARY LAYERS

Having obtained estimates of flows above the bed, the modeler must convert these flows to seabed stresses. This involves consideration of the structure of sea bed boundary layers.

Excellent recent reviews of continental shelf boundary layers are by Bowden[30] and Soulsby.[31] It is not necessary, therefore, to describe again the major features of boundary layers and the commonly accepted methods for estimating seabed stresses. Here we concentrate on those features of the boundary layer which can complicate sediment-transport prediction and which undoubtedly can have a major influence in limiting the accuracy of predictions.

#### 1. Form Drag

Bedforms commonly occur in regions of sediment-transporting flows on the continental shelf. These bed features create an internal boundary layer in the flow, within which there is an acceleration in the water flow over the crest of the bedform and a deceleration near the trough. If the flow in this internal boundary layer is averaged spatially over a bedform, at fixed heights above the bed, the velocity profile is logarithmic, with a slope related to the shear stress at the seabed and a roughness length related to the grain roughness of the bed. At a distance greater than about one tenth of a bedform wavelength away from the boundary, the spatial variability of the internal boundary layer has become smoothed out, the flow is uniform, and the velocity profile is logarithmic at all positions over the bedforms. However, in this upper region the slope of the logarithmic velocity profile is determined by the average shear stress on the bed *plus* the force resulting from form drag on the bedform. In other words, the bedforms appear as roughness elements to the flow in the upper region. In the presence of several scales of bedforms, Smith and McLean[32] found evidence for multiple log layer velocity profiles, each corresponding to a particular bedform scale.

The form drag is the average, over a bedform wavelength, of the normal pressure distribution on the bedform, and it does not contribute to sediment transport. Thus, to estimate sediment motion, it is essential to be able to remove the form drag component from any measurement, or prediction, of total bottom stress. It is clear from the literature that this problem has been underestimated until very recently, and many of the standard assumptions about sediment transport may have to be modified in the light of recent results. Numerical models of flow over ripples, such as those of Smith[33] and Taylor et al.,[34] suggest that for sinusoidal waveforms typically 25 to 33% of the total drag results from form drag. Richards and Taylor[35] developed a numerical model for flow over sand waves and considered two nonsinusoidal waveforms, one with peaked crests and the other asymmetric in shape. Surprisingly, the increase in form drag, compared with sinusoidal waveforms, was not large, typically rising to only about 37% even for a strongly asymmetric shape. However, their model broke down when separation occurred over the bedforms.

Field experiments seem to suggest a much higher proportion for the form drag. Smith[36] found that form drag contributed 82% of the total drag for nonseparating flows and 75% for separating flows over sand waves in the Columbia River. Chriss and Caldwell,[37] measuring the seabed stress directly, in the viscous sublayer, and comparing it with the stress inferred from logarithmic profiles above, found form drag contributions ranging from 74% to as high as 91% on the Oregon Shelf. Although there are too few measurements to be sure that these ratios are characteristic of conditions on continental shelves, clearly form drag can be a significant part of the stress except very close to the bed. Some estimates suggest that, to be within the log layer related to the true seabed stress, measurements must be made within one tenth of a bedform wavelength,[38] others suggest within two or three bedform heights.[39]

The traditional way to treat this problem is to assume that the bedforms are formed by, and in equilibrium with, the ambient flow conditions.[40,41] In this way the magnitude of the form drag is assumed to be a unique function of the flow. Figure 2 shows Fredsoe's plot. This figure shows, in agreement with data from Raudkivi,[42] that the form drag contribution to the total stress is greatest for intermediate flow speeds, falling off at low speeds, where bedload transport is weak or nonexistent, and at high speeds, where sheet flow flattens down the bedforms.

On the continental shelf, where the flow regime can change much more rapidly than the bedforms, the assumption of equilibrium bedforms is unlikely to be generally valid. Dyer[43] gives a nice demonstration, from field data, of the lag between bedforms and the flow conditions in the special case of a tidally dominated environment. He found a hysteresis effect in the apparent bottom roughness and total stress between flood and ebb currents as a result of nonequilibrium ripple sizes.

If bedforms are not in equilibrium with ambient conditions, the form drag must be estimated on the basis of what might be rather scanty data on the size and geometry of the existing bedforms. Smith[36] presented a scheme that can be used when only a single scale

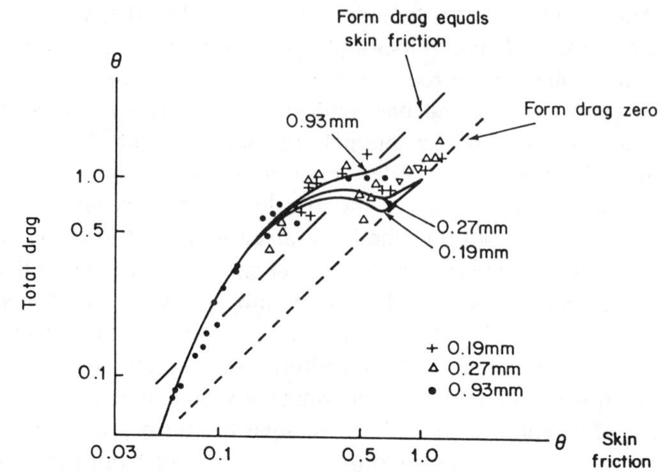


FIGURE 2. Relationship between total friction and skin friction for various grain sizes. Friction is shown in terms of the dimensionless shear stress θ.[41] (Reprinted with the permission of the American Society of Civil Engineers.)

of bedform is present. His scheme depends upon matching the water velocity at the height where the lower log layer meets the upper log layer. There are two parameters which must be estimated in order to calculate the ratio of skin friction to total stress; $z_*$, the height at which the two log layers are matched, and $z_t$, the apparent roughness length in the upper layer. Following Arya,[44] $z_*$ is taken to be the average height to which the internal boundary layer diffuses from one roughness element to the next; an equation for air flow over a warm runway is used to calculate this diffusion distance. This estimate of $z_*$ is most suited to relatively widely spaced bedforms. Again using Arya's measurements of air flow over Arctic pressure ridges, Smith estimated $z_t$ using a drag coefficient which, with the height and wavelength of the bedforms, provides an independent estimate of the form drag. Other estimates of $z_t$ are also available, for example from Wooding et al.[45] or from Lettau.[46] With these estimates, Smith provided an equation, algebraically messy but straightforward to use, for the ratio of skin friction to total stress.

Input parameters for Smith's scheme are the height and wavelength of the bedforms and a drag coefficient, which is presumably dependent on bedform shape. Unfortunately, there are insufficient measurements at present to provide accurate estimates of the drag coefficient. Smith's own measurements give the drag coefficient as 0.15 for separated flow and 0.7 for unseparated flow over his river sand waves. Chriss and Caldwell[37] inverted the procedure to estimate the bedform size which would be responsible for their measured form drag; the results, using a variety of drag coefficients, show that, for a constant form drag, doubling the drag coefficient halves the estimated bedform height.

Smith and McLean[32] extend Smith's procedure to allow for multiple log layers, the effect of suspended sediment load on the velocity profile, and the effect of bedload transport on the roughness length. The result is a complex iterative scheme allowing skin friction to be found from a knowledge of bedform geometry and velocity profile in one of the upper layers. Their theory appears consistent with their field observations, but there are several very poorly known constants involved, which make their scheme hard to use with confidence.

Clearly there is a place for further theoretical and experimental work on this problem. Taylor et al.,[34] Taylor and Dyer,[47] and Richards and Taylor[35] discuss some aspects of modeling the turbulent boundary layer over bedforms. As with all turbulence problems, the major difficulty lies in specifying the turbulence, but there is hope that increasingly realistic models can be developed. Also, the rapidly increasing number of field programs should begin to provide better data against which to assess the theoretical models.

### 2. Combined Wave and Current Effects

A combination of surface wave currents and quasi-steady flow is the most common environment at the seabed on the continental shelf. Field observations (e.g., Butman et al.[48]) show that sediment movement by waves can occur right out to the edge of the shelf. It is therefore somewhat surprising to find that boundary layers for combined wave and current conditions have generally received little attention. Recently, however, a number of theories of wave/current boundary layers have appeared which suggest that wave motion at the seabed can significantly increase the *mean* stress as well as the maximum stress at the seabed. In this section we will briefly review these theories and the field evidence supporting them.

The wave boundary layer is much thinner than the boundary for steady flow. We can therefore distinguish two separate regions in a wave/current boundary layer; a region close to the seabed in which both oscillatory and steady viscous stresses are important, and a region above in which only the steady flow friction is locally important. However, the flow in the upper region is modified appreciably by the presence of the wave boundary layer; in effect, the flow in the upper layer feels an increased boundary resistance due to the presence of the wave-induced turbulence in the lower layer.

Smith[36] and Grant and Madsen[49] independently chose to approach the problem by assuming an eddy viscosity which increases linearly with height above the bed, though they differ in their choice of its magnitude. The results are schemes for calculating seabed stresses and vertical profiles of velocity when wave orbital velocity (outside the wave boundary layer), mean flow speed at some fixed height above the bed, and physical bottom roughness are known. Both theories predict large wave influences on the mean flow. For example, Smith predicts that a wave friction velocity only 25% of the mean flow friction velocity increases the mean boundary shear stress by 56%. An enhancement of the bottom stress of this magnitude is clearly important for sediment transport prediction.

More recent theories use a more complete description of the turbulence in the wave and current boundary layers. Davies et al.,[50] for example, describe a numerical model using turbulent kinetic energy closure. Dyer and Soulsby[51] compare four different theories predicting mean and maximum seabed stresses in wave/current boundary layers, those of Grant and Madsen[49] and Davies et al.,[50] as well as a mixing length theory of Bijker[28] and a momentum deficit model of Fredsoe.[52] Large differences between the theories were found. Bijker's theory consistently gave results significantly larger than the others. Grant and Madsen predictions were 35 to 70% above those of Fredsoe and Davies et al. for mean bed stresses and typically 25 to 40% higher for maximum bed stresses.

Field measurements to test these theories and assess their relevance to real conditions are difficult and costly, and few have been made. Wiberg and Smith[53] analyzed data from the Alaska shelf and found values of friction velocity which were more than twice those predicted for the mean flows alone, but which were in fair agreement with predictions using the theories of both Smith[36] and Grant and Madsen.[49] Grant et al.[54] reported on measurements on the northern California shelf and found again that seabed stresses were typically more than twice values predicted using the mean flows alone, but which were within 10 to 15% of predictions using the Grant and Madsen theory. A third set of observations, from the Nova Scotia shelf,[9] also support the Grant and Madsen predictions, provided that the significant wave velocity is used in the prediction scheme, though the authors point out the large range of potential errors inherent in all the field estimates of wave/current stresses made to date.

The conclusions to be drawn from this survey of our understanding of wave/current boundary layers are, on the one hand, that high-level theoretical models exist to predict boundary layer stresses, but, on the other hand, that the accuracy of these predictions for real field conditions is not well tested. Furthermore, implementation of the more complex of these models within the context of a regional sediment-transport model would probably lead to prohibitively large demands on computer time. There is therefore a clear need for more field measurements against which to test the theories, and further theoretical work to simplify the models while retaining the important physics.

### 3. Other Effects

A number of other phenomena may complicate the prediction of seabed stresses from measurements made above the bed. These include the feedback effects of moving sediment on the water movement, and acceleration and deceleration effects on boundary layer structure.

Suspended sediment in the water column tends to cause density stratification which partially suppresses turbulence which, in turn, modifies the boundary layer velocities and the seabed stresses. The effect appears to be most significant for fine sediments (e.g., the clay suspensions studied by Gust[55]), though McCave[56] found some effect of sediment suspension in velocity profiles over a sand bank in the North Sea. Taylor and Dyer[47] use an advection-diffusion equation to model a coupled sediment-water system, obtaining explicit forms for the sediment and velocity profiles. Smith and McLean[32] consider a similar model but include nonlinear effects which require numerical solutions.

Intense bedload movement of sediment may also influence the boundary layer flows by producing increased values of bottom roughness. Expressions for this effect are given by Smith and McLean[32] and Grant and Madsen[49] but they have not been well tested for field conditions. Gust and Southard[57] suggest major changes to the boundary layer once bedload transport occurs.

## C. THE POTENTIAL INADEQUACY OF STRESS ALONE

The discussion so far has proceeded on the basis that an accurate estimate of seabed stress will allow accurate estimation of sediment transport. There are, however, a number of observations that imply that, at least under some circumstances, this will not be the case. When wave and steady currents flow over seabed topography, and separation of the flow occurs, the resulting vortex patterns can be complex and the implications for sediment transport profound, and often surprising.

Longuet-Higgins[58] modeled the formation of vortices resulting from symmetric oscillatory flow over symmetric ripples and was able to reproduce theoretically many of the observations of Bagnold[59] and others. In this symmetric case there is no net sediment transport. Under waves in the ocean, however, there may be asymmetry of the oscillatory velocity, asymmetry of the bedform profile, and mean flows may also be present, both wave-induced and independently imposed. The laboratory results of Inman and Bowen[60] and Bijker et al.[61] give a flavor of the complexities that can occur in these cases.

Inman and Bowen[60] found, for example, that although for waves alone the transport of sediment and the ripple migration were, as expected, in the direction of wave propagation, a small superimposed mean flow also in the wave-propagation direction could cause *upstream* sediment transport and *upstream* migration of ripples. This surprising result occurred because the added mean flow strengthened the vortex formed on the downstream-facing side of the ripples during the forward flow, and the extra sediment entrained in this vortex was carried upstream during the backward-flow part of the wave cycle. As the authors remark, this phenomenon can only occur for a range of small mean flows and must be reversed as the mean flow increases. However, the magnitude of their mean flows, a few centimeters per second, is comparable with wave-induced flows experienced in the field so their observations may be significant.

Bijker et al.[61] conducted similar experiments but emphasized the importance of wave velocity asymmetry. They also found that the direction of sediment transport can be reversed by a small superimposed mean flow, and they concluded, not unreasonably, that an understanding of vortex formation, as a function of wave and flow parameters, is essential for a proper modeling of sand transport by waves and currents over steep ripples.

These observations clearly have profound significance for the modeling of sediment transport, implying that time lags between hydrodynamics and sediment response must be modeled. For modeling transport of very fine materials, for example in estuaries, the need to consider time lags is well known. However, more measurements, preferably under field conditions, are needed to assess the importance of lags to sediment transport on continental shelves.

## III. THE SEDIMENTARY ENVIRONMENT

The input data on the local environment required for sediment transport modeling can be grouped into four, somewhat overlapping, categories:

1. The grain properties: grain size and density, grain size distribution, settling velocity and its distribution
2. The distribution of grain properties in space: in suspension, moving as bedload, as a function of depth in the bottom sediment, the total depth of sediment
3. The bottom roughness: surface-grain size, local sedimentary structures, biological structures and activity, the appropriate drag coefficient
4. Regional patterns: position relative to large-scale features (bedrock or sedimentary), the regional slope of the sea bed

A knowledge of the grain properties is central to the problem as almost all predictive formulae for sediment movement require a definition of either the sediment grain size and density, or the settling velocity (primarily a function of the grain size and density). These formulae were, however, derived using data from laboratory experiments which were specifically designed to study the behavior of very-well-sorted sand, where a single grain size represents all characteristics of the material; in threshold movement, bedload transport and suspension, grains are moving relative to, and in interaction with, only material of the same size. In a real situation, sediment consists of a mixture of grain sizes and, probably, densities. In the geological literature, the spatial patterns of grain-size distributions are interpreted in terms of hydraulic processes.[2,62] However, most of the sediment-transport formulae cannot logically be recast to include a grain-size distribution. To use these formulae a single mean size must be assumed; Vincent et al.,[29,63] for example, take 530 μm as representative of the inner shelf off Long Island.

Direct observations of material in suspension show that, while the trends predicted by simple models are qualitatively correct, the actual behavior is complex. Lees,[26] for example, notes that "these grain size distributions serve to emphasize the complexities of suspended sediment mechanisms and show in particular that material in suspension does not necessarily bear a simple relationship to that on the seabed beneath it. Furthermore it is not clear what happens to the transparent, subangular grains of quartz ($d_{50}$ = 180 μm) seen moving southwards in the flood tide, but absent in the ebb."

In extreme cases the material in suspension is totally unrelated to the material on the bed. For example, Amos and Greenberg[63] model the transport of silt in the Minas Basin, and include large areas where the actual seabed is a gravel lag overlying mud.

Overall, the choice of a grain-size representation is likely to be a critical decision in any modeling program. Taking the average between a "relict" gravel and a mobile silt does not make much physical sense. Other measures contain similar, but more subtle, difficulties. Bagnold[4] suggests defining the settling velocity of a mix of grain sizes by

$$\bar{w} = \frac{\Sigma p_d w_d}{\Sigma p_d} \tag{1}$$

where $w_d$ is the settling velocity and $p_d$ is the percentage composition of a particular size fraction of diameter $d$. This weights the values of $\bar{w}$ towards large sizes. Equation 1 is easy to apply using actual measurements of the size distribution of the suspended load, but much more complex if the distribution has to be estimated theoretically. Some sort of multiple grain-size model is still needed.

The observed distribution of grain size, item 2, has already been touched upon in the discussion of the gravel-mud layering in the Minas Basin. The depth of material available for movement is clearly an important parameter. Substrates of bedrock, gravel, clay, or indeed any material significantly different from the surface layer, complicates the modeling of erosional sequences by controlling the total sediment supply or the supply of material of a particular size. Most predictive models give only the *potential* for sediment transport, assuming an unlimited sediment supply of a single given size.

The third factor required is information on the bottom roughness. A number of models require this as a specific input. Heathershaw[25] assumes that the bottom velocity profile is logarithmic, and calculates the long-term bottom-stress variations from velocity measurements taken 2 m above the bed. The bottom stress calculated in this way is strongly dependent on the value assumed for $z_0$, the bottom roughness. For a logarithmic velocity profile the equivalent drag coefficient $c_d$ is given in terms of Von Karman's constant $K$ by

$$c_d = \left[\frac{1}{K} Ln \frac{z_{ref}}{z_0}\right]^{-2} \tag{2}$$

from which it is clear that varying the roughness length is equivalent to varying the drag coefficient. Figure 3 indicates typical values used for $z_0$ and the dependence of the value of the threshold velocity for movement of a particular grain size $d$ on the choice of $z_0$. As there is relatively little observational evidence for the systematic variation in $c_d$ implicit in Equation 2 the rationale is, in effect, to partition the total stress into skin friction and form drag, with the value given by this equation being an estimate of the skin friction felt by grains at the boundary.[37,64] Clearly an appropriate drag coefficient, or bottom roughness, will depend upon bedform shape as well as size.

In practice, changes in the first three parameters defining the sedimentary environment are difficult to predict. In the absence of direct observations, simple models will have to assume that the sedimentary parameters are constant and adequately defined by the historical data base.

The regional morphology, item 4, is a quantity that is readily measurable, though often known only in very general terms for large regions of a shelf. There are currently no operational models that update the topography and recompute the sediment fluxes in the new geometry. This would be necessary to follow any changes that significantly alter the local hydrodynamic conditions.

## IV. SEDIMENT-TRANSPORT RELATIONSHIPS

Armed with suitable representations of the hydrodynamic and sedimentary environments, the modeler must next choose a set of equations to predict sediment transport. As we discussed in the opening section, sediment transport calculations generally involve two distinct considerations, the threshold for sediment movement, and the rate of transport once motion begins. While some transport equations do not explicitly include a threshold, most of the commonly used equations result in predictions which depend sensitively upon the choice of threshold condition.

| $z_0$ (cm) | $U_{CR}$ ($z$ = 100 cm) (cm s$^{-1}$) | | | | |
|---|---|---|---|---|---|
| | $d$ (cm) | 0.01 | 0.02 | 0.05 | 0.01 |
| 0.01 | | 30 | 30 | 36 | 50 |
| 0.05 | | 25 | 25 | 30 | 41 |
| 0.10 | | 23 | 23 | 27 | 37 |
| 0.50 | | 17 | 17 | 21 | 29 |

FIGURE 3. Threshold velocities, $U_{CR}$, calculated from Yalin's [78] modified Shields curve for different grain sizes and roughness lengths. ($U_{CR}$ is the critical velocity at 100 cm above the seabed).[25] (Reprinted with permission from Elsevier Scientific Publishing Company.)

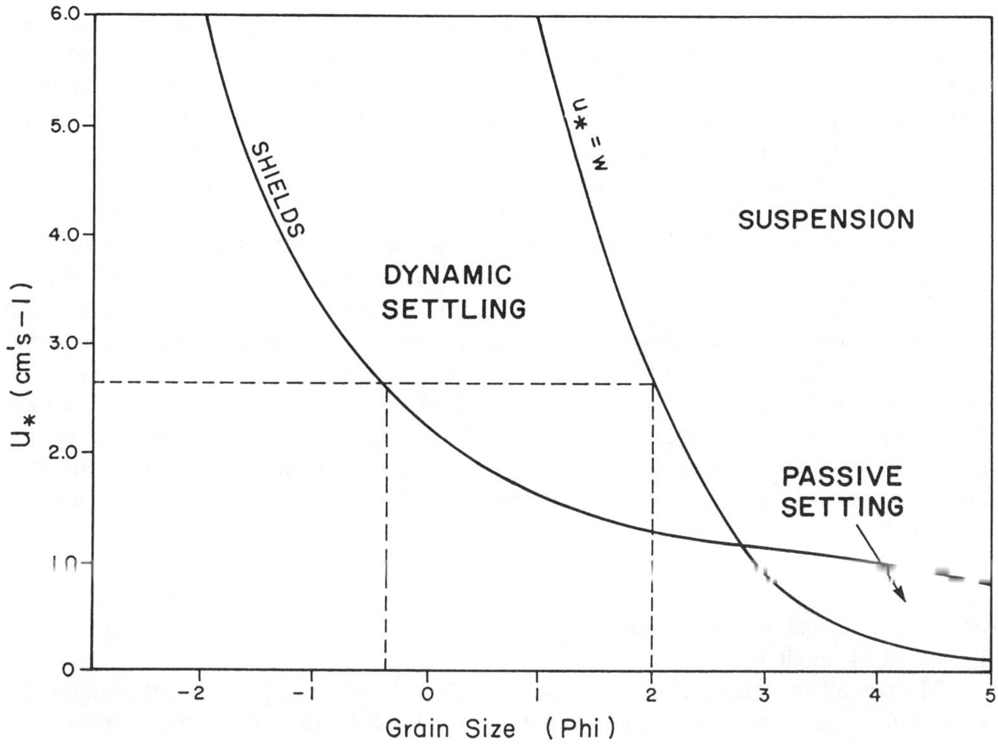


FIGURE 4. The Shields curve for quartz grains in water. Also shown is the curve of friction velocity, $u_* = w$, the sediment-fall velocity. Above the crossover, at $\phi$ just below 3, grains are predicted to go straight into suspension once the stress reaches their threshold.

## A. THRESHOLDS FOR SEDIMENT MOVEMENT

Threshold conditions for sediment movement are generally defined either in terms of a critical stress, or of a critical velocity at a standard height above the bed. Competency curves, of critical stress (or some parameter related to stress) plotted against grain characteristics, have been produced by a number of authors. The competency curves most frequently used are based on those given by Shields.[65] He nondimensionalized the stress acting on a grain by the gravitational force on the immersed weight of the grain, the resulting dimensionless stress being known as the "Shields parameter". This parameter is then plotted against the grain Reynolds number, or a dimensionless grain fall velocity. For quartz grains in water the quantities can be made dimensional, as shown in Figure 4.

There is considerable uncertainty about the shape of the Shields curve for small particle diameters. This uncertainty is discussed further by Miller et al.[66] and Nowell et al.[67] The latter conclude that a critical stress for particle diameters below about 100 μm cannot be predicted with any accuracy. A major part of this uncertainty is due to variable cohesion between particles with small grain sizes.

Application of the Shields curve is strictly only valid for steady flow over a flat bed and for a uniform sediment size. Modification of the Shields curve over seabed topography is complicated by the form drag discussed earlier. Although the form drag does not cause sediment movement, it makes the critical total stress higher over bedforms. Inman[68] shows a modified Shields curve with a branch towards higher threshold stresses for rippled beds. However, as Middleton and Southard[69] pointed out, the critical bed stress over a rippled bed is actually *less* than over a flat bed, once form drag has been removed, because the higher flow velocity over the ripple crests produces locally a higher than average stress. Nevertheless, measurements by Sternberg[70] in tidal channels in Puget Sound suggest that Inman's modified Shields curve predicts sediment thresholds over rippled beds quite well if the total stress (including the form drag) is used. His data also show surprisingly good agreement for thresholds for the sand fraction even when sand is incorporated in a poorly sorted sediment mix including much-larger-sized particles. This indicates that the Shields curve may not be critically dependent on the requirement of a uniform sediment.

The influence of biological activity on sediment-water interaction has been the subject of increasing general interest.[67,71,72] Two opposing influences may occur: binding of surface sediments, for example by mucous coatings, and destabilizing of the sediment by, for example, burrowing, tracking, or the interception of boundary layer flow by animal tubes. Biological effects are very variable and depend, among other factors, on sediment composition and organism density and type. Modifications to Shields diagrams to account for biological activity tend therefore to be site specific and it is as yet difficult to derive general results. For fine to medium sands from an intertidal region, Grant et al.[72] found, in laboratory flume experiments, that although biological cohesion tended to increase the critical stress relative to the Shields diagram, the increase was never more than a factor of two. Since this change is well within the generally accepted level of uncertainty in the Shields curve due to a number of other factors, it is not perhaps of major significance. As expected, biological reworking tends to lower the critical value back to the average Shields curve. Flume studies by Nowell et al.[67] showed that, for fine sand, animal tracks cause significant increases in bed roughness and hence decrease critical stresses. In their experiments the critical stresses decreased by about 40%.

The overall importance of biological activity for sediment transport on continental shelves is difficult to assess, not only because of the potentially conflicting effects and their variability in space and time, but also because of a paucity of quantitative information about levels of benthic biological activity and its patchiness. Nevertheless, the experimental work carried out to date suggests that this merits further investigation.

Madsen and Grant[73] showed laboratory results which suggested that the Shields curve can also be used for oscillatory flows, provided that the stress is calculated using a suitable wave friction factor. Komar and Miller,[74,75] on the other hand, provided threshold equations based on the orbital diameter of the wave motion and the maximum orbital velocity.

The scatter of observed threshold stresses on a Shields curve, from both laboratory and field measurements,[43,76] makes it clear that the threshold stress can at present only be estimated to within a factor of two or three. It is not entirely clear why the values are subject to such variability, though contributing effects must be the existence of varying small-scale roughnesses, and the differing definitions of the threshold of motion used in interpreting results. Since much of the sediment movement on continental shelves probably takes place at stress levels not greatly exceeding threshold levels, the difficulty of choosing a threshold condition will be reflected in the uncertainty of the resulting predictions.

## B. EQUATIONS FOR SEDIMENT TRANSPORT

### 1. Bedload Transport

Many equations have been proposed to predict bedload transport, and the most important of these are described in detail in Graf,[77] Yalin,[78] and Dyer.[79]

Bagnold's equation[59] is probably the most popular, perhaps because it has a more physical basis than other more empirical equations. However, the difficulty with the Bagnold equations is in the choice of an efficiency parameter, $K_b$, linking the energy needed for bedload transport to the energy available from the mean flow. Various forms for $K_b$ have been proposed, for example by Bagnold,[59] Sternberg,[80] and Langhorne,[81] but there is no generally agreed form and $K_b$ is often treated as a variable to be chosen to make predicted transport rates match observed rates. Many variations, based loosely on the Bagnold equation, also exist. Gadd et al.,[27] for example, fitted flume data to an equation containing a threshold velocity and an empirical constant. Its only claim to being related to Bagnold's equation is that it depends upon velocity to the third power.

For oscillatory flows Bagnold[59] argued that the stress caused by the orbital motion of the waves, with no net transport, acts on the seabed to put sediment into motion, and then it can be readily moved by any net current superimposed on the wave motion. The resulting equation forms the basis for a sediment-transport equation for longshore currents in the surf zone, and can be generalized to include suspended-sediment transport, with only a change in the meaning of an efficiency parameter, $K_b$. Once again, however, the value of $K_b$ is hard to estimate. An additional problem is choosing the appropriate forms of wave and mean motions. Bagnold, in effect, used an rms velocity to define the wave stress and a long-term average velocity for the mean flow. However, it is now known that in at least some conditions net sediment transport can occur even when the mean flow is zero, as a result of asymmetry of the wave motion (see, for example, Huntley and Hanes,[82] for field measurements of this effect). With field measurements of the time-varying flow, this effect can be included in a form of Bagnold's equation by combining the *instantaneous* wave stress and the *instantaneous* velocity and calculating the time average of this product. Vincent et al.[63] take this approach, and include a threshold stress. However, in the absence of full measurements of the velocity variations, including wave orbital velocities, it is generally difficult to include this effect in a model of sediment transport.

Another popular bedload transport equation is due to Einstein,[83] often used in a simplified form due to Brown.[5] The starting point for Einstein's equation is an analysis of the forces on an individual sand grain, but the final equation includes a number of empirically adjusted coefficients which cloud the physics implied by the technique. Madsen and Grant[64] found that the Brown version of the Einstein equation worked well for oscillatory flows, provided that, as for the threshold stress, the wave stress is calculated using the wave-friction factors of Jonsson.[84] They also suggested using the Einstein-Brown equation even where bedforms are present if the bed roughness is taken from the grain diameter rather than from the bedform scales. This procedure, they argued, provides the true skin friction and ignores the form drag contribution to the total stress.

Vincent et al.,[85] among others, have pointed out the importance of the different velocity dependencies of the Bagnold and Einstein approaches. Well above the threshold for movement, the Bagnold-type equations have $u^3$ dependence, whereas the Einstein-Brown equation depends upon $u^6$. It might be thought that this widely differing dependence could be readily distinguished in comparing predictions with observations. However, the empirical constants in the equations are generally determined from narrow ranges of velocities and there is, as we shall see later, no consensus as to which dependency is closest to reality.

Other bedload equations are also used. Dyer[43] reviews some of them. Smith,[36] for example, considers the Yalin[86] equation to be the best method of estimating bedload transport if used in conjunction with his method for estimating bed stress from total boundary stress.

## 2. Suspended-Load Transport

There is still substantial uncertainty about the relative importance of bedload and suspended-load transport on continental shelves. For the fine sands of the Washington shelf, Smith and Hopkins[87] and Smith[36] considered bedload transport to be relatively inefficient and suspended transport to be of primary concern, whereas Vincent et al.[63] concluded that bedload transport dominates in the New York Bight, where medium sands cover much of the seabed.

Suspended-transport equations can in principle be derived more rigorously than bedload equations. Dynamical equations for the suspended sediment profile and the velocity profile can be solved, under certain assumptions, either separately[36,88] or as coupled equations in which the sediment profile is allowed to influence the velocity profile and vice versa.[32,47] Numerical solutions to equations with better representation of the turbulence are also being developed.[107] The problem with these approaches lies in determining appropriate boundary conditions at the seabed. Most methods require a value of sediment concentration at the base of the suspended-sediment profile, at an estimated reference height above the bed. Einstein[40] used his bedload equation to estimate this concentration, taking the reference distance as twice the grain diameter. In fact, since the reference concentration so determined is proportional to the bedload transport rate, bedload and suspended load are generally considered together in a single equation when applying Einstein's method. Smith,[36] following Yalin,[86] uses an empirical equation for the reference concentration, based on the excess shear stress and the volume concentration in the bed. Other equations use as a boundary condition the *flux* of sediment through a reference level near the bed, generally through some form of "pickup function" (e,g., Nielsen et al.[89]), but this approach leads to a dependence on poorly understood parameterizations of sediment resuspension and deposition.

An alternative approach has been to use empirical equations, devised for river flow, to estimate the *average* concentration of the suspended sediment, and to multiply this by a representative fluid velocity and water depth to estimate the suspended load transport. Smith and Hopkins[87] used this method to calculate transport based on data from current meters moored 3 m above the bed. Although this crude approach perhaps represents the best that can be done when very little information is available, it is clear that is can only give a rough estimate (Smith and Hopkins suggest they are within a factor of two or three but this is probably optimistic). There is general agreement that inferring suspended load transport rates from measurements at a small mumber of discrete heights provides very unstable results.

The best hope for future refinement of suspended transport equations probably lies in the rapidly improving ability to make accurate field measurements of this transport. Vincent et al.[63] show results from the New York Bight using an acoustic suspended sediment profiler. High-resolution, rapidly varying concentration profiles can be measured which, in conjunction with similar measurements of current, provide direct measurements of suspended-sediment transport rates. This technique is being used in the highly variable nearshore zone[90] and will undoubtedly become more widely used in deeper water as the methods are improved.

# C. FIELD TESTING OF SEDIMENT TRANSPORT EQUATIONS

For all of the transport equations which may be chosen by the modeler, there are empirical coefficients and model assumptions which must be checked against field data. Unfortunately there are as yet very few data sets which can be used for this purpose, and such results as exist are generally confusing and do not obviously point the way towards improving the predictions.

Heathershaw,[25] in the analysis which resulted in Figure 5, considered five equations for sediment transport, three bedload, Einstein,[40,83] Bagnold,[4,59] and Yalin,[86] and two total load, Ackers and White[91] and Engelund and Hansen.[92] At the shallow-water site T1 all models

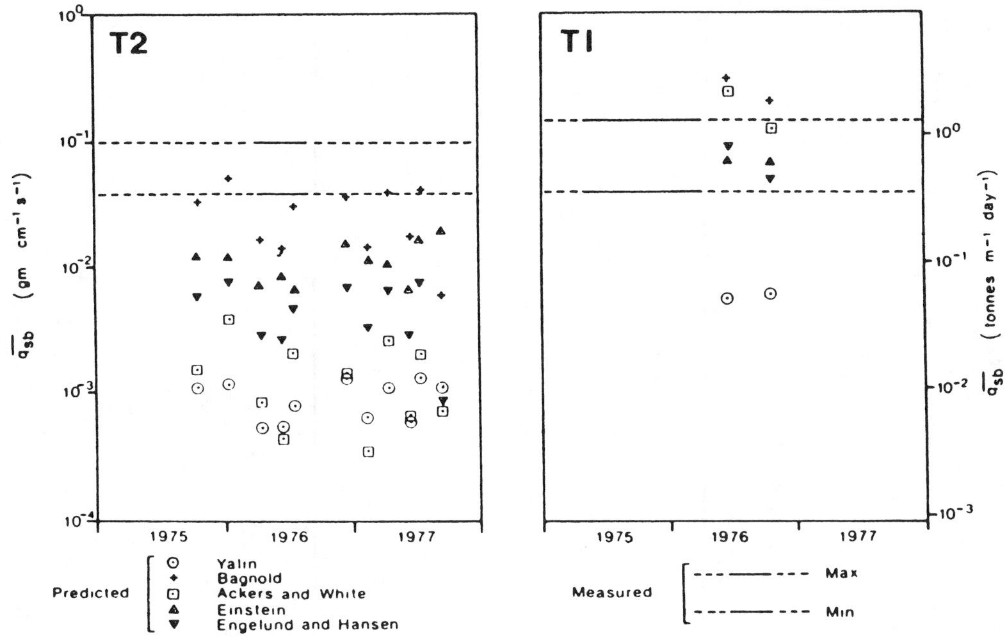


A

FIGURE 5.. Comparison of measured and predicted net bedload transport rates, $q_{ab}$. (A) Maximum and minimum measured rates derived from radioactive tracer experiments at two sites, T1, T2 in Swansea Bay, S. Wales.[25] (Reprinted with permission from Elsevier Scientific Publishing Company.) (B) Comparison of measured net transport with predictions for two stations, N and P in the general area of the tracer experiment [26]. (Reprinted with permission from Blackwell Scientific Publications.)

except Yalin's give similar values, in reasonable agreement with the tracer results (which are *net* transport over a number of tidal cycles). However at T2, the deeper site, the predictions scatter over two orders of magnitude with only Bagnold's prediction being in the same ballpark as the measurements. Heathershaw discusses at some length the accuracy of these predictions, emphasizing the fact that they involve products of the friction velocity and the excess stress, quantities not easy to estimate.

Heathershaw concluded that his work supported the use of a modified Bagnold equation, at least for tidal environments on a shelf. This is consistent with the results of Gadd et al.[27] and would provide a reasonable working hypothesis if Lees,[26] working in the same institute as Heathershaw, on the same type of problem, had not come to very different conclusions. For her measurements, Yalin's model seems to provide the best prediction (Figure 5B).

A more complete method for testing bedload transport formulae against field data is described by Amos et al.[93] Hourly time-lapse photographs of ripple migration on the Nova Scotia shelf over a period of 15 d were used to estimate bedload transport rates, for a range of wave and current conditions measured with bottom-mounted current meters. Ripple heights required for these estimates were not directly measured, so were estimated from observed ripple wavelengths and an assumed ratio of height to wavelength. Ripple type and orientation were found to respond rapidly to the changing hydrodynamic conditions.[94] The estimated transport rates were compared to rates predicted by four equations, including versions of the three bedload equations used by Heathershaw (Einstein, Bagnold, and Yalin) and the Engelund-Hansen total-load equation. In fact, the form of Einstein's equation given by Brown[5] was used. In addition, the bottom stresses used as inputs to these models were calculated using either of two wave/current interaction models reviewed earlier, those of

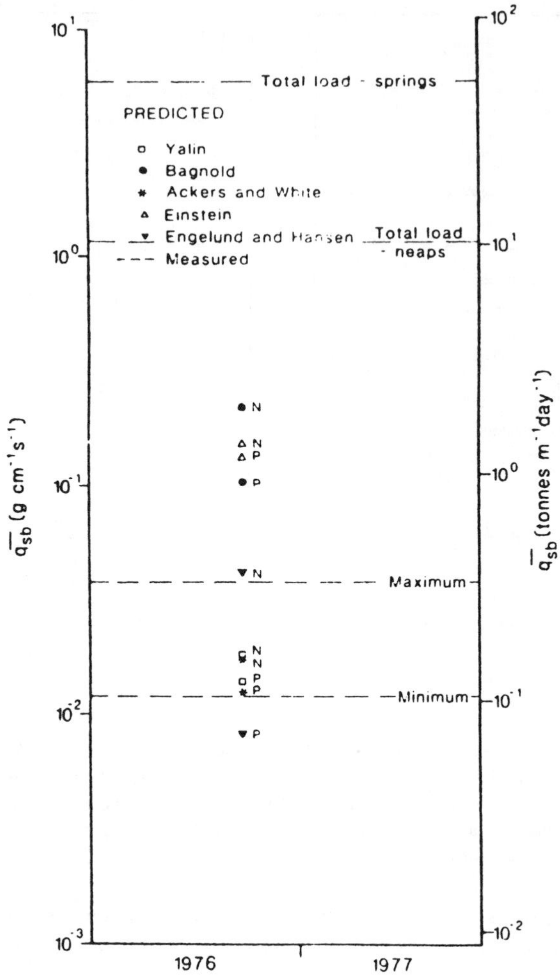


FIGURE 5B

Smith[36] and of Grant and Madsen,[49] the latter modified so that the wave/current friction factor was not allowed to fall below the value expected for waves alone.

Some preliminary results are shown in Figure 6. Though the results show the kind of scatter we have come to accept from field measurements, they clearly allow some assessment of the different prediction techniques. The absolute fit between observations and predictions, primarily dependent on empirical constants, is perhaps of less interest than the *trends*. The Bagnold equations, for example, predicts transport rates which vary much less than the observed rates. The trends appear to favor the Einstein-Brown and Yalin Equations, with the modified Grant-Madsen scheme for bottom stress providing the better absolute fit. It should be emphasized that these results are preliminary, but they are nevertheless encouraging. The technique could obviously be improved upon in further experiments, particularly by measuring ripple heights directly.

In trying to assess the validity of any modeling procedure, one very obvious difficulty is that the description of a model as "Bagnold" or "Einstein" is more a convenience than a definition. Bagnold himself has suggested many versions of his various models. As Heathershaw puts it, his work "supports the use of Gadd et al.'s (1978) recalibrated and restructured version of Bagnold's (1966) equation." The Einstein equation used by Heathershaw was also recalibrated by Gadd et al.[27] using flume data, not field data. The first uncertainty,

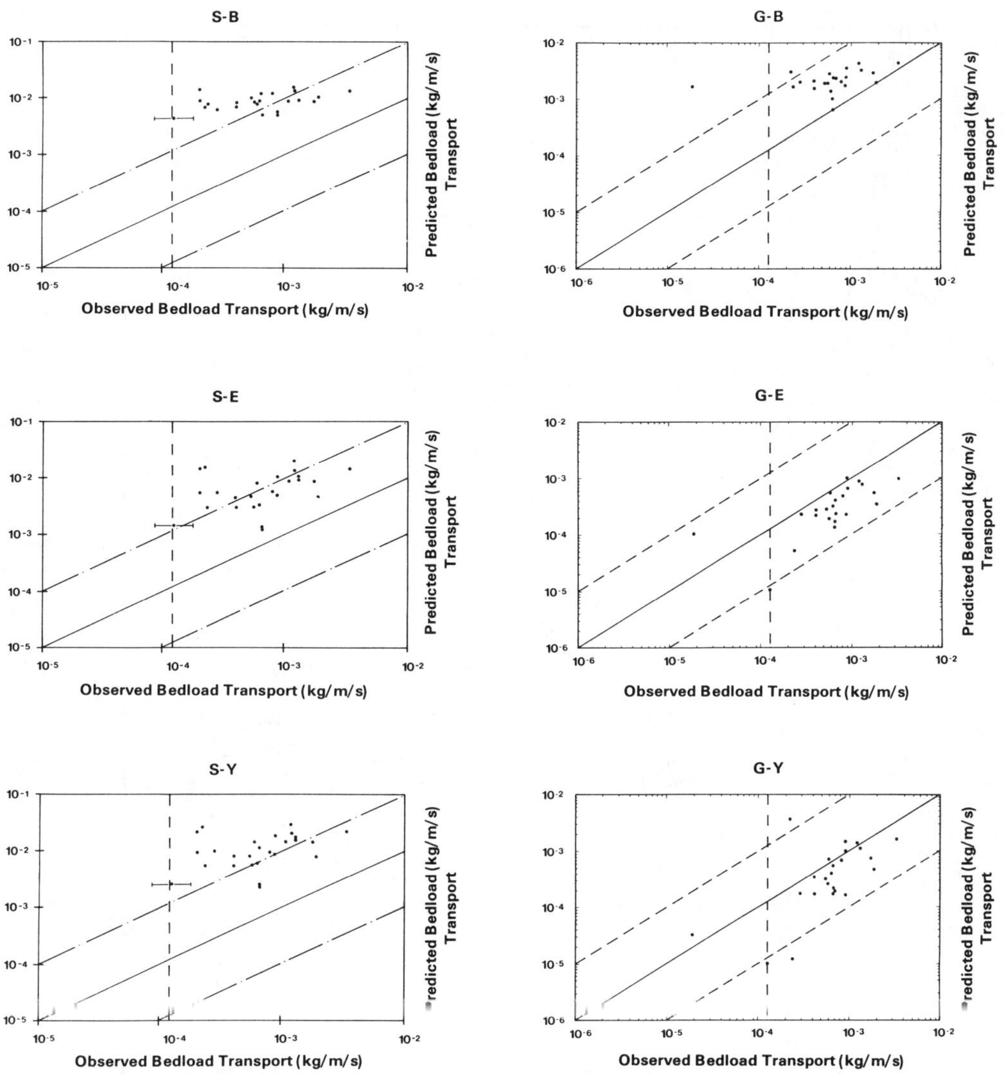


FIGURE 6. Comparisons between measured bedload transport, based on ripple migration, and predicted transport for a range of combined wave and current conditions on the Nova Scotia continental shelf. Predictions are based upon calculations of wave-current stresses at the seabed using the formulae of either (S) Smith,[36] or (G) Grant and Madsen.[49] In the latter case the wave-current friction factor has been limited to a minimum value of $6 \times 10^{-3}$. Bedload-transport rates are calculated using the equations of (B) Bagnold,[4] (E) Einstein-Brown,[5] or (Y) Yalin.[86] Thus, for example, the top left-hand diagram, denoted S-B, is based on Bagnold's theory for bedload transport, with the bottom stress calculated by the wave-current theory of Smith.

therefore, in assessing the adequacy of any formula is the precise nature of the expression, particularly the values of the constants.

The second problem is the assumptions in the hydrodynamic data. Neither Heathershaw nor Lees had wave data, and clearly did not think it was essential to hindcast for the appropriate area. However, Heathershaw has produced a diagram of the expected increase in bottom stress and bedload transport for water depths of 12 m (Figure 7). This very graphically illustrates the critical effect of wave activity on the threshold of movement. It immediately suggests that a shelf model that includes threshold effects but ignores waves is highly suspect. Even at high velocities, wave effects are significant; factors of two appear rather trivial in the immense range of transport rates covered by Figure 7. Note that the

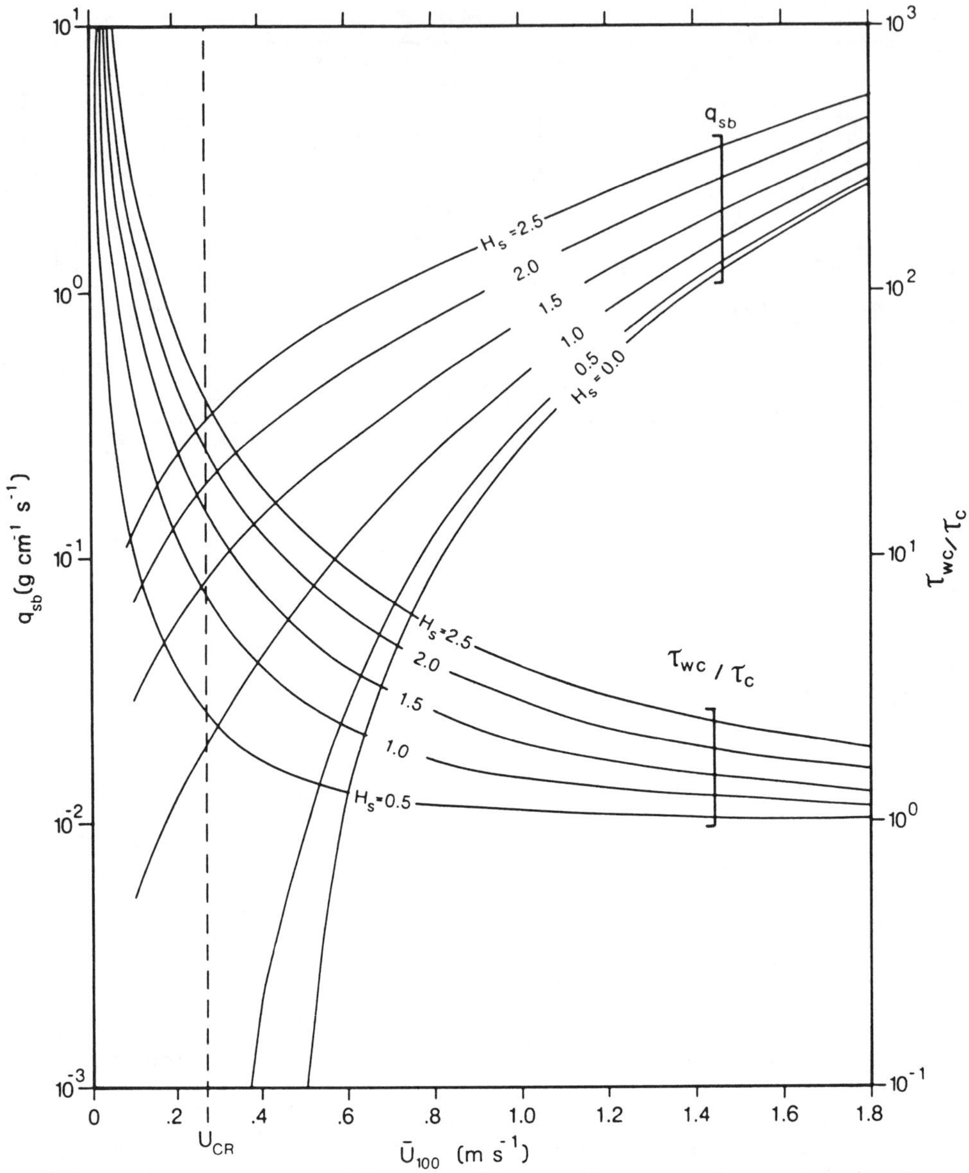


FIGURE 7. Effect of waves on sediment transport, using Bijker's magnification factor, $(\tau_{wO}/\tau_O)$, illustrating the effect of increasing wave height, $H$, on bedload-transport rates, $Q_{ab}$, as a function of current at height 100 cm above the seabed, $U_{100}$. The calculations have assumed waves of 6-s period, water depth of 12 m and a roughness length of 0.05 cm, typical of fine sand.[26] (Reprinted with permission from Blackwell Scientific Publications.)

typical values for the net transport given in Figure 6 are 0.1 gm $cm^{-1}s^{-1}$. These are of the same order as those observed by Gadd et al. on the U.S. East Coast.

These three examples are the only data sets in which tracer experiments provide estimates of sediment transport rate that are independent of other measurements. As none of these experiments obtained wave data, they do not provide a critical test of any model expected to predict sediment transport in extreme conditions.

In energetic environments, material in suspension may be sufficiently dispersed upwards into the water column that it can be sampled with reasonable confidence. Under these conditions the thin layer of slow-moving bedload is assumed to contribute relatively little to the total transport. The product of the observed velocity and concentration may then provide a good estimate of the total sediment transport. Observations are most easily made in strong tidal currents. Figure 8 illustrates two such data sets, which show reasonable agreement.

Lees[26] finds the suspended transport rate $q_{ss}$ to be given empirically by

$$q_{ss} \sim u_*^{2.8} \tag{3}$$

Thorn[1] has data from very shallow water and significant wave activity. Regarding the orbital velocity v as a separate variable he finds

$$q_{ss} \sim u_*^{2.7} \tag{4}$$

where $u_*$ is the friction velocity generated by the tidal stream. This then leads to a multiplier *M* (Figure 8) which accounts for the additional sediment transport associated with wave activity, a result analogous to Figure 7, but purely empirical. As can be seen in Figure 8, in Thorn's environment, the sediment transport may be very strongly enhanced by wave action, with orbital velocities of 0.20 $ms^{-1}$ leading to a 10 to 20 times increase in suspended sediment flux.

These results are extremely valuable and seem to vary consistently within the parameter ranges observed. The difficulty is in extrapolating this type of result to extreme conditions, to small currents and big waves for example.

Another problem is in modeling the intermediate stages of flow, particularly the transition from bedload to suspended load. The central problem seems to be obtaining satisfactory understanding of the factors that control the concentration of the suspended load close to the bed.[36]

The general conclusion, having looked at the bits and pieces available, is that a reasonable first attempt can be made to model movement of a particular grain size in a particular type of environment. Amos and Greenberg's[63] model of the transport of fine material in Minas Basin falls nicely into this category. However, if one needs a model for a shelf, ranging in depth from zero to hundreds of meters, for a range of grain sizes, and a range of hydrodynamic conditions from "normal" to extreme storm, the problem becomes exceedingly complex.

## V. MODELING SEDIMENT TRANSPORT

### A. TWO-DIMENSIONAL MODELS

#### 1. Diagnostic Models

The basic building block of two-dimensional models is the prediction of the sediment transport at a point. Uncertainties here are carried over to any spatial model in which the sediment movement in a region is described in terms of the transport at a number of grid points. However, some consistency is gained by using the same assumptions everywhere and it may be possible to make useful distinctions between the sedimentary response in different areas. In cases where the underlying physical processes are sufficiently dominant, very simple models may be of interest. Pingree[3] in showing that sand banks are associated with eddies in the residual tidal currents avoids the problem of a transport formula entirely. Perhaps more significantly, the model proposed by Huthnance[95,96] for the formation of linear banks by the interaction of the tide with topography uses a simple, Bagnold type, transport relation of the form

$$q_x \sim |u|^n(u + \lambda|u|dh/dx) \tag{5}$$

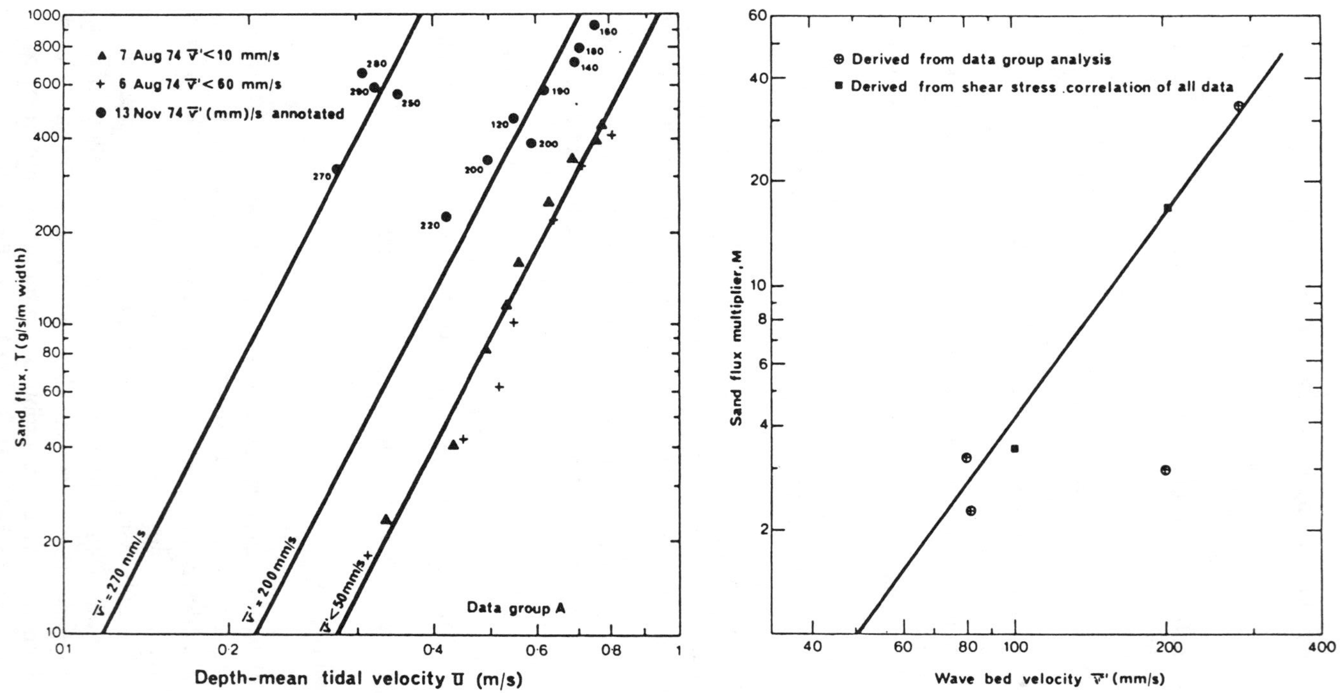


FIGURE 8. Variation of total suspended-sediment transport rate. (A) Transport of sand as a function of tidal velocity ($\bar{u}$) and wave orbital velocity at the bed ($\bar{v}'$).[1] (Reprinted with permission from M. F. C. Thorn.) (B) Suspended transport as a function of $u_*$ in a tidal environment.[26] (Reprinted with permission from Blackwell Scientific Publications.)

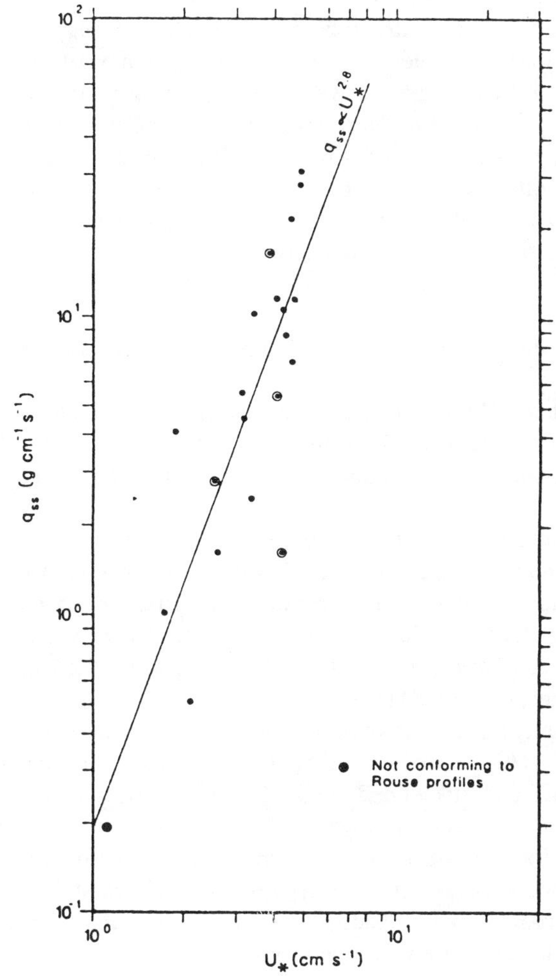


FIGURE 6B

where downslope transport is enhanced by a small factor λ. In terms of Bagnold's original formulation

$$\lambda \frac{\partial h}{\partial x} \sim \beta/\tan\phi \tag{6}$$

with β the bottom slope and ϕ the friction angle. This is generally a small quantity as tanϕ is normally assumed to be about 0.6, essentially the angle of repose of the material. Although the downslope term is normally small it plays a significant role because the tidal oscillations from ebb to flood result in small *net* transport rates. This is illustrated in Figure 6, where Lees shows estimates of maximum transport rates two orders of magnitude larger than the net transport averaged over a longer time scale.

A major shortcoming of all existing sediment-transport formulae except Bagnold's, is the omission of any dependence on bottom slope. As these are empirical functions there is no logical way to include slope effects.[64] It is tempting to assume that bottom slope is not an essential parameter. However, models for tidal generation of sand waves[97,98] (and sand banks[95]) strongly suggest that an important balance is made possible by including an equilibrium slope which reduces the net transport to zero. A similar result, used to investigate beach equilibrium[99] and complex bar structures on beaches,[100] suggests that relatively small slopes may play a significant role in balancing the tendency of surface waves to move material onshore. Recent experimental work on slope effects on sediment thresholds and sediment transport rates reported by Hardisty and Whitehouse[101,102] confirm the importance of slopes. In fact, their results for wind-driven sand movement suggest that transport rates are related to the seventh power of the bedslope, much higher than predicted by Bagnold.

A model of a shelf on which banks, sand waves, and beaches are of central interest may have to include a bottom slope dependence.

**2. Predictive Models**

The ideal two-dimensional model for sediment transport would step forward in time by:

1. Computing typical orbital velocities for a particular time step over the spatial grid. This would require a wave-forecasting model, a wave refraction/diffraction model with a breaking criterion, and wave theories appropriate for a range of heights, periods, and depths.
2. Computing the current velocity at the sea bed over the grid. Requirements include wind and tidal models of the area, possibly three dimensional to include stratification, the input of statistical parameters (means and variances) to represent processes not specifically reproduced by these models, and a model for wave-driven currents.
3. Combining wave, current, and bottom roughness to predict the parameters required for sediment transport formulae, particularly $u_*$, $z_0$.
4. Computing sediment transport vectors $q$ over the grid with some control on the grain-size distribution. This requires a sediment transport model incorporating threshold of motion, bedload, and suspended load, the influence of bottom slope, a grain-size differentiation, and checks on the state of the sedimentary environment.
5. Computing erosion/deposition, through spatial gradients of transport rates, over the grid. This may not be necessary for every time step. The model would check availability of material for erosion, material added or removed at boundaries, and, possibly, update the topography in light of erosion/deposition.

Figure 9, taken from Sundermann and Klocker,[103] shows the mean sediment vector over a tidal cycle and the predicted erosion and accretion patterns for a grain size of 0.5 mm. The grid resolution is 400 m. This is one of the more sophisticated models available and it is therefore interesting to see how it compares to the "ideal" model above. In this case,

1. There is no wave input.
2. Tidal currents are modeled in a two-dimensional numerical scheme into which some wind forcing could probably be included without serious difficulty.
3. The shear velocity is expressed as the depth mean velocity divided by an empirical function involving the total water depth (a poor representation of the bottom boundary layer flow).
4. A set of empirical equations for threshold, bedload, and suspended load that are in the author's words "based essentially on hydraulic model experiments, their extensibility for natural waters being not yet proved" (no effect of bottom slope is mentioned).
5. As the authors also point out "the figure shows only possible sediment movements because the model does not check the actual covering of the bottom".

There is clearly some way to go to reach a practical prediction algorithm.

### B. DEVELOPMENT OF A MORE COMPLETE MODEL

#### 1. Existing Models

The models that have been most widely used to predict sediment transport on a shelf have not included wave effects. It is consequently not clear whether the scatter of predictions relative to the tracer data in Figure 5 indicates a fundamental prediction problem. Simple models suggest that wave effects may be very significant (Figure 7) and Thorn's data[1] shows that, at least in his particular environment, orbital velocities of 0.20 $ms^{-1}$ increase the suspended transport rate by as much as a factor of twenty.

Except in some sheltered regions, dominated by tidal currents, sediment transport on continental shelves is generally storm dominated, and wave effects on the sea bed are of central interest. Engineering interest in extreme events further reinforces the need for a wave/current model for the shelf.

For the Nova Scotia continental shelf this requirement led to the development of a model to predict the stress and sediment transport at a point as a function of both waves and currents.[104] The basic concept follows the work of Grant and Madsen.[49] The sediment transport model used was the modified Einstein-Brown equation which Madsen and Grant[64] suggest using even in the presence of bedforms. However, provision was made in the model to use other transport formulae, the Engelund and Hansen[92] total-load equation and the Bagnold[4] formulae for bedload and suspended load. All the equations were defined in relatively simple and unmodified forms, as opposed to the recalibrated and restructured versions of Gadd et al.[27]

The input parameters required are

1. The wave height, frequency, and direction and the water depth (the parameters necessary to define the wave orbital velocity above the sea bed)
2. The mean current (or tidal current) vector above the seabed
3. The grain size and bottom roughness

In fact, some provision was included to estimate waves (or currents) using the local wind conditions. However, these are of very little value for a real shelf where conditions experienced at a point depend strongly on large-scale patterns of forcing and topography.

The advantage of such a model is that it provides an opportunity to explore the implications of a rather complex formulation. Conditions for sediment movement in combinations of currents, wave conditions, and water depths can be examined. The difficulty, inevitably, is the lack of data against which critically to test the results.

#### 2. A Two-Dimensional Model

In a second report on transport modeling on the Scotian Shelf, Martec[105] have proposed an approach close to that outlined. The structure of the model is shown schematically in Figure 10.

The hydrodynamic input proposed is much more complete than in most previous models. The model would:

1. Hindcast the wave field, including refraction and breaking
2. Predict tides and wind-driven circulation in a two-dimensional barotropic model
3. Compute bottom stress as described in the single point model[104]
4. Compute transport
5. Determine accretion or erosion between each grid point over the period of interest

A particularly innovative idea is the use of a sediment-size distribution function defined by 10 size intervals of width 0.5$\phi$ in the range $-0.75\phi$ to 3.75$\phi$, the observed range for Sable Island Bank. The total sediment-transport rate is then defined by

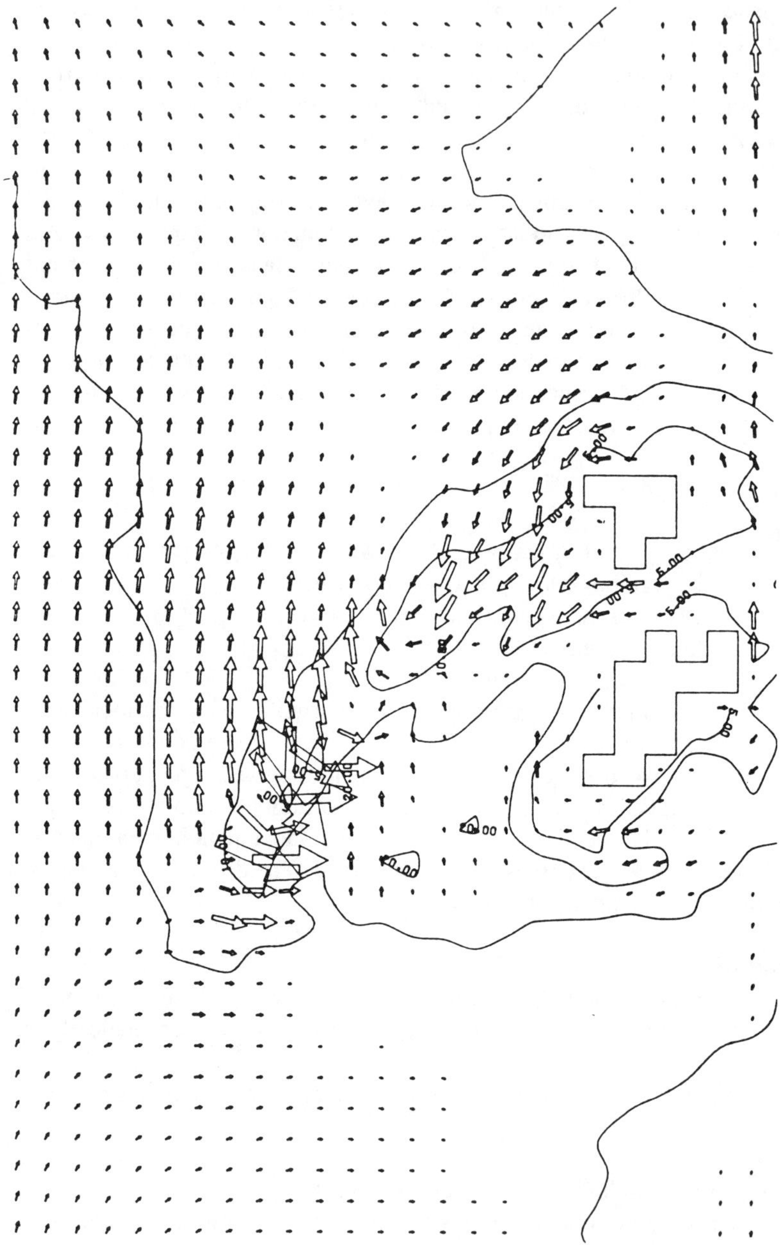

A

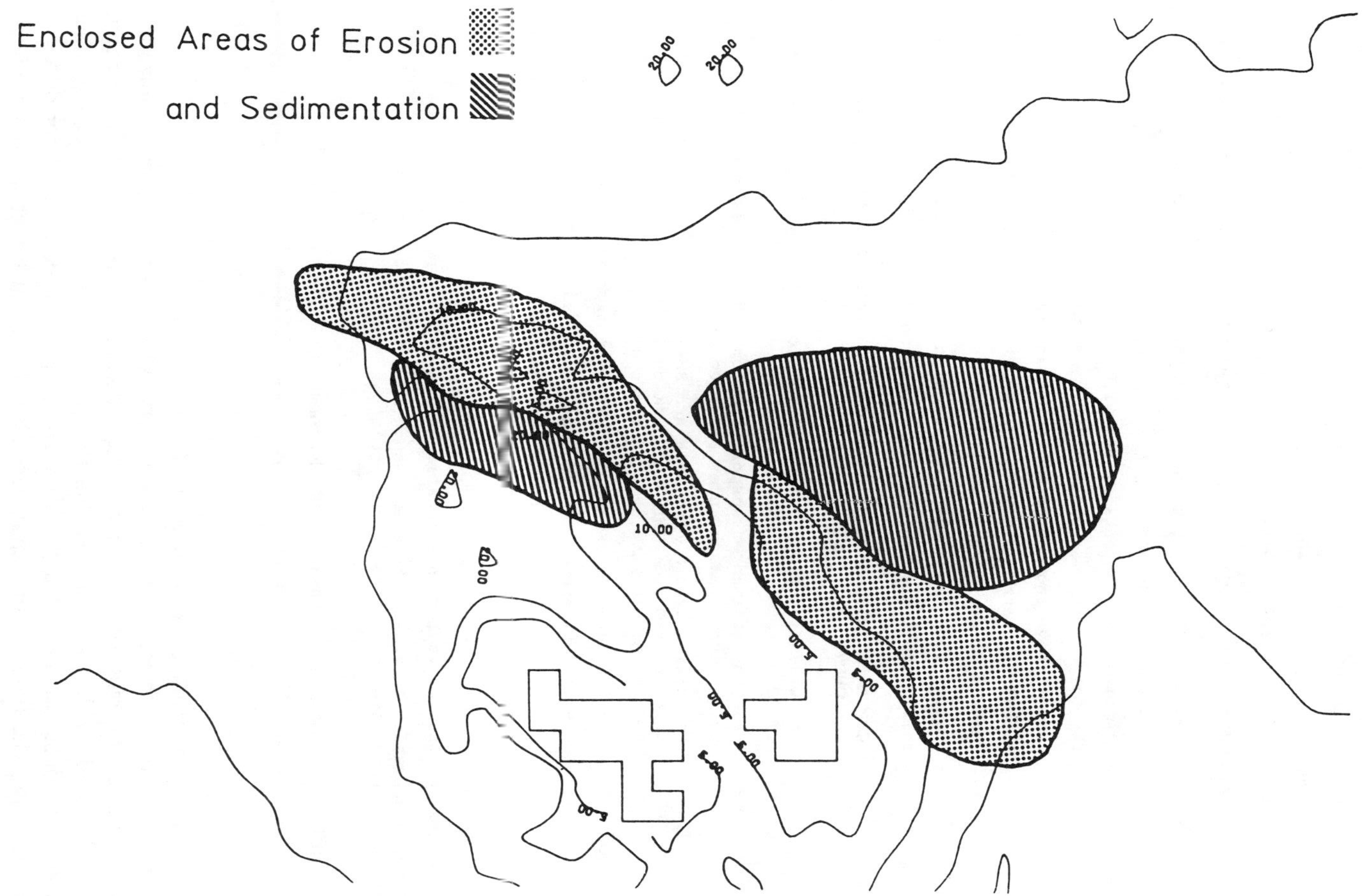


B

FIGURE 9. (A) Transport vectors round Heligoland, grain diameter 0.5 mm. (B) Divergence of sediment transport expressed as areas of significant erosion or deposition.[103] (Reprinted with permission from Springer-Verlag.)

```
DETERMINE SPACIAL DISTRIBUTION OF ACCRETION OR EROSION

    CALCULATE NET SEDIMENT TRANSPORT FOR SELECTED GRID POINTS

    For columns from IF to IL by increment of II, and rows from JF to JL
    by increments of JI:

      - Read grid point data

      - Compute grain size distribution

      - Initialize net transports to zero

         CALCULATE THE NET TRANSPORT AT EACH TIME STEP DURING A STORM

         For times from ITF to ITL by increments of ITI times three hours:

           - Read storm conditions

           - Compute tidal currents UT and VT.

           - Compute wind-driven currents UWI and VWI

           - Compute wave induced currents UWA and VWA

           - Add all currents, U = UT + UWI + UWA and V = VT + VWI + VWA

           - Compute direction and magnitude of total current

           - Compute wave height/period/direction distribution

           - Initialize instantaneous transports to zero

              CUMULATE INSTANTANEOUS TRANSPORT FOR EACH WAVE AND SEDIMENT
              COMPONENT

              For wave directions D from 11.25° to 78.75° by increments
              of 22.5°:

              Read the sediment transport look-up table corresponding
              to the current magnitude, and wave direction

              For grain sizes 5 from SMIN and SMAX of NS, wave orbital
              velocities O from OMIN and OMAX by increments of NO, periods
              P from PMIN to PMAX by increments of NP, and

                - Cumulate instantaneous transport from each component

         - Compute direction and magnitude of instantaneous transport

         - Convert transport to North and East components

         - Write instantaneous transports rate

         - Cumulate net transports

    - Write net transports for this location

- Determine accretion or erosion between each grid point by taking the
  difference in net transport at adjacent locations
```

FIGURE 10. Schematic outline of a model for computing sediment transport.[105]

$$q = \Sigma p_d \cdot q_d \tag{7}$$

where $q_d$ is the transport of material of diameter $d$.

Some caution is needed in estimating $q_d$. As we have seen, most transport formulae are not readily generalized to multiple grain sizes. The problem is illustrated by Sundermann and Klocker,[103] who show diagrams of potential transport, such as Figure 9, for a number of grain sizes. This does not imply that such transport rates can be attained for different sizes simultaneously. The physical limitations are clearly seen in Bagnold's development based on energy arguments. Bagnold suggests that there is a fixed maximum work rate which must be shared by the various grain sizes. Despite these difficulties, the development of realistic algorithms to handle grain-size distributions is an essential step which will eventually bring the model much closer to the geological view of sedimentary dynamics and consequently provide new possibilities for the critical evaluation of predictive models. However, in the short term, the inclusion of multiple grain sizes will emphasize the uncertainty associated with the transport relations for both bedload and suspended load.

In the present formulation the sediment distribution and sedimentary environment are assumed to be time invariant; this greatly reduces the advantages of a multiple-grain-size approach. For example, an interesting experiment would be to impose a spatially uniform grain-size distribution and model its evolution in time. The patterns of size distribution created would provide a further test of the validity of the size-dependent terms in any transport model.

An immediate need is therefore an algorithm for the conservation of sediment within each grain-size category. This, in turn, leads to two further problems, an adequate definition of the initial sedimentary conditions (grain-size distribution as a function of depth, perhaps) and appropriate boundary conditions for the study region.

In many ways the Martec prototype is a significant step towards the ideal model. However, the practical limitations are severe. The grid scale proposed is relatively coarse (7.4 × 5.4 km) and would not resolve the North Sea banks that Huthnance[96] has been modeling. The 3 h time step suggested will have to be shortened significantly to adequately describe tidal currents. However, the computing requirements for even this coarse resolution are formidable.

One of the central modeling problems is that practical constraints on computer time seriously reduce the possibility of exploring the parametric dependence of the model. In practice these models involve three types of parametric input.

**Parameters associated with the basic physical equations** — For example, tidal models must usually be calibrated by adjusting both the bottom friction coefficient and the tidal input at open boundaries. For the modeling of sediment transport there is the complication of the choice of possible equations, and various versions of these formulations, each containing some further parameters requiring definition (drag coefficient, bottom roughness, critical stress for movement, etc.).

**Parameters defining the forcing, initial conditions and boundary conditions for a model run** — There is a natural tendency in discussing the modeling of a specific area such as the Scotian Shelf to think in terms of realistic inputs. However, this may not be the best approach to understanding the implications of the modeling procedure. The real structure is extremely complex and it may be difficult, both conceptually and computationally, to perform a reasonable sensitivity analysis. Model runs with a simplified topography, hydrodynamic forcing, or sediment distribution may provide far more insight into the model as a system. The use of a multiple-grain-size distribution which is time invariant,[105] indicates the danger of getting the worst of all possible worlds; a complex model that contains internal inconsistencies. However, resolving these inconsistencies by keeping track of all grain sizes again leads to very severe computational requirements.

**The last set of parameters define the modeling process itself, particularly the grid spacing and time step** — The obvious danger is that the practicality of actually performing the computations may impose a minimum grid spacing and, in doing so, redefine the actual problems to be addressed. Ideally, the maximum spacing should be limited by the assumption, implicit in the model, that each grid point is truly representative of conditions within a region defined by the grid spacing. In practice, it seems likely that models of sufficiently small grid scale to define the wave conditions will provide reasonable insight into the spatial variation of sediment transport. A sensible strategy may be to have tidal- and wind-driven models of relatively coarse resolution, a wave refraction/diffraction model of necessarily finer scale and a sediment transport model of variable scale. It is worth noting that there are no dynamical constraints on the spacing of the sediment transport calculations themselves. However, rather severe difficulties emerge in following sediment fluxes through a region. The use of flux-corrected methods for such calculations is discussed by Amos and Greenberg.[63]

The disadvantage of proceeding with further development of a large-scaled two-dimensional model is that relevant field data on sediment distribution and movement are not available. There is, therefore, the pragmatic problem of determining an optimal sediment-transport formulation from among those shown in Figure 5. The development of an optimal formula will almost certainly depend on comparison with field data from only one or two points in the region of interest. The next step, in fact, is likely to be a refinement of the single-point models such as that of Martec,[104] as new data become available.

# VI. DISCUSSION AND CONCLUSIONS

"In view of the variety of processes involved, the insufficient theories and the lack of adequately observed data, modeling of sediment transport for natural areas is still in its infancy . . . A main problem of modeling is the lack of data from nature . . . a direct quantitative or even qualitative verification is impossible . . . "[103]

Having said that, perhaps one has said it all! It is clear that extremely sophisticated models, computationally very demanding, can be developed whose relevance to the real world is virtually unknown.

The specific needs for improving the detailed understanding of sediment transport processes include further research into transport thresholds and the many boundary-layer problems discussed earlier. Major advances have been made in recent years in field instrumentation, laboratory studies, and theories relating to sediment transport prediction. However, this development is proving slow to provide a clear discrimination between the various sediment transport formulae despite the obvious differences of their dependence on key parameters such as the velocity $u$ (or friction velocity $u_*$). We need to know more about the physics of bottom boundary layers, particularly if we are primarily concerned with the prediction of extreme conditions. For this purpose, it is essential to know whether the transport varies as approximately $u_*^3$, as in Figure 7, or as $u_*^7$.

In view of the uncertainty of these predictive formulae for sediment transport, it is easy to overlook the importance of the accurate prediction of the hydrodynamic processes. The dependence of the sediment transport on high powers of the velocity makes transport calculations very sensitive to errors in the definition of hydrodynamic conditions. Fortunately, tidal currents are often a significant component of the velocity field and are reasonably predictable. Unfortunately, they are oscillatory and the net transport may be the result of a small difference between two large transport rates (as in Figure 5). For this reason, the prediction of large, short-term transport rates may provide a more rational first objective than the prediction of net rates over a long time scale. Unfortunately, tracer experiments tend to be representative of the longer time averages. An obvious windfall would be a tracer experiment during storm conditions with a strong preferred-flow direction.

The numerical models for calculating waves, tidal currents, and wind-driven currents also need field calibration. This is relatively straightforward for the tides, and an interesting but tractable problem for the wind-driven currents (such models have been under development for a considerable time, initially for storm surge forecasting), but a major problem for the wave predictions. Wave models, rather like sediment transport models, have reached a very sophisticated stage on a very narrow data base. Field programs to measure directional spectra are needed both to calibrate and validate the model results. Our particular interest in extreme conditions also focuses attention on the current field driven by wave dissipation and breaking which may be significant over the shallower banks on an exposed continental shelf.[106]

Such overall needs for data suggest an extensive field program. However, most of the hydrodynamic problems are being tackled for reasons that are not primarily related to sediment transport. The central requirement at the present time is an experimental program focused on field measurement of sediment transport which includes measurements of all the primary inputs. This would include the following.

**Hydrodynamics** — Measurements of the wave orbital velocity, tidal, and mean currents at some reference height above the bed. Measurements of the detailed velocity structure of the boundary layer, with estimates of skin friction and form drag.[37] Regional measurements of the velocity field to examine horizontal coherence.

**Sedimentary environment** —Bottom topography on all length scales, grain-size distribution of suspended, surficial, and subsurface sediment. Time-lapse photography of bottom boundary conditions. Assessment of biological activity.

**Sediment transport** — Tracer experiments (with careful consideration of the size and size distribution to be employed). Suspended sediment sampling, optical, and acoustic monitoring of the suspension.

There are other requirements, direct measurement of bedload for example, for which methodology or instrumentation is not available. Indeed, some of the requirements already included are difficult to achieve in practice, suspended sediment sampling in extreme conditions being one example. Such a program would provide an important focus for the wide-ranging interests in sediment transport and sedimentation. There are clearly models waiting to be tested; further model development, other than to improve the efficiency of the actual computation, is pointless without improvements in our basic understanding of these processes or comprehensive data sets specifically designed to provide all necessary model parameters.

## REFERENCES


1. **Thorn, M. F. C.,** The effect of waves on the tidal transport of sand, Hydraulic Res. Station, Wallingford, U. K., Notes 21, 4, 1979.
2. **Hill, P. R. and Bowen, A. J.,** Modern sediment dynamics at the shelf-slope boundary layer off Nova Scotia, in *The Shelfbreak: Critical Interface on Continental Margins,* Stanley, D. J. and Moore, G. T., Eds., SEPM Spec. Pub. 33, 1983, 265.
3. **Pingree, R. D.,** The formation of the Shambles and other banks by tidal stirring of the seas, *J. Mar. Biol. Assoc. U.K.*, 58, 211, 1978.
4. **Bagnold, R. A.,** An approach to the sediment transport problem from general physics, U.S. Geol. Survey Prof. Paper 422-I, 1966, 37
5. **Brown, C. B.,** Sediment Transportation, in *Engineering Hydraulics,* Rouse, H., Ed., John Wiley & Sons, New York, 1973.
6. **Heathershaw, A. D. and Simpson, J. H.,** The sampling variability of Reynolds stress and its relation to boundary shear stress and drag coefficient measurements, *Estuarine Coastal Mar. Sci.,* 6, 263, 1978.
7. **Bowden, K. F. and Ferguson, S. R.,** Variation with height of the turbulence in a tidally-induced bottom boundary layer, in *Marine Turbulence,* Nihoul, J. C. J., Ed., Elsevier Oceanography Series No. 28, Elsevier, Amsterdam, 1980.
8. **Grant, W. D., Williams, A. J., III, and Glenn, S. M.,** Bottom stress estimates and their prediction on the Northern Californian continental shelf during CODE-1: the importance of wave-current interaction, *J. Phys. Oceanogr.,* 14, 506, 1984.
9. **Huntley, D. A. and Hazen, D. G.,** Seabed stresses in combined wave and steady flow conditions on the Nova Scotia continental shelf; field measurements and predictions, *J. Phys. Oceanogr.*, 18, 347, 1988.
10. **Greenberg, D. A.,** Modeling the mean barotropic circulation in the Bay of Fundy and Gulf of Maine, *J. Phys. Oceanogr.*, 13, 886, 1983.
11. **Davies, A. M.,** Application of a three-dimensional shelf model to the calculation of North Sea Currents, in *North Sea Dynamics,* Sundermann, J. and Lenz, W., Eds., Springer-Verlag, Berlin, 1983, 44.
12. **Davies, A. M.,** A three-dimensional modal model of wind-induced flow in a sea region, *Progr. Oceanogr.*, 15, 71, 1985.
13. **Amos, C. L. and Greenberg, D. A.,** A simulation of suspended particular matter in the Minas Basin, Bay of Fundy -- a region of potential tidal power, Proc. Canadian Coastal Conference 1980, Burlington, Ontario, NRC. 2, 1980.
14. **Tryggestad, S., Selange, K. A., Mathisen, J. P., and Johansen, O.,** Extreme bottom currents in the North Sea, in *North Sea Dynamics,* Sundermann, J. and Lenz, W., Eds., Springer-Verlag, Berlin 1983, 148.

15. **Davies, A. M. and Flather, R. A.,** On computing extreme meteorologically induced currents, with application to the north-west European continental shelf, *Cont. Shelf. Res.*, 7, 643, 1987.
16. **Clarke, A. J. and Brink, K. H.,** The response of stratified frictional flow of shelf and slope waters to fluctuating, large-scale, low-frequency wind forcing, *J. Phys. Oceanogr.*, 15, 439, 1985.
17. **Huthnance, J. M.** Tidal current assymetries over the Norfolk Sandbanks, *Estuarine Coastal Mar. Sci.* 1, 89, 1973.
18. **Huthnance, J. M.,** On mass transports generated by tides and long waves, *J. Fluid Mech.*, 102, 367, 1981.
19. **Loder, J. W.,** Topographic rectification of tidal currents on the side of Georges Bank, *J. Phys. Oceanogr.*, 10, 1399, 1980.
20. **Darbyshire, M. and Draper, L.,** Forecasting wind-generated sea waves, *Dock Harbour Eng.*, 195, 482, 1963.
21. **Bretschneider, C. L.,** Prediction of waves and currents, Look Laboratory Report, Vol. 3, No. 1, 1973, 1.
22. **Hasselman, D. E., Dunkel, M., and Ewing, J. A.,** Directional spectra observed during JONSWAP, *J. Phys. Oceanogr.*, 10, 1264, 1973.
23. **Komen, G. J.,** Activities of the WAM (Wave Modelling) Group, in *Advances in Underwater Technology, Ocean Science and Offshore Engineering*, 6, Oceanology; Graham & Trotman, London 1986, 486.
24. **Bowen, A. J., Chartrand, D. M., Daniel, P. E., Glodowski, C. W., Piper, D. J. W., Readshaw, J. S., Thibault, J., and Willis, D. H.,** Canadian Coastal Sediment Study, Hydraulics Lab. Tech. Rep. No. TR-HY-013, Final Report of the Steering Committee, Div. Mech. Eng., National Research Council of Canada, 1986, 96.
25. **Heathershaw, A. D.,** Comparisons of measured and predicted sediment transport rates in tidal currents, *Mar. Geol.*, 42, 75, 1981.
26. **Lees, B. J.,** The relationship of sediment transport rates and paths to sandbanks in a tidally dominated area off the coast of East Anglia, U.K., *Sedimentology*, 30, 461, 1983.
27. **Gadd, P. E., Lavelle, J. W., and Swift, D. J. P.,** Estimates of sand transport on the New York shelf using near-bottom current meter observations, *J. Sediment Petrol.*, 48, 239, 1978.
28. **Bijker, E. W.,** Some considerations about scales for coastal models with moveable beds, Delft Hydraulics Lab. Rep. No. 50, 1967, 142.
29. **Vincent, C. E., Young, R. A., and Swift, D. J. P.,** Sediment transport on the Long Island shoreface, North American Atlantic Shelf: role of waves and currents in shoreface maintenance, *Cont. Shelf Res.*, 2, 163, 1983.
30. **Bowden, K. F.,** Physical problems of the benthic boundary layer, *Geophys. Surveys*, 3, 255, 1978.
31. **Soulsby, R. L.,** The bottom boundary layer of shelf seas, in *Physical Oceanography of Coastal and Shelf Seas*, Johns, B., Ed., Elsevier, Amsterdam, 1983, 470.
32. **Smith, J. D. and McLean, S. R.,** Spatially averaged flow over a wavy surface, *J. Geophys. Res.*, 82, 1735, 1977.
33. **Smith, J. D.,** Investigations of turbulent boundary layers and sediment transport phenomena as related to shallow marine environments. II. Studies of non-uniform boundary layer flows, Report A 69-7, Department of Oceanography, University of Washington, Seattle, 1969.
34. **Taylor, P. A., Gent, P. R., and Keen, J. M.,** Some numerical solutions for turbulent boundary-layer flow above fixed, rough, wavy surfaces, *Geophys. J. R. Astron. Soc.*, 44, 177, 1976.
35. **Richards, K. J. and Taylor, P. A.,** A numerical model of flow over sand waves in water of finite depth, *Geophys. J. R. Astron. Soc.*, 65, 103, 1981.
36. **Smith, J. D.,** Modeling of sediment transport on continental shelves, in *The Sea*, Vol. 6, Goldberg, E. D., McCave, I. N., O'Brien, J. J., and Steele, J. H., Eds., Wiley-Interscience, New York, 1977, 539.
37. **Chriss, T. M. and Caldwell, D. R.,** Evidence for the influence of form drag on bottom boundary layer flow, *J. Geophys. Res.*, 87, 4148, 1982.
38. **SANDS,** Shelf sediment dynamics: a national overview, Rep. Workshop Vail, Colorado, November 1976, Gorsline, D. S. and Swift, D. J. P., Eds., NSF (IODE), 1977.
39. **Krugermeyer, L. and Grunwald, M.,** The influence of sea waves on the wind profile, *Boundary Layer Meteorol.*, 10, 403, 1978.
40. **Einstein, H. A.,** River sedimentation, in *Handbook of Applied Hydrology*, Section 17--11, Te Chow, V. Ed., McGraw-Hill, New York, 1964, 35.
41. **Fredsoe, J.,** Shape and dimensions of stationary dunes in rivers, *J. Hydraul. Div. ASCE*, 108 (HY8), 932, 1982.
42. **Raudkivi, A. J.,** *Loose Boundary Hydraulics*, Pergamon Press, Oxford, 1976.
43. **Dyer, K. R.,** Velocity profiles over a rippled bed and the threshold of movement of sand, *Estuarine Coastal Mar. Sci.*, 10, 181, 1980.
44. **Arya, S. P. S.,** A drag partition theory for determining the large scale roughness parameter and wind stress on the Arctic pack ice, *J. Goephys. Res.*, 80, 3447, 1975.

45. **Wooding, R. A., Bradley, E. F., and Marshall, J. K.,** Drag due to regular arrays of roughness elements of varying geometry, *Boundary Layer Meteorol.*, 5, 285, 1973.
46. **Lettau, H.,** Note on aerodynamic roughness — parameter estimation on the basis of roughness element description, *J. Appl. Meteorol.*, 8, 828, 1969.
47. **Taylor, P. A. and Dyer, K. R.,** Theoretical models of flow near the bed and their implications for sediment transport, in *The Sea,* Vol. 6, Goldberg, E. D., McCave, I. N., O'Brien, J. J., and Steele, J. H., Eds., Wiley-Interscience, New York, 1977, 579.
48. **Butman, B., Noble, M., and Folger, D. W.,** Long-term observations of bottom current and bottom sediment movement on the mid-Atlantic continental shelf, *J. Geophys. Res.*, 84, 1187, 1979.
49. **Grant, W. D. and Madsen, O. S.,** Combined wave and current interaction with a rough bottom, *J. Geophys., Res.*, 84, 1797, 1979.
50. **Davies, A. G., Soulsby, R. L., and King, H. L.,** A numerical model of the combined wave and current bottom boundary layer, *J. Geophys. Res.*, 93, 491, 1987.
51. **Dyer, K. R. and Soulsby, R. L.,** Sand transport on the continental shelf, *Annu., Rev. Fluid Mech.*, 20, 295, 1988.
52. **Fredsoe, J.,** Sediment transport in currents and waves, Ser. Pap. 35, Inst. Hydrodyn. Hydraul. Eng., Technical University of Denmark, Lyngby, 1984, 37.
53. **Wiberg, P. and Smith, W. D.,** A comparison of field data and theoretical models for wave-current interactions at the bed on the continental shelf, *Contin. Shelf Res.*, 2, 147, 1983.
54. **Grant, W. D., Williams, A. J., III, Glenn, S. M., Cacchione, D. A., and Drake, D. W.,** High frequency bottom stress variability and its prediction in the CODE region, CODE Tech. Rep. no. 15, Woods Hole Oceanographic Institution, Woods Hole, MA, 1984.
55. **Gust, G.,** Observations on turbulent drag reduction in a dilute suspension of clay in sea-water, *J. Fluid Mech.*, 75, 29, 1976.
56. **McCave, I. N.,** Some boundary layer characteristics of tidal currents bearing sand in suspension, *Mem. Soc. R. Sci. Liege,* 6(6), 107.
57. **Gust, G. and Southard, J. B.,** Effects of weak bedload on the universal law of the wall, *J. Geophys. Res.*, 88, 5939, 1983.
58. **Longuet-Higgins, M. S.,** Oscillatory flow over steep sand ripples, *J. Fluid Mech.*, 107, 1, 1981.
59. **Bagnold, R. A.,** Mechanics of marine sedimentation, in *The Sea,* Vol. 3, Hill, M. N., Ed., Wiley-Interscience, New York, 1963.
60. **Inman, D. L. and Bowen, A. J.,** Flume experiments on sand transport by waves and currents, *Proc. 8th Conf. Coastal Engineering,* American Society of Civil Engineers, New York, 1963, 137.
61. **Bijker, E. W., Hijum, E. V., and Vellinga, P.,** Sand transport by waves, in *Proc. 15th Coastal Engineering Conf. Hawaii,* American Society of Civil Engineers, New York, 1976, 1149.
62. **Middleton, G. V.,** Hydraulic interpretation of sand size distributions, *J. Geol.*, 84, 405, 1981.
63. **Vincent, C. E., Swift, D. J. P., and Hillard, B.,** Sediment transport in the New York Bight, North American Atlantic Shelf, *Mar. Geol.*, 42, 369, 1981.
64. **Madsen, O. S. and Grant, W. D.,** Quantitative description of sediment transport by waves, *Proc. 15th Coastal Engineering Conf.*, American Society of Civil Engineers, New York, 1977, 1093.
65. **Shields, A.,** Anwendung der Ahnlich Keitsmechanik und Turbulenzforschung auf Geschielsebewegung, Mitteil. Preuss. Versuchsanst; Wasser, Erd, Schiffsbau, Berlin, No. 26, 1936.
66. **Miller, M. C., McCave, I. N., and Komar, P. D.,** Threshold of sediment motion under unidirectional currents, *Sedimentology,* 24, 507, 1977.
67. **Nowell, A. R. M., Jumars, P. A., and Eckman, J. E.,** Effects of biological activity on the entrainment of marine sediments, *Mar. Geol.*, 42, 133, 1981.
68. **Inman, D. L.,** Sediments: physical properties and mechanics of sedimentation, in *Submarine Geology,* 2nd ed., Shepard, F. P., Ed., Harper & Row, New York, 1963, 963.
69. **Middleton, G. V. and Southard, J. B.,** Mechanics of sediment movement, SEPM Short Course No. 3, Tulsa, Oklahoma, Society of Economic Palaeotologists and Mineralogists, 1977.
70. **Sternberg, R. W.,** Measurements of incipient motion of sediment particles in the marine environment, *Mar. Geol.*, 10, 113, 1971.
71. **Rhoads, D. C.,** Organism-sediment relations on the muddy sea floor, *Oceanogr. Mar. Biol. Annu. Rev.*, 12, 263, 1974.
72. **Grant, W. D., Boyer, L. F. and Sandford, L. P.,** The effect of bioturbation on the initiation of motion of intertidal sands, *J. Mar. Res.*, 40, 659, 1982.
73. **Madsen, O. S. and Grant, W. D.,** The threshold of sediment movement under oscillatory water waves: a discussion, *J. Sediment. Petrol.*, 45, 360, 1975.
74. **Komar, P. D. and Miller, M. C.,** The threshold of sediment movement under oscillatory water waves, *J. Sediment. Petrol.*, 43, 1101, 1973.
75. **Komar, P. D. and Miller, M. C.,** On the comparison between the threshold of sediment motion under waves and unidirectional currents with a discussion of the practical evaluation of the threshold, *J. Sediment. Petrol.*, 45, 362, 1975.

76. **Larsen, L. H., Sternberg, R. W., Shi, N. C., Marsden, M. A. H., and Thomas, L.,** Field investigations of the threshold of grain motion by ocean waves and currents, *Mar. Geol.,* 42, 105, 1981.
77. **Graf, W. H.,** *Hydraulics of Sediment Transport,* McGraw-Hill, New York, 1971, 513.
78. **Yalin, M. S.,** *Mechanics of Sediment Transport,* 2nd ed., Pergamon Press, Elmsford, NY, 1977, 238.
79. **Dyer, K. R.,** *Coastal and Estuarine Sediment Dynamics,* Wiley, Chichester, 1986, 342.
80. **Sternberg, R. W.,** Predicting initial motion and bedload transport of sediment particles in the shallow marine environment, in *Shelf Sediment Transport,* Swift, D. J. P., Duane, D. B., and Pilkey, O. H., Eds., Dowden, Hutchinson and Ross, Stroudsberg, PA, 1972, 61.
81. **Langhorne, D. N.,** An evaluation of Bagnold's dimensionless coefficient of proportionality using measurements of sand-wave movement, *Mar. Geol.,* 43, 49, 1981.
82. **Huntley, D. A. and Hanes, D. M.,** Direct measurement of suspended sediment transport, *Coastal Sediments '87,* American Society of Civil Engineers, New York, 1987, 723.
83. **Einstein, H. A.,** The bedload function for sediment transportation in open channel flows, Tech. Bull. 1026, Soil Conservation Service, U. S. Department of Agriculture, 1988, 78.
84. **Jonsson, I. G.,** Wave boundary layers and friction factors, *Proc. 10th Coastal Engineering Conf.,* American Society of Civil Engineers, New York, 1966, 127.
85. **Vincent, C. E., Young, R. A., and Swift, D. J. P.,** Bedload transport under waves and currents, *Mar. Geol.,* 39, M71, 1981.
86. **Yalin, M. S.,** An expression for bed load transportation, *Proc. Am. Soc. Civ. Eng., J. Hydraul. Div.,* 89 (NY3), 221, 1963.
87. **Smith, J. D. and Hopkins, T. S.,** Sediment transport on the continental shelf off Washington and Oregon in light of recent current measurements, in *Shelf Sediment Transport,* Swift, D. J. P., Duane, D. B., and Pikley, O. H., Eds., Dowden, Hutchinson and Ross, Stroudsburg, PA, 143, 1972.
88. **Rouse, H.,** Modern conceptions of the mechanics of turbulence, *Trans. Am. Soc. Civ. Eng.,* 102, 436, 1937.
89. **Nielsen, P., Svendsen, I. A., and Stanb, C.,** Onshore-offshore sediment transport on a beach, *Proc. 16th Int. Conf. Coastal Engineering, Hanburg, American Society of Civil Engineering, New York, 1978, 1475.*
90. **Hanes, D. M., Vincent, C. E., Huntley, D. A., and Clarke, T. L.,** Acoustic measurements of suspended sand concentration in the $C^2S^2$ experiment at Stanhope Lane, Prince Edward Island, *Mar. Geol.,* 81, 185, 1988.
91. **Ackers, P. and White, W. R.,** Sediment transport: new approach and analysis, *Proc. Am. Soc. Civ. Eng., J. Hydraul. Div.,* HY 11, 2041, 1973.
92. **Engelund, F. and Hansen, T.,** A monograph of sediment transport in alluvial streams, Technisk-Vorlag, Copenhagen, 1967, 62.
93. **Amos, C. L., Bowen, A. J., Huntley, D. A., and Judge, T.,** A comparison between observed and predicted sand transport on the Eastern Canadian Continental Shelf, in preparation.
94. **Amos, C. L., Bowen, A. J., Huntley, D. A., and Lewis, C. F. M.,** Ripple generation under the combined influences of waves and currents on the Canadian Continental Shelf, *Cont. Shelf Res.,* 8(10), 1129, 1988.
95. **Huthnance, J. M.,** On one mechanism forming linear sand banks, *Estuarine Coast. Shelf Sci.,* 14, 79, 1982.
96. **Huthnance, J. M.,** On the formation of sand banks of finite extent, *Estuarine Coast. Shelf Sci.,* 15, 277, 1982.
97. **Cartwright, D. E.,** On submarine sand-waves and tidal lee waves, *Proc. R. Soc. London Ser. A,* 253, 218, 1959.
98. **Richards, K. J.,** The formation of ripples and dunes on an erodable bed, *J. Fluid Mech.,* 99, 597, 1981.
99. **Bowen, A. J.,** Simple models of nearshore sedimentation; beach profiles and longshore bars, in *The Coastline of Canada,* McCann, S. B., Ed., Geological Survey of Canada, Ottawa, 1980, 1.
100. **Holman, R. A. and Bowen, A. J.,** Bars, bumps and holes: models for the generation of complex beach topography, *J Geophys. Res.,* 87, 457, 1982.
101. **Hardisty, J. and Whitehouse, R. J. S.,** Evidence for a new sand transport process from experiments on Saharan dunes, *Nature,* 443, 532, 1988.
102. **Whitehouse, R. J. S. and Hardisty, J.,** Experimental assessment of two theories for the effect of bedslope on the threshold of bedload transport, *Mar. Geol.,* 79, 135, 1988.
103. **Sundermann, J. and Klocker, R.,** Sediment transport modeling with applications to the North Sea, in *North Sea Dynamics,* Sundermann, J. and Lenz, W., Eds., Springer-Verlag, Berlin, 1983.
104. **Martec Limited,** Sediment transport on a continental shelf, Atlantic Geoscience Center, Energy, Mines and Resources Canada, 1982, 60.
105. **Martec Limited,** A 2-D sediment transport model for continental shelves, Department of Supply and Services, Canada 1983.
106. **Dolata, L. F. and Rosenthal, W.,** Wave set-up and wave-induced currents in coastal zones, *J. Geophys. Res.,* 89, 1973, 1984.
107. **Davies, A. G.,** personal communication.

Chapter 11

# QUALITATIVE ASPECTS OF TOPOGRAPHIC WAVES IN CLOSED BASINS, GULFS, AND CHANNELS

**Thomas F. Stocker and Kolumban Hutter**

## TABLE OF CONTENTS

# I. ABSTRACT

After presenting the governing equation for topographic waves and discussing its properties and some known solutions in bounded and unbounded domains we point out an interesting and potentially useful connection between the spectrum of the topographic wave operator in a semi-infinite channel and that of the Schrödinger equation of an electron subject to the potential well. We then show that in closed basins there are three types of modal structures: global, basin-wide; small scale, basin-filled; localized. Effects of the variation of the topography on the dispersion relation are discussed and the influence of the curvature of an elongated basin on the dispersion relation and on the modal structure is studied. To further aid in the identification of the individual mode types in rectangular basins the current ellipses and the Stokes drift vectors are computed. A preliminary analysis of a double trench finally demonstrates how shelf waves may, through resonance, excite topographic waves in fjords or estuarine channels.

# II. INTRODUCTION

Wave phenomena are among the distinctive features that can be observed in the velocity and temperature records of instruments which are moored in the ocean or in lakes. A large amount of them manifest themselves as barotropic or baroclinic gravity waves and, in enclosed basins, give rise to external or internal seiches with usually periods of a few hours and at most 1 to 2 d. Long periodic processes can, in general be attributed to the vorticity nature of the motion. Existence of these second-class waves is due to the rotation of the Earth and the variation of the bathymetry. In the ocean these so-called shelf waves have widely been identified as vorticity-dominated motions that are essentially trapped along the shores; in enclosed basins they form the topographic waves and enjoy a particularly rich structure. Their occurrence has been observationally corroborated in only a few individual cases (Lake Michigan, Lake Ontario, Lake of Lugano and Zurich, see Chapter 1 in Stocker and Hutter[1]), but the interpretation is partly controversial because of the scantiness of the data that were collected.

The aim of this study is to describe the structure of topographic waves in infinite and semi-infinite channels and in closed basins which in plan view form a rectangular region or a sector of an annulus and have a trough-like topography.

We first point out an interesting and potentially useful connection between the spectrum of the topographic wave operator in a semi-infinite channel and that of the Schrödinger equation subject to the potential well. It is then shown in detail that the topographic wave operator possesses a rich and dense spectrum with essentially three types of modal structures: global basin-wide modes, small-scale modes which fill the entire basin, and localized bay modes. Each of these modes has its distinctive structure; however, the modes cannot be ordered according to frequency, and often a mode with localized small-scale structure is hardly separated from another that "fills the entire basin". This aggravates identification of individual modes by observation. Dense nets of current meters and detailed analyses of time series that include rotary spectra and drift-current estimates may be necessary to arrive at reliable conclusions regarding mode identification. It is shown that drift currents of different modes are significantly distinct and may aid substantially in the interpretation of complicated data analyses. A preliminary analysis of a double trench finally demonstrates how shelf waves may excite topographic waves near river mouths or fjords. It also naturally indicates in which direction future research in topographic waves is likely to advance.

Consider the barotropic shallow-water equations subject to the rigid-lid assumption. Let $\psi$ be the mass-transport stream function according to which

$$Hu = -\frac{\partial \psi}{\partial y}, \qquad Hv = \frac{\partial \psi}{\partial x} \tag{1}$$

$H(x,y) > 0$ is the water depth, and u,v are the vertically averaged velocity components in the x- and y- directions, respectively. The evolution equation for $\psi$ evolves from the conservation law of potential vorticity and yields in general a first order hyperbolic partial differential equation that must be solved in the lake domain $\mathcal{D}$, subject to the boundary condition of no flow through the shore boundary $\partial\mathcal{D}$. Thus, the following boundary value problem emerges for $\psi$:

$$\begin{aligned} \tilde{\Pi}\psi &= 0, \qquad (x,y) \in \mathcal{D}, \\ \psi &= 0, \qquad (x,y) \in \partial\mathcal{D} \end{aligned} \tag{2}$$

Here $\tilde{\Pi}$ is the time-dependent wave operator

$$\begin{aligned} \tilde{\Pi} &= \frac{\partial}{\partial t} \nabla \cdot (H^{-1}\nabla(\cdot)) - \hat{z} \cdot \nabla(fH^{-1}) \times \nabla(\cdot) \\ &= \frac{\partial}{\partial t} E[\cdot] - J[fH^{-1},\cdot] \end{aligned} \tag{3}$$

in which f is the Coriolis parameter, $\hat{z}$ is a unit vector pointing in the direction opposite to the gravity vector, and $\nabla$ is the horizontal gradient operator. Equation 3, when subject to f = constant, is called the *topographic wave operator*; it enjoys the following properties.[1]

- Whenever $J \equiv 0$, no waves can propagate. Any nonsteady solution is due to the presence of the rotation of the Earth and the variation of the bathymetric profile.
- The boundary value problem (Equation 2) is scale invariant, i.e., changing x,y by the scale [L] and the depth by the scale [H], leaves Equation 2 unchanged. This property is due to the rigid-lid assumption. Scrutiny indicates that it holds as long as [L] is small in comparison to the external Rossby radius

$$R = \frac{\sqrt{g[H]}}{f} \sim 500 \text{ km} \tag{4}$$

- Most significant, the topographic wave equation is invariant under conformal mappings. Indeed, in a general orthogonal curvilinear coordinate system $\xi = \xi(x,y)$, $\eta = \eta(x,y)$ the operators E and J read

$$\begin{aligned} E[\cdot] &= \frac{\partial}{\partial \xi}\left(\frac{J_2}{J_1}\frac{1}{H}\frac{\partial(\cdot)}{\partial \xi}\right) + \frac{\partial}{\partial \eta}\left(\frac{J_1}{J_2}\frac{1}{H}\frac{\partial(\cdot)}{\partial \eta}\right), \\ J[fH^{-1},\cdot] &= \frac{\partial}{\partial \xi}(fH^{-1})\frac{\partial(\cdot)}{\partial \eta} - \frac{\partial}{\partial \eta}(fH^{-1})\frac{\partial(\cdot)}{\partial \xi}, \end{aligned} \tag{5}$$

where $J_1$ and $J_2$ are the scale factors

$$J_1 = \left|\frac{\partial \xi}{\partial x}\right|, \qquad J_2 = \left|\frac{\partial \eta}{\partial y}\right|$$

Since under conformal mappings $J_1 = J_2$, Equation 5 reveals the stated invariance property. This property was pointed out and used by Johnson.[2,3] It was implicitly used also by Mysak.[4]

The invariance property of the topographic wave operator under conformal mappings is a useful tool, as a wealth of solutions can be generated from known solutions to special geometries. Thus Stocker and Hutter[1] have indicated, how shelf wave solutions along a hyperbolic shoreline can be obtained from Buchwald and Adams[5] shelf wave analysis for straight shorelines; and they show how straight-channel solutions can be used to construct topographic wave modes in hyperbolic channels. More important, Johnson[3] generates analytic topographic wave solutions in a semi-infinite channel by conformally mapping the infinite strip into a semi-infinite strip having a cut.

The conformal mapping *property* of the topographic wave operator has vastly enlarged the set of configurations for which topographic wave solutions are known, but the conformal mapping *technique* does not provide us with a tool to generate characteristically new solutions. A mode with a basin-wide structure in one domain will be stretched and torted in the transformed domain but it will still be basin wide. This is perhaps a limitation of its usefulness, as is the fact that bathymetric contour lines are transformed with the mapping so that topographies in the transformed domains may be very special and perhaps unrealistic.

Exact solutions to the boundary value problem (Equation 2) have been constructed in simple *bounded* and in special *infinite* domains. A summary of the existing solutions is given in Chapter 3 of Stocker and Hutter.[1] Lamb[6] is the first to present the topographic wave solution in a circular basin with parabolic bottom profile. Saylor, et al.[7] find exact formulas for modes in circular basins whose bottom profile varies radially according to a power law. Elliptical basins with parabolic bottom were treated by Ball,[8] and elliptical basins whose depth curves follow confocal ellipses are studied by Mysak,[4] Mysak et al.,[9] and Johnson.[2] All these solutions enjoy the property of having quantized real frequencies and wave numbers. Associated stream functions fill the entire basin and thus exhibit basin-wide structure. Figure 1 illustrates this for the three lowest modes in an elliptical lake with an exponential bottom profile.

The literature on topographic waves in unbounded domains is far more extensive, and it is impossible to do justice to all the relevant works. We thus limit our remarks to the essential properties and direct the reader to the reviews by Mysak[10] and Stocker and Hutter.[1,12] The configurations are generally domains that are infinitely long in the x-direction with one (shelf) or two (channel) shore boundaries parallel to it and with a topographic profile that varies only with y, the direction perpendicular to the x-direction. The following are characteristic properties of the solutions.[19]

- The spectrum, i.e., the dispersion relation is continuous: to each real frequency there is at least one real wave number. For a bounded slope parameter $S = |\partial H/\partial y|/H$ there is a countably infinite number of shelf modes and the dispersion relation is double valued, as shown in Figure 2 for exponential shelf waves.[5] Accordingly, long shelf waves are nondispersive, i.e., as $k \rightarrow 0$, $c_{gr} = \partial\sigma/\partial k = c = \sigma/k$. Phase and group velocities are the same. Moreover, for $0 < |k| < |k_0|$ the group velocity is positive, $c_{gr} > 0$, and energy and phase propagate in the same direction. For $|k| > k_0$, however, $c_{gr} < 0$ and the energy propagates against the phase. As $|k| \rightarrow \infty$ $\sigma \rightarrow 0$, $\sigma(k_0) = \sigma_0$ is called *critical frequency* or *cut-off frequency*.
- When S is unbounded at a point, $\sigma(k)$ is in general monotone with $\sigma \rightarrow \sigma_{max}$ as $k \rightarrow \infty$. The dispersion relation is now single valued.
- In doubly connected domains (topographic waves around islands) the above statements remain valid, but the spectrum is again quantized due to the $2\pi$-periodicity of the x-coordinate.

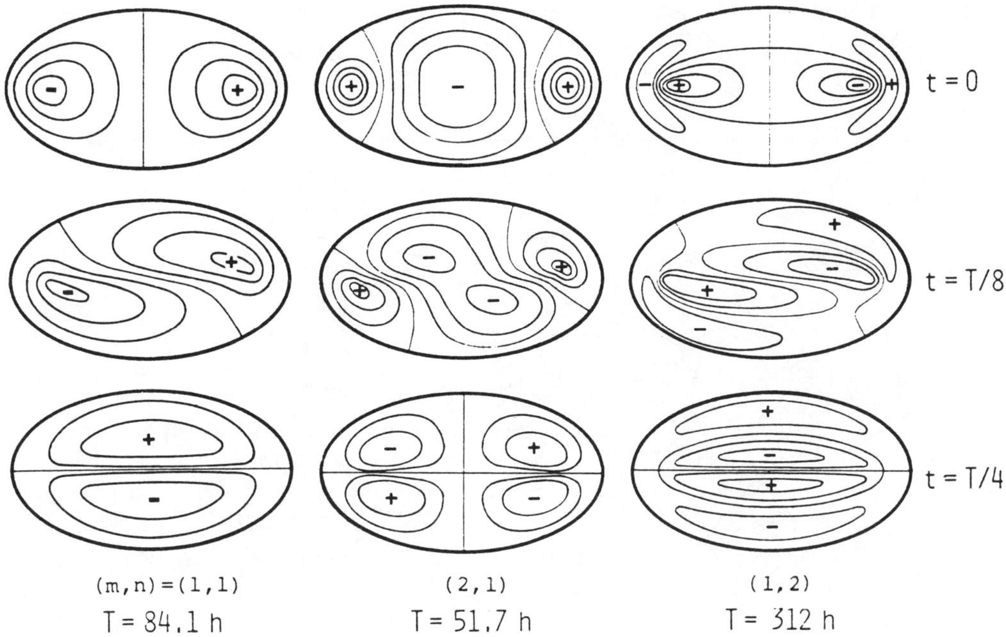


FIGURE 1. Stream line contours of the three lowest modes in an elliptical lake with exponential bottom. (From Johnson, E. R., *Geophys. Astrophys. Fluid Dyn.*, 37, 279, 1987. With permission.)

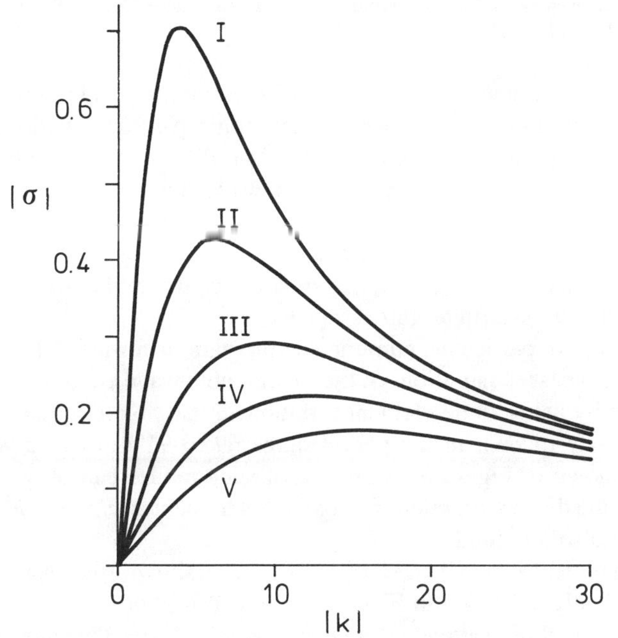


FIGURE 2. Dispersion relation σ(k) for the first five modes of the topographic wave equation on an exponential shelf with H (y) = ϵ exp (by), b = 5.4. (From Buchwald, V. T. and Adams, J. T., *Proc. R. Soc. London Ser. A*, 305, 235, 1968. With permission.)

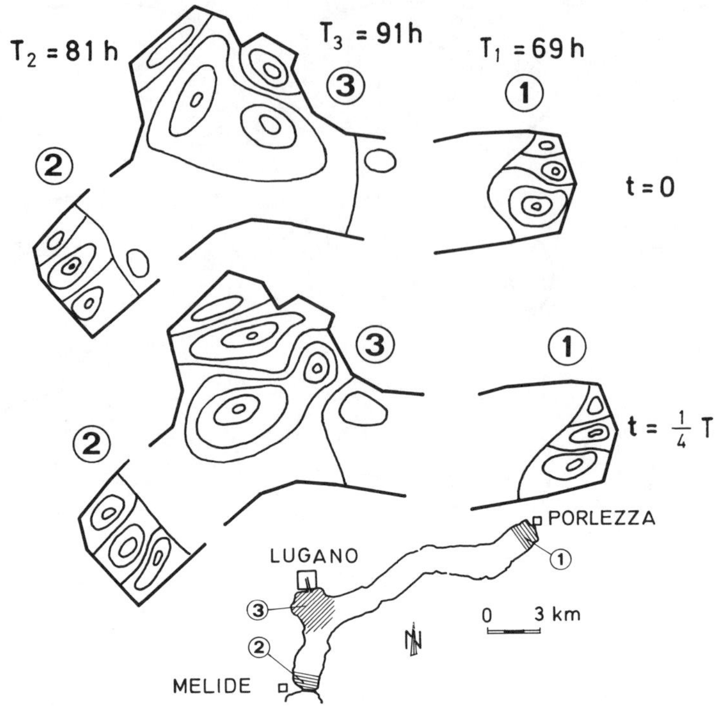


FIGURE 3. Three modes of long periodic waves in Lake of Lugano obtained by finite element technique. (From Stocker, T. and Hutter, K., *J. Fluid Mech.*, 185, 107, 1987. With permission.)

To our knowledge no analytical solutions of the boundary value problem (Equation 2) are known for domains and bathymetries H when non-separable equations emerge or equations cannot be transformed to special forms like in Ball's case of the ellipse with parabolic bottom. Neither have cases been studied where wave numbers would be complex. Such solutions are, however of significance as they would demonstrate existence of solutions of Equation 2 which show substantial wave activity in a relatively narrow region and die out as one moves away from these regions. The situation is not unlike that of Taylor reflections of Kelvin waves in a semi-infinite gulf.

Observations and, in particular, numerical computations for the Lake of Lugano indicate that such spatially evanescent solutions of the topographic wave equation do exist (see Figure 3). Trösch[13] found by using finite element techniques, that among other basin-wide modes, three independent bay modes, in the period range 69 to 91 h, existed with wave activity only in the bays shown in Figure 3. Existence of such modes can only be demonstrated if the *real branch* of the dispersion relation is extended to the *complex branch,* i.e., if complex wave numbers are also admitted.

In our semianalytical channel model[1,11,12] we have shown that dispersion relations *for each mode* are typically as shown in Figure 4. The real part of it (for $\sigma < \sigma_0$) is the typical shelf-wave-dispersion relation already shown in Figure 2. Above the cutoff frequency follows a complex domain, i.e., for $\sigma_0 < \sigma < \sigma_1$ to each $\sigma$ there exist four complex wave numbers, two of which are conjugate complex. Above $\sigma = \sigma_1$ the wave number is purely imaginary. A complete system consists of a countably infinite number of such *mode units*. If $\sigma$ is plotted against the modulus of k (for simplicity), then the dispersion relation for three modes is as shown in Figure 5. These special properties will be needed in the sequel.

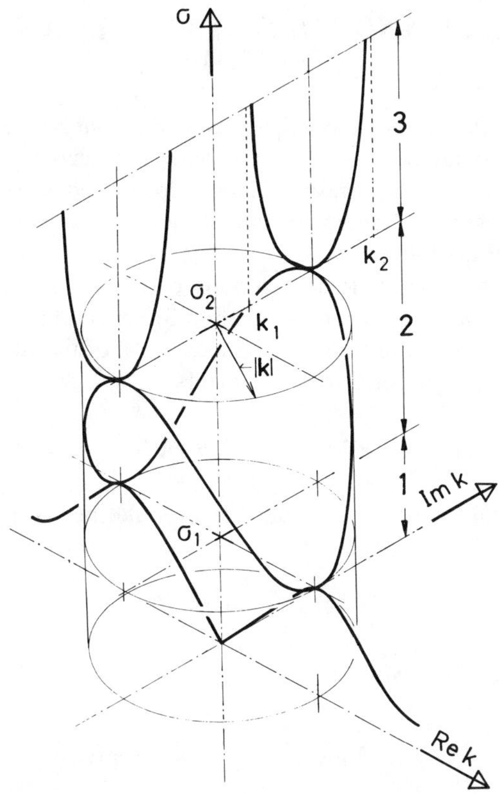


FIGURE 4. Schematic plot of the complex dispersion relation $\sigma(k)$ for an infinite channel. In regime 1, k is real; in regime 2, it is complex with the constant modulus $|k|$; and in regime 3, k is purely imaginary, taking asymptotic values $k_1$ and $k_2$ for large $\sigma$. (From Stocker, T. and Hutter, K., *J. Fluid Mech.*, 170, 435, 1986. With permission.)

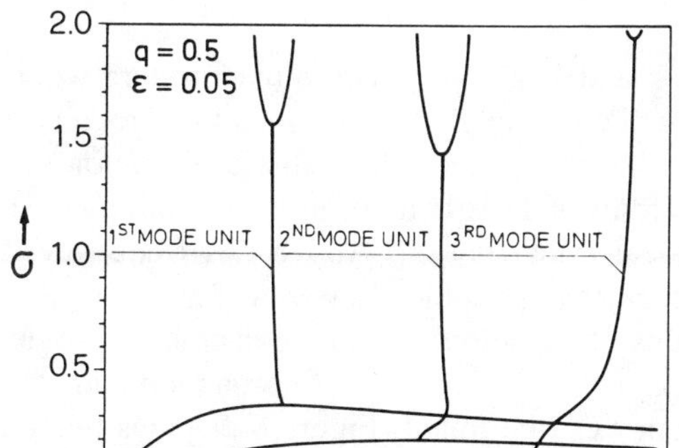


FIGURE 5. Modulus k of the third-order disperision relation for an infinite channel. (From Stocker, T. and Hutter, K., *J. Fluid Mech.*, 170, 435, 1986. With permission.)

# III. BOUND STATES AND FREE STATES IN A GEOPHYSICAL SYSTEM

In this section a conspicuous correspondence is worked out between the structure of the solution space of the topographic wave equation and an other well-known and important problem in quantum physics. This correspondence helps to understand the problem of topographic waves in a more general framework and provides further important questions which remain to be answered in the future.

Topographic waves arise as periodic solutions of the linear, homogeneous boundary value problem (Equation 2). The operator (Equation 3) acts on the complex mass-transport stream function $\psi$, which must vanish on the boundary $\partial\mathcal{D}$ of the two-dimensional domain. Introduction of solutions proportional to $e^{-i\omega t}$, yields

$$\overline{\Pi} = -i\sigma\nabla \cdot H^{-1}\nabla - \hat{\mathbf{z}} \cdot \nabla H^{-1} \times \nabla \tag{6}$$

where the nondimensional frequency $\sigma \equiv \omega/f$ has been defined. Equation 6 can be written as

$$\Pi = -\Delta + \mathbf{V} \cdot \nabla \tag{7}$$

in which the vector function $\mathbf{V}$ is given by

$$\mathbf{V} = \nabla \log H - \frac{1}{i\sigma}\hat{\mathbf{z}} \times \nabla \log H \tag{8}$$

and contains the eigenvalue $\sigma$. Equation 8 gives the decomposition of the vector field $\underset{\sim}{V}$ into an irrotational and a solenoidal part.

An open domain $\mathcal{D}$ where the depth function H is smooth and satisfies the conditions

$$\begin{aligned} \frac{\partial H}{\partial x} &= 0, \qquad x_0 < x < \infty \\ |\nabla H|^2 &> 0, \qquad 0 < x < x_0 \end{aligned} \tag{9}$$

describes a semi-infinite channel in a (x,y)-coordinate system whose topography varies in both directions within the shore zone $0 < x < x_0$ and has a constant thalweg depth for $x > x_0$. By studying a semi-infinite channel that satisfies Equation 9, it was demonstrated in Stocker,[14,15] that the spectrum of $\Pi$ is composed of a continuous and a discrete part. The two parts join at the critical frequency $\sigma_0$. Above this frequency, free topographic wave propagation for $x > x_0$ becomes impossible. However, for $\sigma > \sigma_0$ the spectrum is not empty; it is discrete, i.e., solutions of Equation 2 can be constructed provided the frequency takes the special values $\sigma = \sigma_i$ $(i = 1, 2, \ldots)$. Alternatively, in the interval $0 < \sigma < \sigma_0$ solutions with *any* frequency can be found. Figure 6 displays two typical modes belonging to the discrete and the continuous spectrum, respectively. Evidently, there is a distinct difference between these two representations.

Modes which belong to the discrete spectrum are *bound states* of the system. This is clear, because $\sigma > \sigma_0$ implies evanescent wave activity in the zone $x > x_0$ (see Figure 6). Although no wave energy is fed in from infinity, trapped wave energy can be found in the shore zone at *distinct* frequencies. Likewise, due to the existence of a cutoff frequency, energy cannot radiate away from the *open* shore zone. During one cycle vortices are formed close to the upper shore line. Their phase propagates in a right-bounded fashion along the

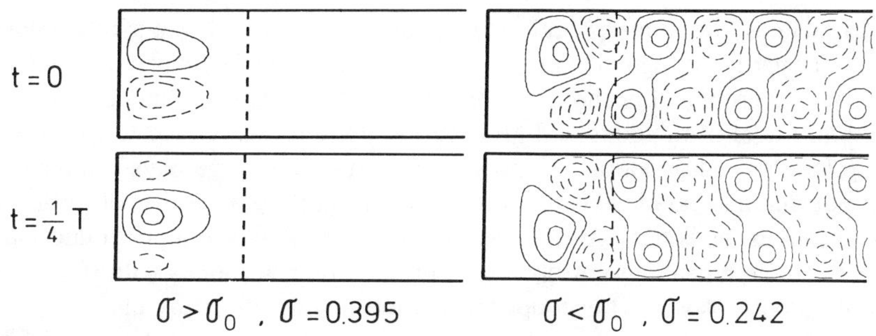


FIGURE 6. Solutions belonging to the discrete ($\sigma > \sigma_0$) and the continuous ($\sigma < \sigma_0$) spectrum. The discrete spectrum contains bound states: the stream function is spatially evanescent as one moves away from the bay zone, and the vortices remain trapped. Free states arise in the continuous spectrum: a wave incident from infinity is reflected in the bay. (From Stocker, T. and Hutter, K., *J. Fluid Mech.*, 185, 107, 1987. With permission.)

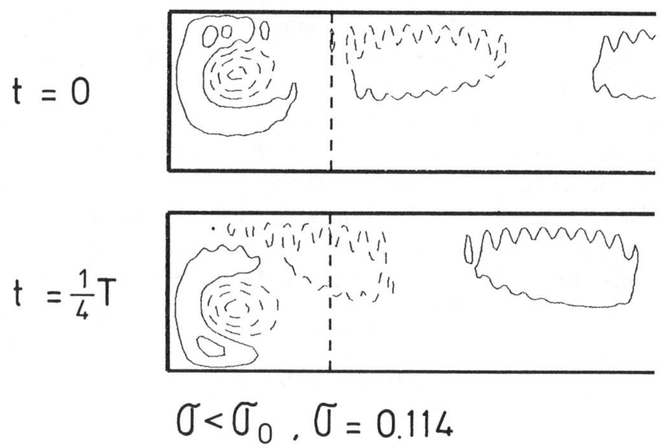


FIGURE 7. Resonance within the continuous spectrum. An incident wave excites a higher order bay mode.

channel endwall, where they evanesce at the opposite shore line. The continuous spectrum, on the other hand, shows *free states* of the system. Wave energy which is incident from infinity propagates along the upper boundary. The topography H which is acting like a repelling force in the shore zone $0 < x < x_0$ due to $\partial H/\partial x \neq 0$ induces a reflection of this energy flux and forces the vortices to follow the lines of constant f/H. It is this mechanism which produces the reflected wave of equal wave length which propagates towards infinity along the opposite channel shore. A characteristic quality of topographic waves is that to each frequency there always exists a short and a long wave (see Figure 4). Thus, wave energy could be distributed onto different modes. The study in Stocker[14,15] has shown, however, that for frequencies not close to $\sigma_0$ most of the reflected energy lies in the mode with the same wave number. Only close to $\sigma_0$ is the reflected energy contained in the mode with the second wave number. The continuous spectrum also contains *resonances*. These are bound states of higher transverse order. Since for the lowest order transverse mode periodic wave motion with $\sigma < \sigma_0$ is possible, trapped modes of higher order can be excited. A weak energy flux from infinity may then cause strong wave activity in the shore zone (Figure 7).

The above theoretical results still await verification in nature via carefully designed mooring campaigns, or reinterpretation of measurements of long periodic signals which exhibited so far partly ambiguous portions. Nevertheless, the findings are of interest to both numerical and experimental oceanographers. They tell us that, in a numerical model of the ocean, not only tidal oscillations and propagating vorticity waves are to be expected when considering long periodic motion. Trapped vortex modes in large bays can also occur. We also point out that a separation of the motion due to gravity from that of vortex stretching (topographic waves) as is done in Equation 1 can not be achieved a priori under all circumstances. For a large basin or semi-open domain bay trapped modes with $\sigma > \sigma_0$ may well interact with gravity waves. This coupling was left undetermined by us.

Let us return to the conspicuous structure of the solution space of Equation 7 for a semi-infinite channel. This is a geophysical system which may allow trapped, free, and resonant solutions. For vorticity waves this fact was first demonstrated by Stocker and Hutter.[1] Other geophysical configurations have shown similar compound spectra: e.g., irrotational flow in a channel containing a cylindrical obstacle.[16] It must be emphasized, however, that our results are based on a semianalytic method. In Stocker and Johnson[17] the spectrum of the topographic wave operator is investigated for a simpler and more basic geometry of a semi-infinite channel. It is demonstrated that the number of both bay modes and resonances is governed by the bay topography. Further, estimates of the frequencies and the number of bay modes are deduced.

The topographic wave problem and its special solution space in the geophysical system of a semi-infinite channel enjoys a remarkable similarity with a well-known and thoroughly studied problem in modern physics. This correspondence is discussed here in order to provide a guide line for future theoretical and basic investigation concerned with the boundary value problem (Equation 2). The related problem is the stationary motion of an electron in the force field of a still proton. It can be understood by solving the Schrödinger Equation in space representation

$$\mathcal{H}\phi = E\phi,$$
$$\mathcal{H} = \Delta + V \tag{10}$$

where $\mathcal{H}$ is the scaled Hamiltonian operator, V is a potential describing the attractive Coulomb force field of the still proton, E is the energy of the electron and $\phi$ is a complex probability function for the position of the electron. Ignoring the spherical geometry of the problem, one can study solutions in one single dimension x with a simple potential well of the form

$$V(x) = \begin{cases} 0, & |x| > x_0, \\ -V, & |x| < x_0, \end{cases}$$
$$V > 0 \tag{11}$$

The structure of the solution, i.e., the probability of the particle position crucially depends on the eigenvalue E, the particle energy. Considering the potential (Equation 11), two cases must be distinguished.

For a particle with the energy E lower than the "well-energy" only bound states are allowed. This means that the electron is trapped within the potential well. Moreover, bound states are only possible for distinct energies $E = E_i$, $(i = 1, 2, \ldots)$. Consequently, for $E < 0$ there is a discrete spectrum containing solutions that are localized in the well. It is significant that both the values $E_i$ and the number of possible trapped or bound states depend on the structure of the potential, i.e., on V and $x_0$.

Particles with arbitrary energy $E > 0$ can propagate outside the potential well. The classical theory predicts the complete transmission of a particle approaching the potential

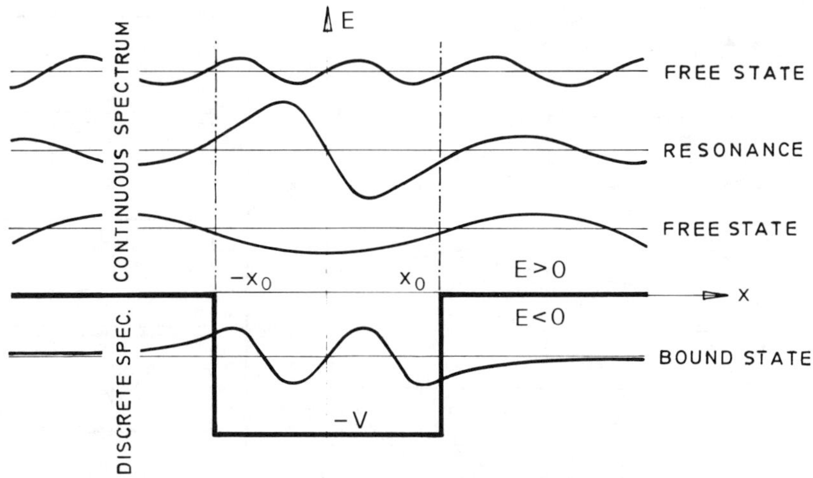


FIGURE 8. Model for an electron in the attractive force field (well potential) of a proton. For discrete E < 0 bound states prevail and the particles form a stable hydrogen atom. For E > 0 the electron is scattered by the proton; these are free states and occur for any E > 0. Resonances arise in the continuous spectrum.

well. Quantum theory, on the other hand, demonstrates that the particle is partly reflected and partly transmitted. At distinct energies the probability density within the potential well assumes a maximum. This is a resonant state where the residence time of the particle in the well is enlarged. Figure 8 summarizes these three qualitatively different solutions.

Comparing it with Figure 9, it is evident that the two solution spaces enjoy very similar properties. Both problems equally lead to three physically different states of the particle or of the topographic wave pattern. For large wave frequencies or negative particle energies only bound states are possible. Both particle and baymodes are not able to escape the well or the bay. In the opposite case when the frequencies do not exceed the critical cutoff, or the particle energies are positive, free states occur. Just as the particle is able to override the potential well and is not caught by it, the topographic wave impinging from infinity is reflected at the shore zone. Moreover, this reflection/transmission process depends on the frequency or energy of the wave or particle. For certain values the system exhibits a resonant state. This consists for the particle in an enlarged probability density to find it in the well. It amounts to a longer residence time of the free particle within the well. Equally, the water body in the open shore zone can be brought into resonance when a topographic wave with a certain frequency propagates towards the bay. Table 1 illustrates further formal correspondences of these two physically different problems. The remarkable correspondence of these two problems arises due to their similar mathematical description. This table therefore strongly suggests that the methods of mathematical physics which have been successfully applied to particle problems in quantum mechanics, could readily be employed for the topographic wave problem. There exists a vast literature on general principles of equations of the form Equation 10. These principles allow important estimates on the structure of the spectrum (e.g., is there a discrete part, etc.), on the behavior of solutions, on bounds for the eigenvalues, and many more. Further, criteria can be formulated which relate the structure of the spectrum to the coupling functions V or **V**. We encourage the mathematical oceanographer to tackle this challenging problem; the number of corresponding facts to a well-known problem may be promising.

We add some further speculations: the tunnel effect is an important mechanism in quantum physics. It predicts that a particle can "tunnel" through a potential barrier which his classical "brother" could never vanquish; an effect which was to experience a vast

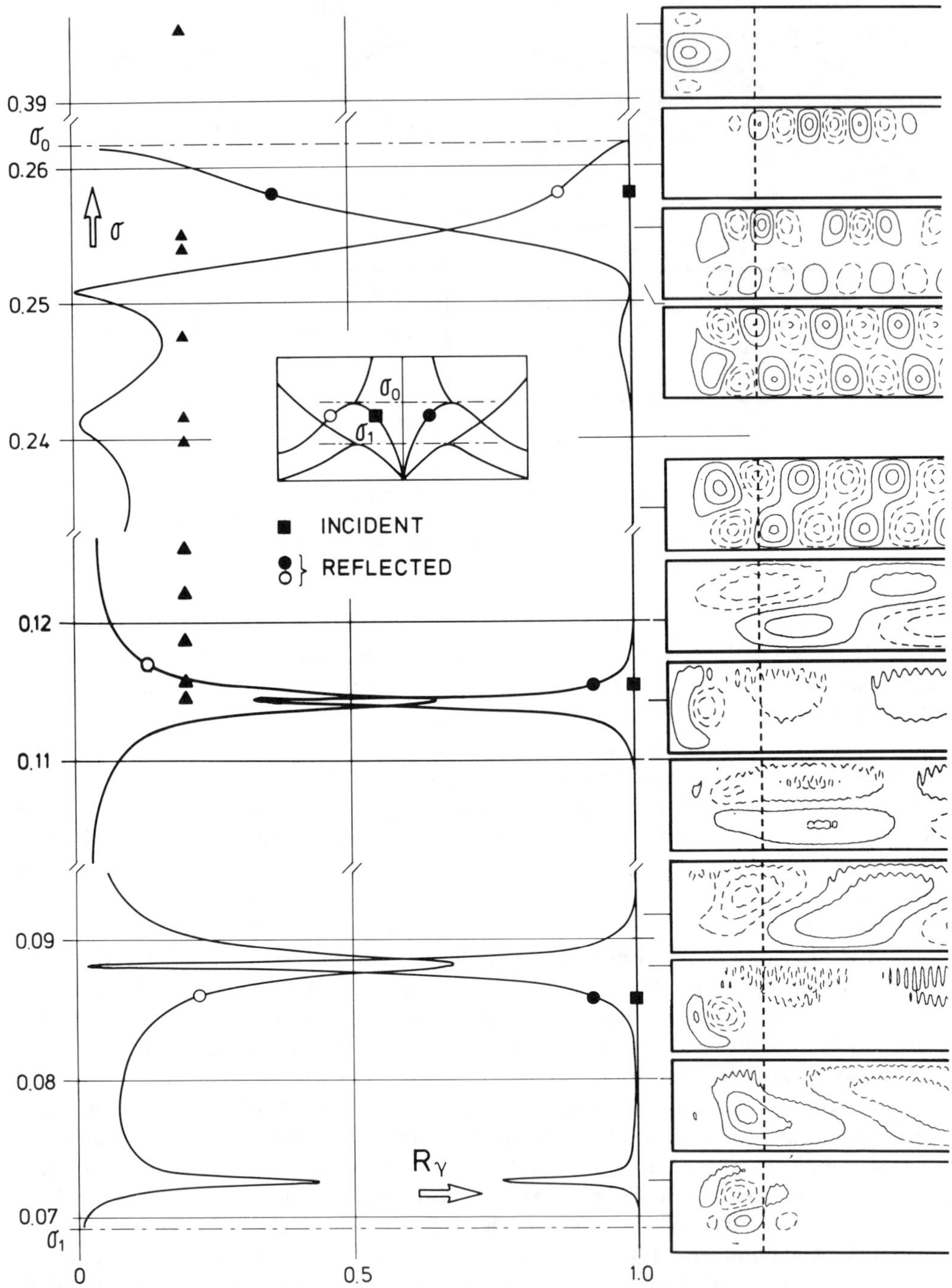


FIGURE 9. Spectrum of topographic waves in a semi-infinite channel consisting of a discrete ($\sigma > \sigma_0$) and a continuous part ($\sigma < \sigma_0$). Bound states exist in the discrete spectrum, free and resonant states occur in the continuous spectrum.[14,15]

technical and industrial application. Topographic wave energy likewise seems to exhibit the ability of tunneling[20]. For instance, imagine a continental shelf and an estuarine domain which is connected to the former. It remains to be verified, both theoretically and practically, that shelf wave energy could tunnel through an estuarine domain and excite at its end a resonant bay mode. Consideration of bathymetries which communicate with the open ocean

**TABLE 1**
**Collection of Correspondences between a Well-Known Problem in Quantum Mechanics and Topographic Waves in a Semi-Infinite Channel with Equations 12 and 13**

| | **Example in quantum mechanics** | **Topographic waves** |
|---|---|---|
| Physical process described | Electron in an attractive Coulomb force field produced by a still proton | Topographic waves in an open domain of varying water depth |
| Equation considered | Schrödinger equation | Potential vorticity equation |
| Quantity | Probability density of particle location, scalar $\phi$ | Mass-transport stream function, scalar $\psi$ |
| Differential operator | $\mathcal{H} \equiv -\Delta + (V - E)$ | $\Pi \equiv -\Delta + \mathbf{V}(\sigma)\cdot\nabla$ |
| Differential equation | $\mathcal{H}\,\phi = 0$ | $\Pi\psi = 0$ |
| Properties of equation | Partial, linear, second order, homogeneous | |
| Coupling to physical configuration | V, scalar potential of the Coulomb conservative force field | **V**, vector function of the gradient of the logarithmic depth field; **V** is not irrotational |
| Eigenvalue | E, particle energy | $\sigma$, wave frequency |
| Domain of integration | $\mathbb{R}^1$: $-\infty < r < \infty$ or $\mathbb{R}^3$: $0 < r < \infty$, $0 \le \varphi \le 2\pi$; $-\pi/2 < \theta < \pi/2$ | $\mathbb{R}^2$: $0 < x < \infty$; $-B/2 \le y \le B/2$ Semi-infinite channel |
| Boundary conditions | $\phi$ finite or 0 for $\lvert r \rvert \to \infty$ | $\psi$ finite or 0 for $x \to \infty$ $\psi = 0$ for $x = 0$, $\lvert y \rvert = B/2$ |
| Structure of spectrum | Spectrum consists of discrete and continuous part | |
| Discrete spectrum | For $E < 0$, countable set, finite or infinite number of eigenvalues dependent on V; $\phi \to 0$ for $\lvert r \rvert \to \infty$; Bound states, electron, and proton form a stable hydrogen atom | For $\sigma > \sigma_0$, countable set? Number of eigenvalues? Dependence on **V**?; $\psi \to 0$ for $x \to \infty$; Bound states, bay modes are trapped in the shore zone |
| Continuous spectrum | For $E > 0$, free states, electron is scattered by the proton | For $0 < \sigma < \sigma_0$ Free states, topographic wave is reflected in the shore zone |
| Resonances | For $E_i < 0$, $i = 1,2,\ldots,$ Particle resides for a longer time close to the proton | For $0 < \sigma_i < \sigma_0$; $i = 1,2,\ldots$ Higher order bay mode is excited by incident wave |

therefore must be viewed as being a *part* of a more complex water system. Interaction with oceanic vorticity wave motion should not be ignored a priori.

# IV. GEOMETRY AND TOPOGRAPHY EFFECTS

In this section the influence of variation in topography and geometry of the basin is studied. It essentially is a summary of earlier work by Stocker and Hutter[1,11,12] and Stocker.[14,15]

## A. VARIATION OF THE TOPOGRAPHY

Consider a semi-infinite channel with the transverse profile

$$h(x,y) = h_0(x)\left(1 + \epsilon - \left|\frac{2y}{B}\right|^q\right) \tag{12}$$

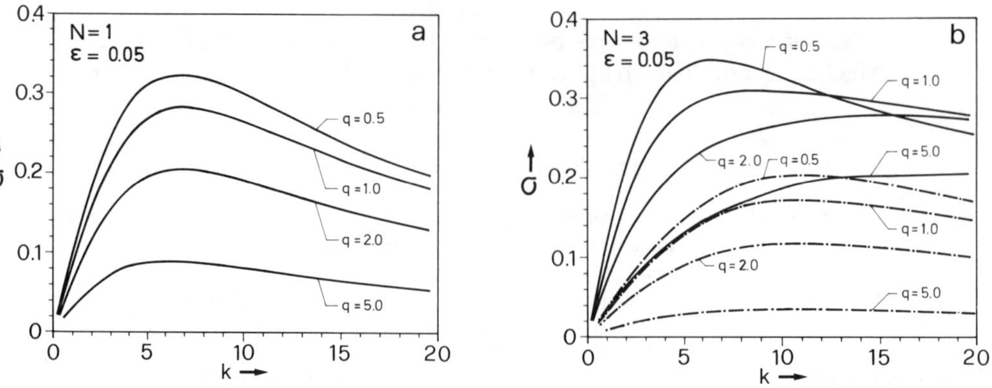


FIGURE 10. Effect of topography on the dispersion relation of the first mode (solid) and the second mode (dashed-dotted) in a channel. (a) N = 1, (b) N = 3. (From Stocker, T. and Hutter, K., *J. Fluid Mech.*, 170, 435, 1986. With permission.)

where $h_0$ (x) is the thalweg depth, $\epsilon$ a small sidewall parameter, B the channel width; $q < 1$ describes locally convex, $q > 1$ locally concave profiles. For $x > 0$ we define

$$h_0(x) = \begin{cases} \eta + \sin^p(\pi x/2x_0), & 0 < x < x_0, \quad p \geq 2, \\ \eta + 1 - \sin^{-p}(\pi(x_0 - x)/2x_0), & 0 < x < x_0, \quad p \leq -2, \\ \eta + 1, & x > x_0 \end{cases} \tag{13}$$

$\eta$ is again a small shore depth parameter, p measures the "steepness" of the along-axis depth and $x_0$ determines the extent of the bay zone.

Numerical solutions of the topographic wave equation with the specifications of Equations 12 and 13 were constructed. Their frequencies depend quantitatively on the parameters introduced above. The value of the cutoff frequency for free wave propagation in $x > x_0$ depends only on the parameters q (and $\epsilon$) which govern the global topography variations. They determine the exact form of the dispersion relation of which a parameter study of the real branches is shown in Figure 10 (a and b). Hence, q (and $\epsilon$) determine the frequency range where free wave propagation in the far field is possible (namely for $\sigma < \sigma_0$). The q modifies also the transverse depth profile in the shore zone and therefore influences the frequencies of the bay modes. Figure 11 shows that this dependence is comparatively weak as values of $\sigma$ only vary less than 10%.

The longitudinal depth variations can be modeled by the parameter p in Equation 13. It is clear that mostly modes confined to the shore zone, where the effect of p is experienced, will be affected by variation of p. Figure 12 gives the frequencies of the bay modes as functions of p. Whereas for $p < -2$ changes are moderate, they are significant for $p > 2$.

## B. VARIATION OF THE GEOMETRY

The semi analytic method introduced by Stocker and Hutter,[1] permits a straightforward extension of the solution technique used for straight geometries to curved channels with constant curvature. This section closely follows Stocker.[14] Consider, therefore, basins whose plan view are sectors of annuli. The thalweg line is defined by the middle radius ($1/\kappa$). The assumption of constant $\kappa$ and width B leads to a domain which has the shape of a ring. Consequently, the continuous dispersion relation would have to be subject to a periodicity condition. This quantization will not be imposed in the figures, simply because the full curve will provide a clearer understanding of the qualitative effects of curvature.

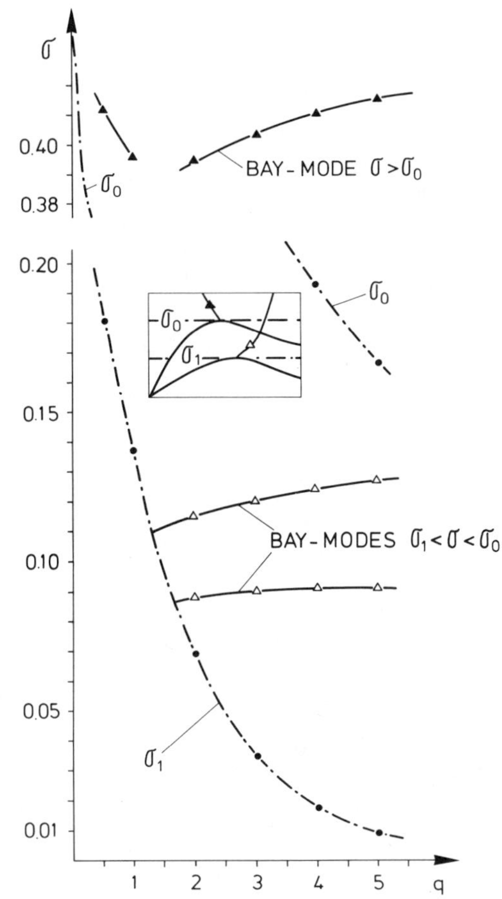


FIGURE 11. Frequencies of the resonances or bay modes in $\sigma_1 < \sigma < \sigma_0$ and a bay mode with $\sigma > \sigma_0$ as functions of the transverse topography parameter q; $N = 2$, $\epsilon = 0.05$, $p = 2$, $\eta = 0.01$. The cutoff frequencies of the two mode units are indicated.[15]

Figure 13 shows the dispersion relation $\sigma$ (k) of a third order model for the first two mode units. For $\kappa = 0$ the symmetry with respect to the vertical is visible, whereas for $\kappa > 0$ it is broken. For a given frequency all wave numbers are shifted to the right, which implies that curvature shortens waves for $k > 0$, whereas they become longer for $k < 0$. Recalling that the solutions are freely propagating waves (i.e., proportional to $e^{i(kx - \sigma t)}$), which are right bounded, it follows that the waves traveling along the inner (outer) shore line are longer (shorter) than in the case $\kappa = 0$. Further, the critical point $(k_0^+, \sigma_0^+)$ in the domain $k > 0$ is translated to lower frequencies and larger wave numbers, and the opposite is true for $(k_0^-, \sigma_0^-)$ in the domain $k < 0$. Consequently, there exists a frequency range $\sigma_0^+ < \sigma < \sigma_0^-$, where only waves with $k < 0$ can propagate. These are trapped along the inner shore line. If there existed eigenfrequencies in this range for a closed basin, their modal structure would exhibit a particular pattern with wave motion primarily at the inner shore line. This is discussed later. Table 2 lists the boundaries of these frequency ranges. For increasing curvature the values of $\sigma_0^-$ and $\sigma_0^+$ lie farther and farther apart; this effect is weak for steep topographies. Note that from an observational point of view the difference is very small, e.g., $T_0^- \approx 57$ h and $T_0^+ = 63$ h for $q = 2$ and $\kappa = 0.2$, a difference that is unlikely to be detectable by field observations.

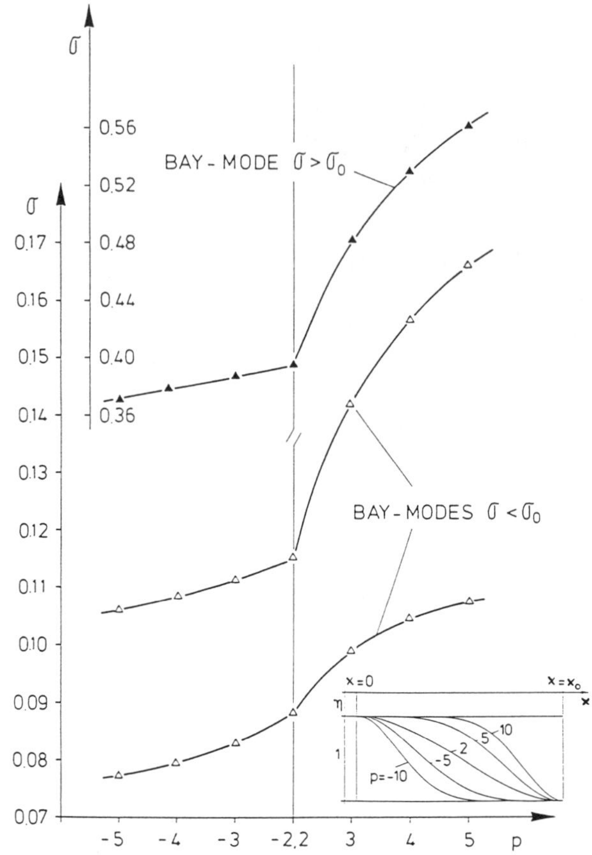


FIGURE 12. Frequencies of the resonances or bay modes in $\sigma_1 < \sigma < \sigma_0$ and a bay mode with $\sigma > \sigma_0$ as functions of the longitudinal topography parameter p; N = 2, q = 2, $\epsilon$ = 0.05, $\eta$ = 0.01.[15]

On the other hand, for increasing κ and q the difference of the wave numbers increases. These properties are also displayed in Figure 14. Two mode units of the dispersion relation are given for different values of the topography and curvature parameters. The results support the findings listed in Tables 2 and 3.

So far we abstained from giving a physical explanation of the curvature effect and only listed and discussed the various alterations that arose in the dispersion relation because of the presence of curvature. In order to understand the physical mechanism which produces the asymmetry of σ(k) we now work out the characteristic difference between the equations formulated in the natural coordinate system and in the Cartesian system. Topographic waves are described by the conservation of potential vorticity. This quantity contains three contributions: the curl of the velocity field representing the relative vorticity, the Earth's rotation as an additional vorticity, and the bottom topography. It is the relative vorticity which is primarily influenced by curvature. We give a rough analysis.

The potential vorticity Equation (2) in cylindrical coordinates (r, φ) describes topographic waves in a domain of constant curvature. Using Equation 5 and $J_1 = 1$, $J_2 = r$ for cylindrical coordinates implies

$$\frac{\partial}{\partial t}\left[\frac{1}{r}\frac{\partial}{\partial r}\left(\frac{r}{H}\frac{\partial\psi}{\partial r}\right) + \frac{1}{r}\frac{\partial}{\partial\varphi}\left(\frac{1}{rH}\frac{\partial\psi}{\partial\varphi}\right)\right] + \frac{\partial\psi}{\partial r}\cdot\frac{1}{r}\frac{\partial}{\partial\varphi}\left(\frac{1}{H}\right) - \frac{1}{r}\frac{\partial\psi}{\partial\varphi}\frac{\partial}{\partial r}\left(\frac{1}{H}\right) = 0 \quad (14)$$

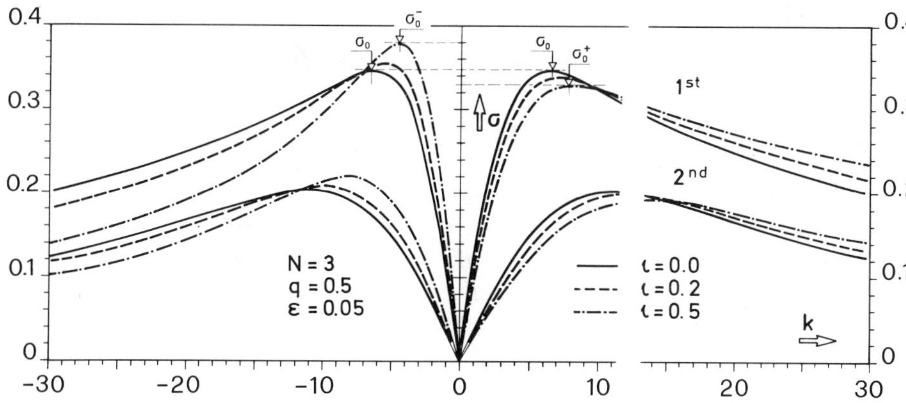


FIGURE 13. Dispersion relation of a third order model for various values of curvature. The parameters are N = 3, q = 0.5, $\epsilon$ = 0.05.[14]

**TABLE 2**
**Frequencies $\sigma_0^{\pm}$ for k $\gtrless$ 0 where $\partial\sigma/\partial k$ = 0 for Different Values of Curvature $\kappa$ and Topography q**

| q | | $\kappa$ = 0 | $\kappa$ = 2.0 | $\kappa$ = 0.5 |
|---|---|---|---|---|
| | $\sigma_0^-$ | 0.2745 | 0.2799 | 0.2947 |
| 2.0 | $\sigma_0^+$ | 0.2745 | 0.2708 | 0.2669 |
| | $\Delta\sigma$ | 0 | 0.0091 | 0.0278 |
| | $\sigma_0^-$ | 0.2081 | 0.2111 | 0.2188 |
| 5.0 | $\sigma_0^+$ | 0.2081 | 0.2059 | 0.2033 |
| | $\Delta\sigma$ | 0 | 0.0052 | 0.0155 |

*Note:* The parameters are N = 3, $\epsilon$ = 0.05, first mode unit.

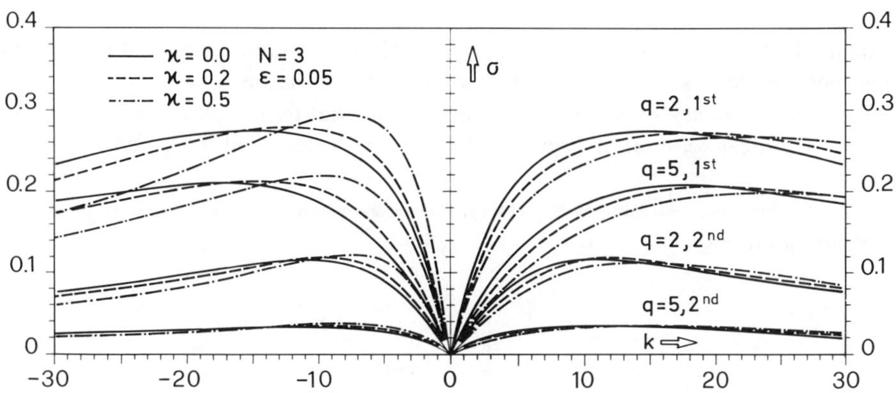


FIGURE 14. Curvature and topography effect on the dispersion relation.[14]

**TABLE 3**
**Wave Numbers k± Corresponding to Table 2**

| q | | κ = 0 | κ = 0.2 | κ = 0.5 |
|---|---|---|---|---|
| | $k_0^-$ | −14.9 | −12.1 | −7.83 |
| 2.0 | $k_0^+$ | 14.9 | 17.8 | 21.9 |
| | $k_0^+ - \lvert k_0^- \rvert$ | 0 | 5.7 | 14.1 |
| | $k_0^-$ | −17.7 | −14.6 | −9.83 |
| 5.0 | $k_0^+$ | 17.7 | 20.7 | 25.1 |
| | $k_0^+ - \lvert k_0^- \rvert$ | 0 | 6.1 | 15.3 |

Consider periodic solutions proportional to $e^{-i\sigma t}$ and domains where the depth profiles are azimuthally constant. Equation 14 then reduces to

$$\left[\psi_{rr} + \frac{1}{r^2}\psi_{\varphi\varphi} - \left(\frac{H_r}{H}\right)\psi_r + \frac{1}{r}\psi_r\right] + \frac{i}{\sigma}\left(\frac{H_r}{H}\right)\frac{1}{r}\psi_\varphi = 0 \tag{15}$$

In order to investigate the main effects of curvature on the wave properties it suffices to assume that the channel is only weakly curved. Such a domain has a radius of curvature $1/\kappa$, which is large compared with the channel width B. The arc length s is approximately given by

$$s = \frac{1}{\kappa}\varphi \qquad \frac{\partial}{\partial\varphi} = \frac{1}{\kappa}\frac{\partial}{\partial s} \tag{16}$$

whereby the varying position r was replaced by the constant radius of curvature $1/\kappa$. The same approximation is introduced in Equation 15 yielding

$$\left[\psi_{rr} + \psi_{ss} - \left(\frac{H_r}{H}\right)\psi_r + \kappa\psi_r\right] + \frac{i}{\sigma}\left(\frac{H_r}{H}\right)\psi_s = 0 \tag{17}$$

Equation 17 is the potential vorticity equation in Cartesian coordinates with the extra term $\kappa\psi_r$ incorporating to first order the effects of constant curvature.

The simple geometry of a weakly curved exponential trench with $1/\kappa \leq r \leq 1/\kappa + w$ allows the analytical solution of Equation 17. Let the water depth be given by

$$H(r) = e^{-2br}$$

and look for solutions

$$\psi = e^{iks} \cdot e^{-\left(b+\frac{\kappa}{2}\right)r} \cdot \sin\frac{n\pi\left(r - \frac{1}{\kappa}\right)}{B} \tag{18}$$

Inserting Equation 18 into 17 requires

$$\sigma = \frac{2bk}{k^2 + \left(b + \frac{\kappa}{2}\right)^2 + \left(\frac{n\kappa}{B}\right)^2} \tag{19}$$

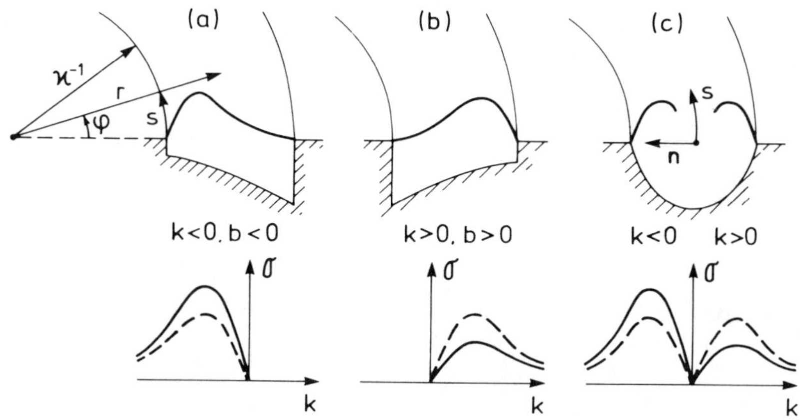


FIGURE 15. Two trench profiles (a) and (b) can model the effect of curvature in the natural coordinate system of a curved channel with parabolic bottom profile (c) by a "first order perturbation" of the TW-equation. The effects on the dispersion relations are indicated (dashed for $\kappa = 0$, solid for $\kappa \neq 0$) below.[14]

This is the dispersion relation of topographic waves in a bend. Figure 15 illustrates how the two curved trenches with $b < 0$, (Figure 15a) and with $b > 0$ (Figure 15b) can model the curved parabolic channel displayed in Figure 15c. On the northern hemisphere the phase of these waves propagates with the shallow water to its right. Hence, for $b < 0$ we must select $k < 0$ since $\sigma > 0$, and the wave travels along the inner curved shoreline at $r = 1/\kappa$. In this situation, $\kappa > 0$ generally increases frequencies, as can be inferred from Equation 19. A wave is trapped along the opposite outer shoreline if $b > 0$, i.e., the water is shallow towards $r = 1/\kappa + B$ and $k > 0$. Curvature $\kappa > 0$ then decreases frequencies.

The dispersion curves given below the two trenches explain the effect of curvature and are qualitatively in agreement with the behavior observed in Figures 13 and 14. The cutoff frequencies of waves traveling around convex bends are increased, whereas those of waves around concave bends are decreased. These fundamental alterations due to the introduction of curvature lead to an understanding of the modal structure of the eigenmodes in closed curved basins.

Figure 16 shows corresponding eigenmodes for different values of the curvature ordered according to frequency. Generally, curvature does not alter the eigenfrequency very much. Deviations of the eigenfrequencies for a strongly curved basin ($\kappa = 0.5$) from the values in the straight lake are throughout less than 5%. Eigenfrequencies decrease with increasing curvature. There is little hope to detect experimentally any effect of curvature on the eigenfrequency. The stream-function patterns, on the other hand, show more pronounced modifications.

The Ball-modes (basin wide, large scale) clearly demonstrate the influence of the shift of the wave numbers along either shore line. Although the total number of vortices remains constant when increasing $\kappa$, the number of gyres along the inner shore decreases in favor of that along the outer. Along with this, the inner vortices become larger.

As could be expected in advance, the stream function of the bay modes ($\sigma = 0.395$ and $\sigma = 0.115$) is hardly altered in the curved basin. This is rather obvious, because the bay does not "see" much of the curved basin. Mainly modes which consist of wave motion over the whole curved domain will be influenced by this change of geometry.

The channel modes (basin wide, small scale) demonstrate remarkable changes. By increasing the curvature, wave motion is significantly attenuated in the region towards the center of curvature. For $\kappa = 0.5$ (an extreme case) the eigenmode only consists of a trail of waves trapped to the outer shore line.

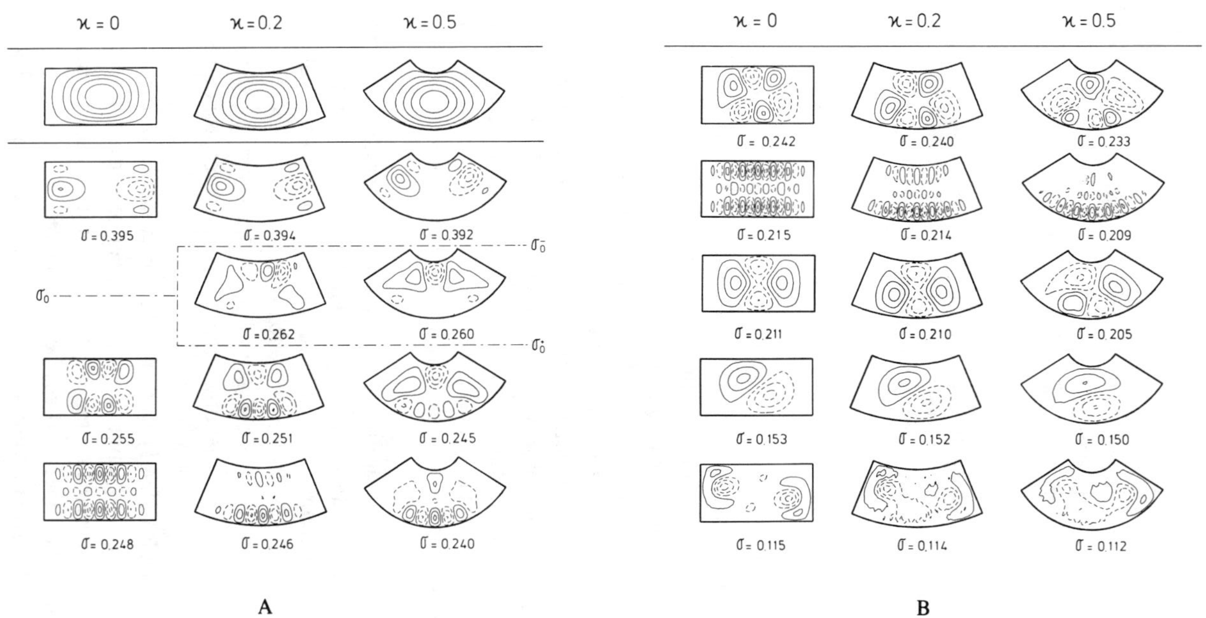


FIGURE 16. Comparison of eigenmodes in straight and curved basins; their lines of constant depth are shown in the top row. The modes are ordered for decreasing frequency. The cutoff frequency $\sigma_0$ splits in the case of curvature giving an interval $[\sigma_0^+, \sigma_0^-]$ in which the modes show unusual patterns.[14]

In the critical interval $I \equiv [\sigma_0^+, \sigma_0^-]$ there are indeed eigenfrequencies which exhibit the conjectured structure. Few large-scale vortices are trapped to the inner boundary of the basin. Solutions for $\kappa > 0$ with $\sigma \in I$ are structurally new. In the course of one cycle the vortices do not propagate around the basin but rather remain trapped in a domain close to the inner boundary. This mode, however is not a true bay mode since wave motion contains a contribution from the real wave number $k < 0$.

The above results suggest that with the exception of frequencies close to the cutoff, curvature effects can safely be ignored, since frequency shifts are small and most likely below the resolution of available instruments. On the other hand, Figure 16 indicates that detection of subtleties in the model structure would require a very high number of moored instruments in the transverse direction of the basin.

## V. CURRENT FIELD

In this chapter the current structure in closed rectangular basins will be analyzed by constructing current ellipses and co-range and co-tidal lines and by evaluating drift currents for the three mode types. These quantities facilitate interpretation of the mode structures that are associated with detected frequencies via field observations.

### A. CURRENT ELLIPSES AND TIDAL LINES

Consider the evolution of the transport vector $\hat{\mathbf{z}} \times \nabla\psi$ and of the velocity field $(\hat{\mathbf{z}} \times \nabla\psi)/H$ at fixed positions within the basin. An outstanding property of topographic waves is the fact that the field vector rotates either cyclonically or anticyclonically at a fixed position. Established statistical methods exist that permit calculation of the orientations of the rotation and of associated eigenfrequencies from periodograms of time series of velocity or transport field components. These rotary spectra are described in Gonella.[18]

In the present study all fields have harmonic time dependence, and, hence, the tip of a field vector will describe an ellipse. This ellipse is called transport ellipse (for the transport vector $\hat{\mathbf{z}} \times \nabla\psi$) or current ellipse (for the velocity vector $[\hat{\mathbf{z}} \times \nabla\psi)]/H$).

The qualitative structure of the stream function does not strongly depend on the values of the geometric and bathymetric parameters. It therefore suffices to examine one particular case. Figures 17 to 19 display eigenmodes of three examples of each type and their transport patterns. Some 63 positions within the rectangular basin are evaluated and the transport ellipse is plotted as a dashed (solid) curve if the transport vector rotates in the (counter) clockwise direction.

The first Ball-modes are characterized by a central area with counterclockwise rotation, (Figure 17). Its size depends on the mode, and we note that only the linear Ball-mode has a nonvanishing current vector in the center. This central area is surrounded by a region of clockwise and weak rotation. However, only Ball-modes with no "radial" node are given here; others emerge at much smaller frequencies, i.e., below the cutoff frequencies of the next mode units and would consist of more interlocking areas with different senses of rotation.

Figure 18 shows a selection of three bay modes. At the lake ends, wave activity is observed in the form of clockwise rotating currents. For higher bay modes these transport ellipses degenerate to nearly linear motion. It follows a zone of strong counterclockwise rotation, roughly at the position where the slope of the thalweg has a maximum. Beyond it, closer to the center, wave motion rapidly decreases and dies out.

The stream functions and transport ellipses of three channel modes are plotted in Figure 19. It is typical and could already be inferred from the stream-function plots, that there are bands along the long side of the basin. Close to the shoreline clockwise rotation is observed, being strongest in the middle of the elongated lake but weak at the long ends. Parallel to it follows a band where the current vectors rotate counterclockwise. Along the whole basin in the neighborhood of the thalweg line nearly no wave activity is experienced.

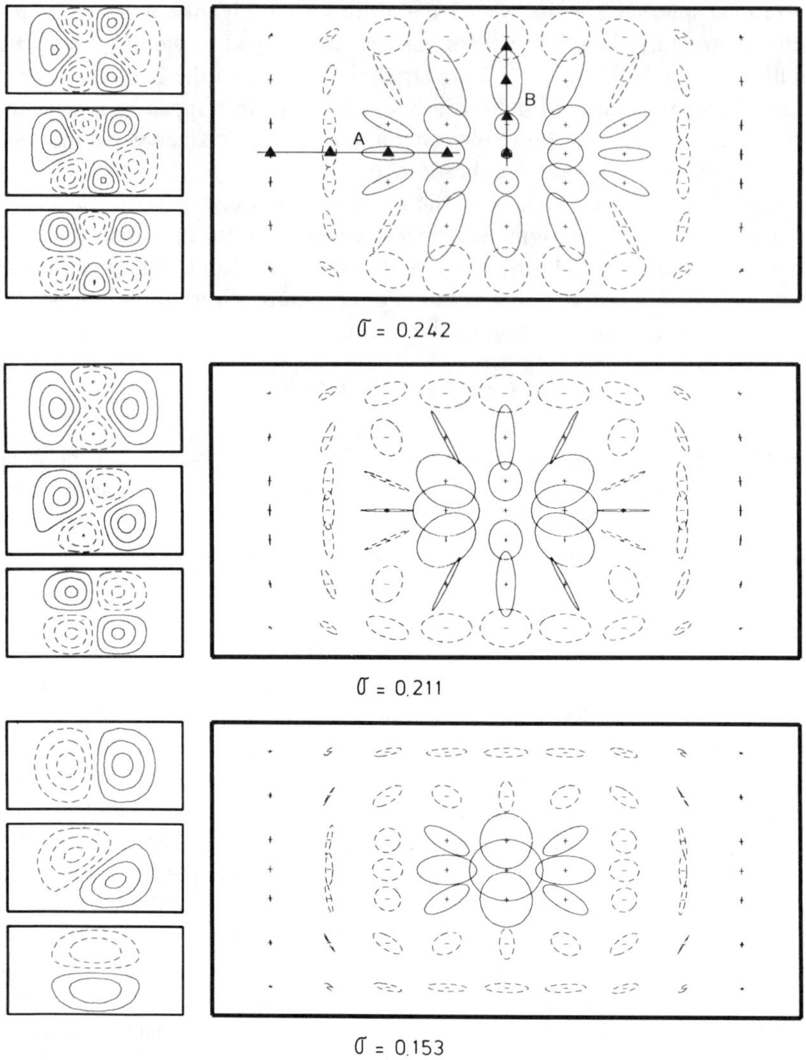


FIGURE 17. Stream function and transport ellipses for three Ball-modes. The small frames show contours of the stream function for t = 0 (bottom), t = T/8, t = T/4 (top). Transport ellipses are dashed (solid) for (counter) clockwise rotating transport vectors. ▲ indicate mooring sites on the linear arrays A and B, respectively.

The spatial and temporal interrelation between the fixed positions where individual mooring measurements can be taken must also be discussed. This can be readily achieved by calculating lines of constant phase and lines of constant amplitude of the transport stream function and the velocity vector fields.

Figures 20 to 22 display co-tidal and co-range lines of all important fields, i.e., the stream-function (scalar) and the transport field (in components). The linear Ball-mode (Figure 20) has the simplest structure. Co-range lines of the stream-function field are mainly circular, giving rise to a positive amphidromic point in the center of the lake. The co-tidal lines join at this point. The transport field (related to $\nabla\psi$) has now two amphidromies located at conjugate positions for both components; each exhibits a positive sense. Note that the transport field is weak towards the lake ends.

Figure 21 gives the quantities for the next higher Ball-mode (quadratic). The co-tidal

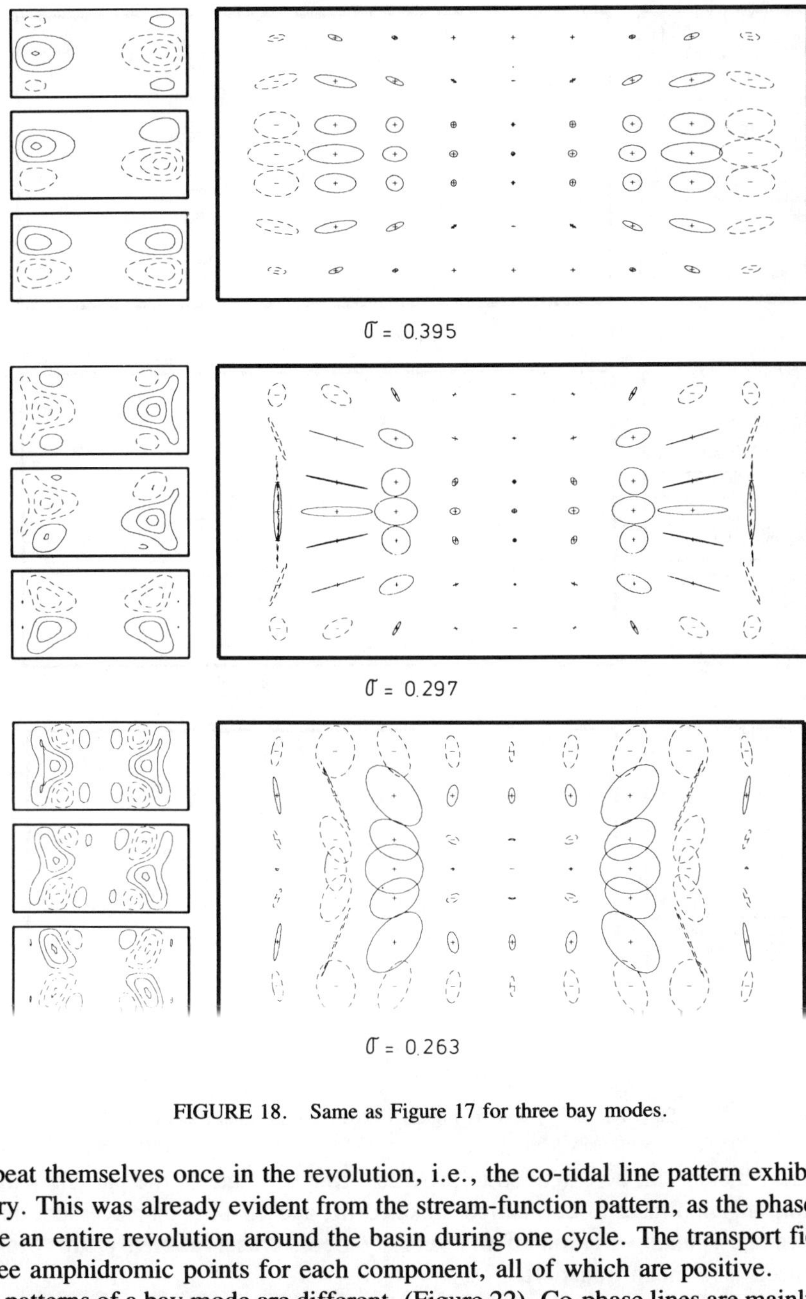


FIGURE 18. Same as Figure 17 for three bay modes.

lines repeat themselves once in the revolution, i.e., the co-tidal line pattern exhibits central symmetry. This was already evident from the stream-function pattern, as the phase does not complete an entire revolution around the basin during one cycle. The transport field shows now three amphidromic points for each component, all of which are positive.

The patterns of a bay mode are different, (Figure 22). Co-phase lines are mainly straight, joining at one or two positive amphidromic points at the center of the basin. The v-component of the transport has a nodal line across the basin near the lake end. In view of the smaller spatial scale of the vortices the channel modes (not shown here) have much more complicated patterns.

## B. PARTICLE PATHS

A consistent linear wave theory predicts closed particle paths. Within this linear approximation, the path line represents the scaled trace of the tip of the Eulerian velocity vector, the current ellipse. Figures 17 to 19 give an impression of the shape and orientation of these particle paths. Note, however, that they apply for transport ellipses and therefore depth-integrated path lines.

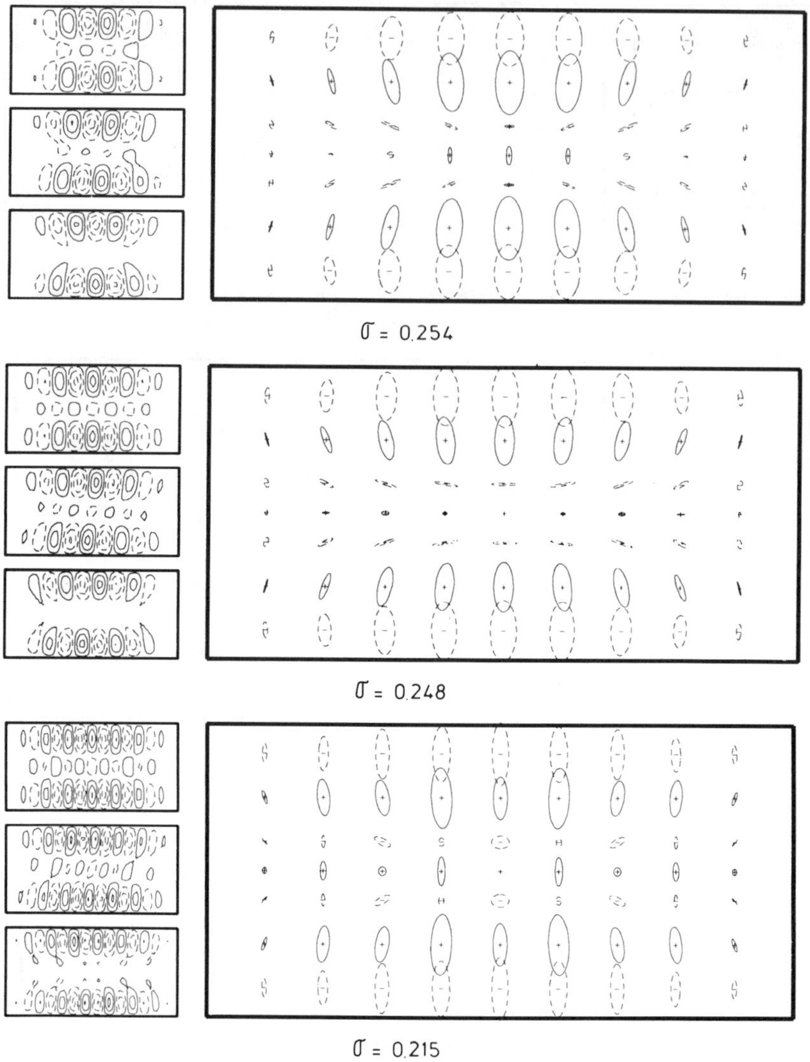


FIGURE 19. Same as Figure 17 for three channel modes.

The linear theory can be extended to provide a first estimate of the particle path caused by advection; this nonlinear effect, called Stokes drift, causes a net displacement during one cycle due to the spatial variation of the velocity field. Observationally, the Stokes drift manifests itself as a mean transport.

Figures 23 to 25 show the depth-integrated Stokes drift and correspond to Figures 17 to 19. It is a property of the Ball-modes (Figure 23) that the counterclockwise propagating gyres produce a net drift in the clockwise direction. This induced circulation is closer to the boundaries the higher the Ball-modes are. For the quadratic and the higher Ball-mode a weak cyclonic rotation in the lake center is observed. At the lake ends there is no drift.

The bay modes in Figure 24 exhibit significant transport at the lake ends, each in a different fashion. The mode with $\sigma = 0.395$ has a pronounced drift along the short sides of the basin against the direction of phase propagation. The lower left and top right corners act as sources; the others are sinks of the drift current. The next bay mode shows a pair of lateral gyres and the structure of the stream function of the mode with $\sigma = 0.263$ implies a drift along the ends of the long sides. The role of the sources and sinks is now interchanged.

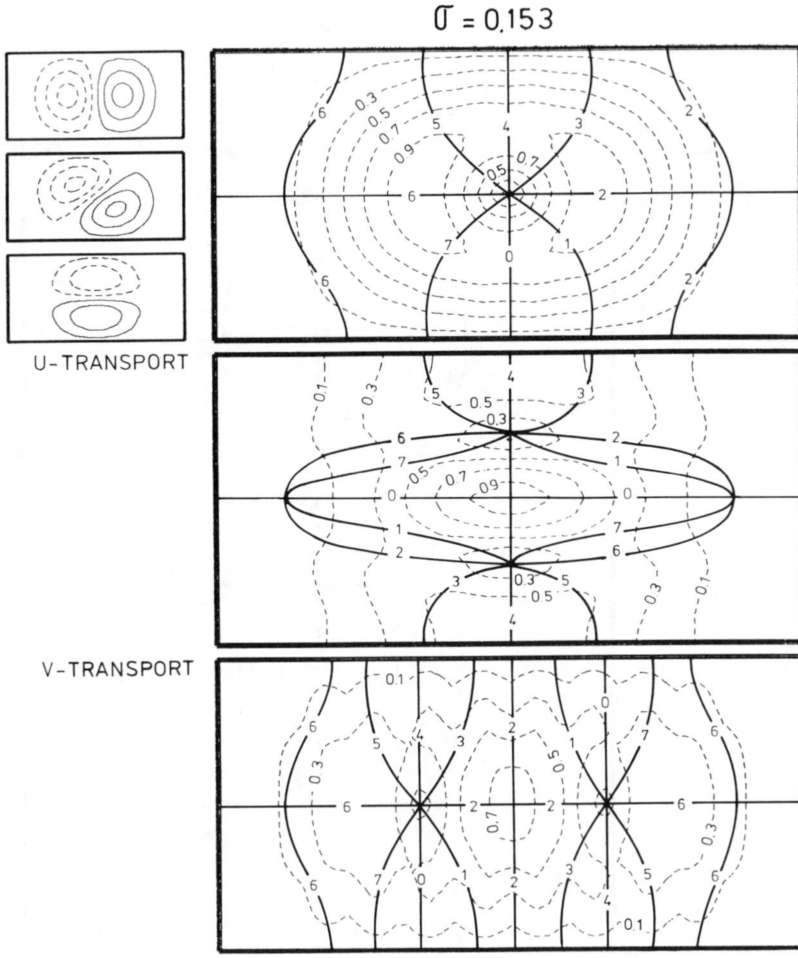


FIGURE 20. Lines of constant phase (solid; co-tidal lines) and lines of constant amplitude (dashed; co-range lines) of the stream function, the u-component (along thalweg axis) of the transport and the v-component for the linear Ball-mode. The numbers labeling lines of constant phase correspond to multiples of an eighth of a period; those of constant amplitude indicate the fraction of the maximum amplitude.[14]

Figure 25, finally, shows drift vectors for the channel modes. They hardly differ from each other, and the pattern consists of two near-shore drifts in opposite direction to phase propagation. The drift is experienced only close to the long sides of the basin and most of the lake has no transport at all. Opposite corners again act as sources and sinks, respectively. The Stokes drift has a particular sense of direction which is mainly clockwise around the basin (on the northern hemisphere). Nevertheless, the three mode types have different transport properties: Ball-type modes exhibit a circular pattern confined to the center of the lake. Bay and channel-type modes show a straight drift current along the short and long sides of the basin, respectively. This again underlines that knowledge of the modal type is of particular importance.

## VI. THE DOUBLE TRENCH

In this section a crude model is presented which shall give an idea of the behavior of shelf waves near river mouths, fjords, or estuarine domains. The simplicity of the model,

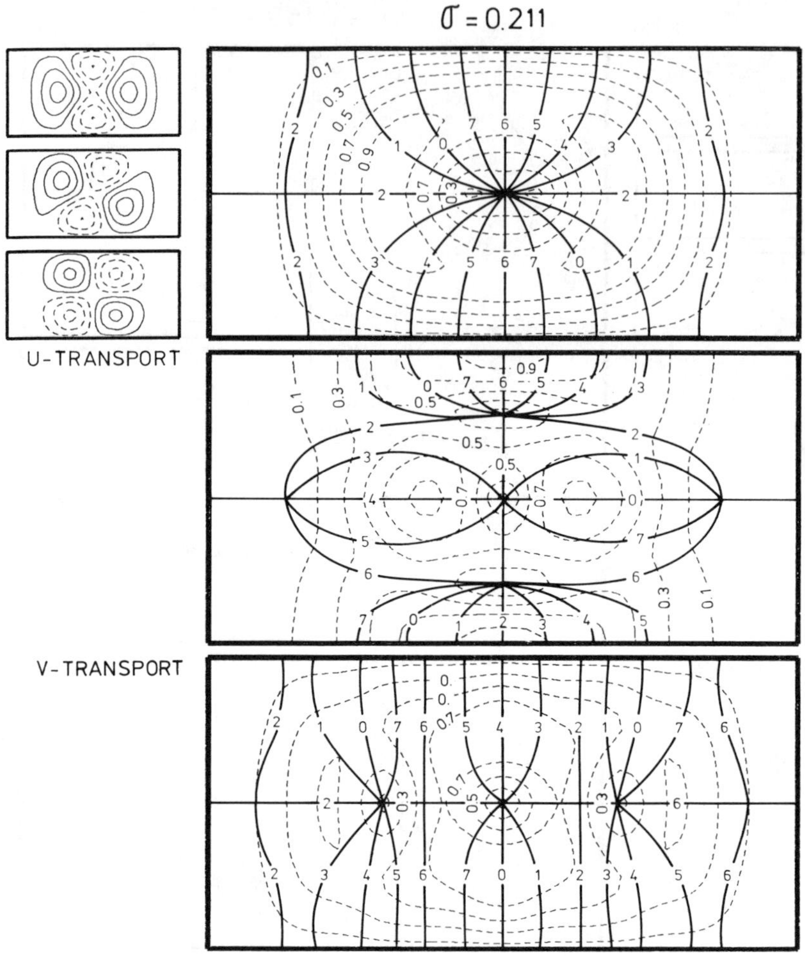


FIGURE 21. Same as Figure 20 for the quadratic Ball-mode.

however, does not lead to resonant bay modes as it only provides a first impression of the possible behavior of the composed geometry. Figure 26 illustrates the configuration in a (x,y)-coordinate system. An infinite trench as an idealized model for a continental shelf of width s communicates with a closed smaller trench of width r and length $\ell$; this appended gulf accounts for the estuarine region. The presence of vertical walls at $x = 0$, $y = 0$ and $x = \ell + s$ is a crucial drawback of this model, particularly because topographic waves are studied. It was shown in Stocker and Hutter[1] that presence of discontinuities in the isobaths affects the propagation mechanism of topographic waves considerably. The reason why the trench is used is its analytical simplicity: the dispersion relation $f(\sigma,k) = 0$ is a simple quadratic equation in k and thus, real and complex wave numbers can easily be determined. The latter are known to play a significant role but are cumbersome to calculate for less restrictive geometries.

The trench profiles in the two domains are given by

$$h^e(x,y) = h_0 e^{by}, \qquad 0 \leq y \leq r \tag{20a}$$

for the estuarine channel and

$$h^s(x,y) = h_0 e^{cx}, \qquad \ell \leq x \leq \ell + s \tag{20b}$$

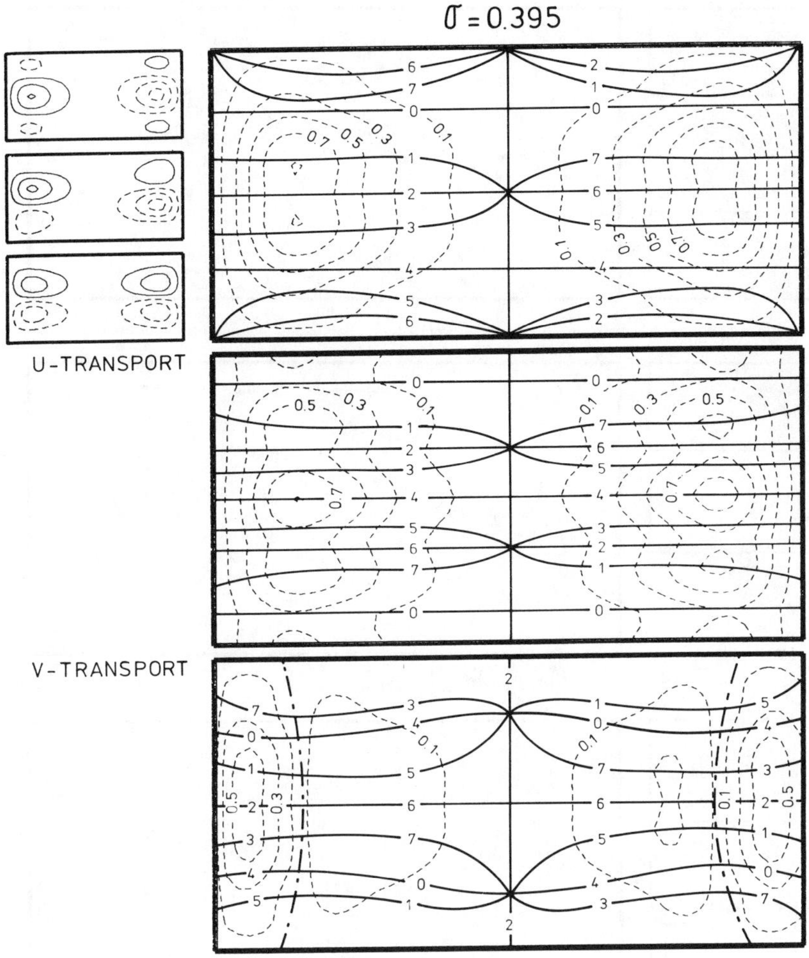


FIGURE [illegible]. Same as Figure 20 for a bay mode. The nodal line (dashed-dotted) is indicated.

for the infinite trench modeling the continental shelf. As demonstrated in Stocker and Hutter[1] the solutions in the respective domains read

$$\psi_n^e = e^{-i\sigma t} e^{(b/2)y} \sin\left(\frac{n\pi y}{r}\right) \exp(i k_n^e x) \tag{21a}$$

$$\psi_m^s = e^{-i\sigma t} e^{(c/2)(x-\ell)} \sin\left(\frac{m\pi(x-\ell)}{s}\right) \exp(i k_m^s y) \tag{21b}$$

and the wave numbers $k_n^e$ and $k_m^s$ of the nth and the mth transverse mode satisfy the dispersion relations.

$$(k_n^e)^2 - \frac{b}{\sigma} k_n^e + \left[\frac{b^2}{4} + \left(\frac{n\pi}{r}\right)^2\right] = 0 \tag{22a}$$

$$(k_m^s)^2 - \frac{c}{\sigma} k_m^s + \left[\frac{c^2}{4} + \left(\frac{m\pi}{s}\right)^2\right] = 0 \tag{22b}$$

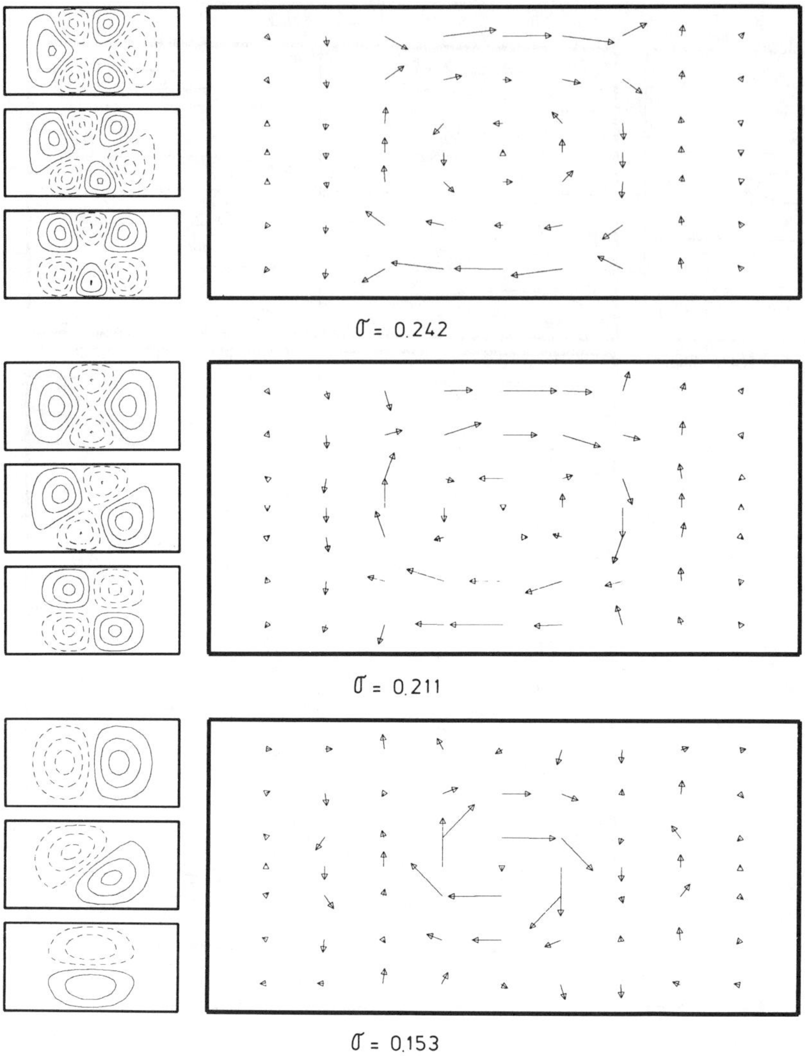


FIGURE 23. Depth-integrated Stokes drift during one period for three Ball-modes.

From these, cutoff frequencies for shelf and estuarine waves can easily be determined. It can be verified that the solutions (Equation 21) satisfy the no flux condition $\psi^e_n = 0$ on the lines $y = 0$, $y = r$, and $\psi^s_m = 0$ on $x = \ell$, $x = \ell + s$. We further require $\psi^e_n$ on $x = 0$ because the estuary is assumed to be closed. In order to achieve no flux across $x = 0$ the stream function $\psi^e_n$ must be a superposition of solutions with the two possible wave numbers $k^e_n$ of the nth mode which satisfy Equation 22a. Equation 21a, therefore, is replaced by

$$\psi^e_n = e^{-i\sigma t} e^{(b/2)y} \sin\left(\frac{n\pi y}{r}\right)[\exp(ik^e_{m}x) - \exp(ik^e_{2n}x)] \tag{23}$$

describing a standing wave in $0 < x < \ell$. It can be seen that the semi-infinite trench cannot sustain a bay mode which is evanescent away from the wall. We are therefore restricted in this simple model to consider only periodic wave motion (superposed evanescent motion) in the estuary. This implies that the frequency must not exceed the cutoff of the fundamental mode in the estuary, i.e.,

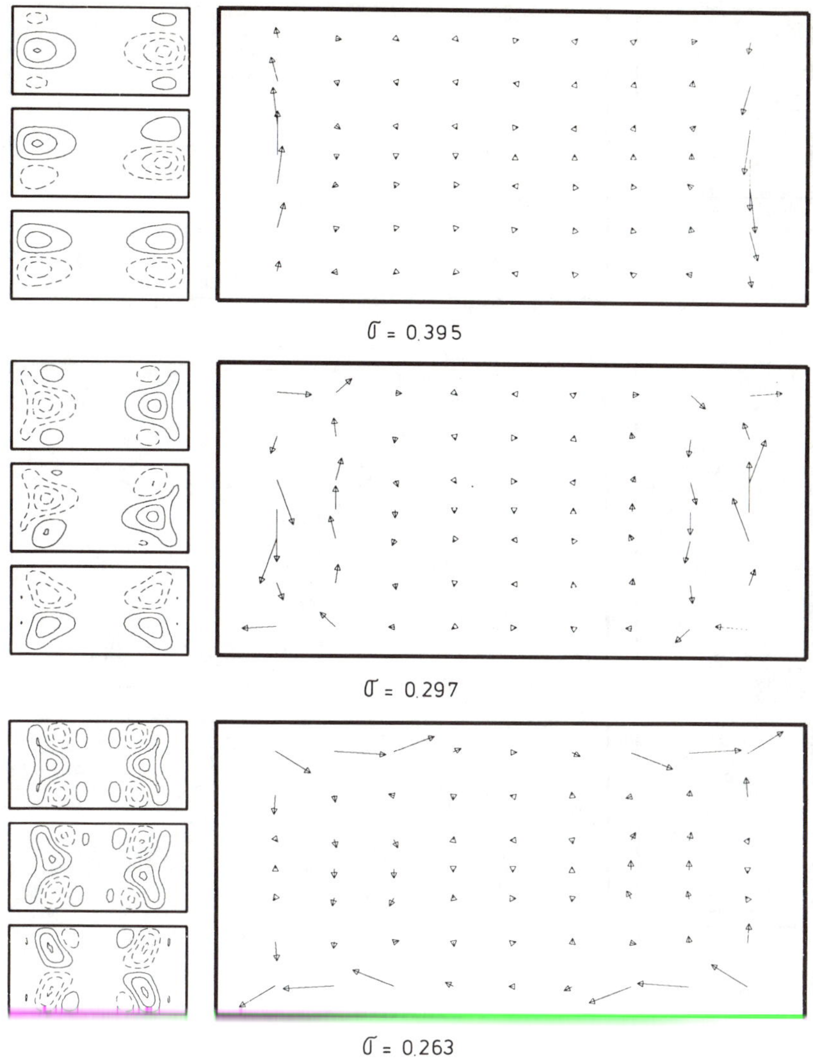


FIGURE 24. Same as Figure 23 for three bay modes.

$$\sigma < \frac{b}{\sqrt{b^2 + (2\pi/r)^2}}$$

The solutions in the respective domains must now be patched together. The junction in the neighborhood of $x = \ell$ has a peculiar structure in that there exists a vertical, roughly triangular wall at $y = r$, $\ell \leqslant x \leqslant \ell + br/c$. $\Gamma_2$ is its surface projection and $\Gamma_1$ is the surface projection of $h^e = h^s$. Aware of the imposed idealizations, we require

$$\left.\begin{aligned} \psi^e &= \psi^s \\ \frac{\partial \psi^e}{\partial y} &= \frac{\partial \psi^s}{\partial y} \end{aligned}\right\} \quad \text{on} \quad \Gamma = \Gamma_1 \cup \Gamma_2 \tag{24}$$

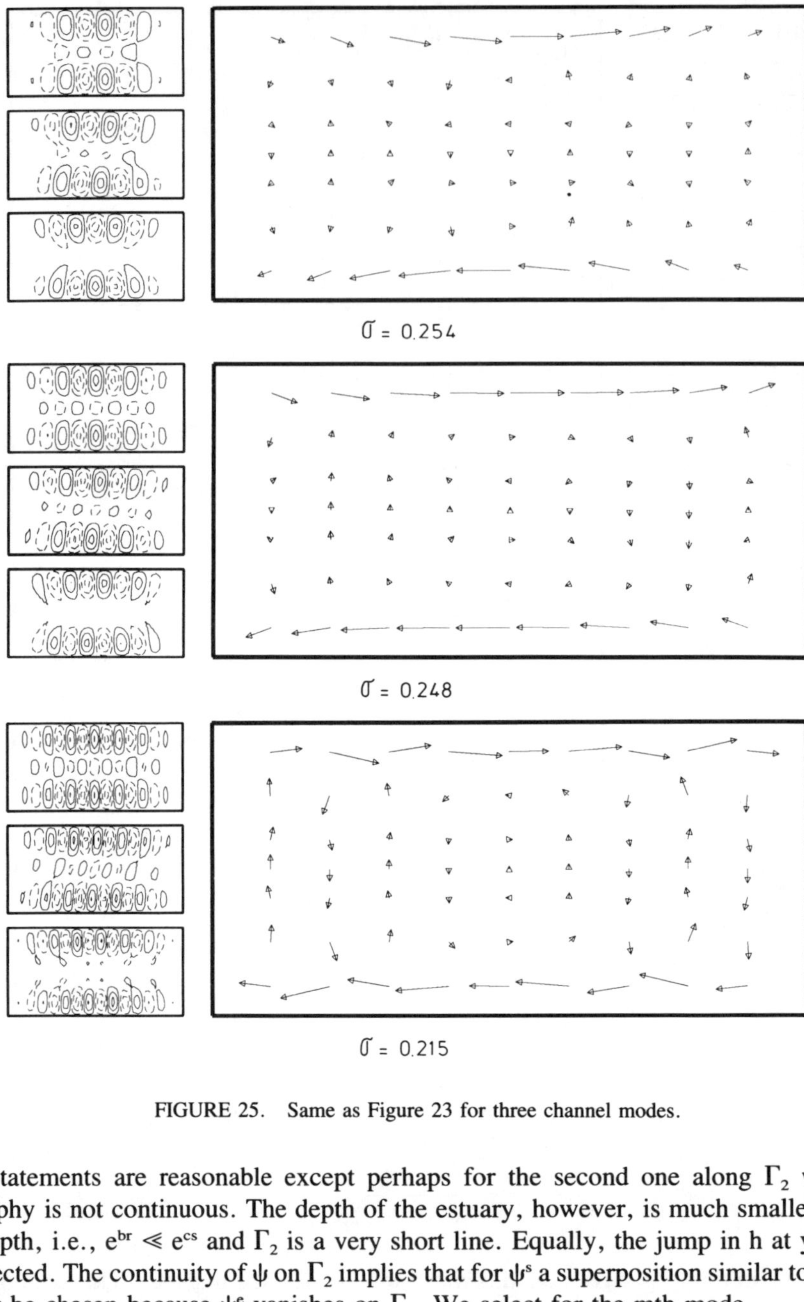


FIGURE 25. Same as Figure 23 for three channel modes.

These statements are reasonable except perhaps for the second one along $\Gamma_2$ where the topography is not continuous. The depth of the estuary, however, is much smaller than the shelf depth, i.e., $e^{br} \ll e^{cs}$ and $\Gamma_2$ is a very short line. Equally, the jump in h at $y = r$ can be neglected. The continuity of $\psi$ on $\Gamma_2$ implies that for $\psi^s$ a superposition similar to Equation 23 must be chosen because $\psi^e$ vanishes on $\Gamma_2$. We select for the mth mode

$$\psi^s_m = e^{-i\sigma t}e^{(c/2)x}\sin\left(\frac{m\pi(x - \ell)}{s}\right)\,[\exp(ik^s_{1m}(y-r)) - \exp(ik^s_{2m}(y-r))]$$

An exact satisfaction of Equation 24 amounts to the superposition of an infinite number of modes. Let us restrict considerations to one pair (i.e., $k^s_{11}$ and $k^s_{21}$) of shelf waves of the first transverse mode (i.e., $m = 1$); trapped shelf modes must not be included. The wave field in the estuary, on the other hand, shall consist of the fundamental pair (Equation 23) with $n = 1$ and a series $n = 2,3,\ldots$ of evanescent solutions of the form Equation 21a. These are needed to fulfil Equation 24 and are assumed to decay exponentially away from $\Gamma_1$ towards $x = 0$, i.e., $\text{Im}k^e_n < 0$, $n = 2,3,\ldots$ . These preliminaries lead to the solution

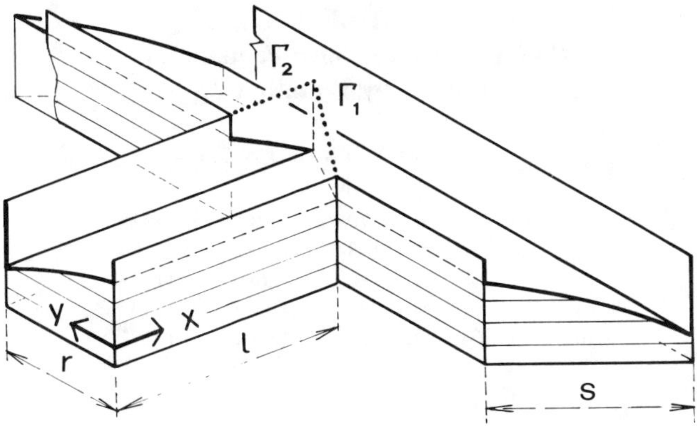


FIGURE 26. Double trench as a crude model of an estuary in connection with a continental shelf.

$$\psi^e = e^{-i\sigma t}e^{by/2}$$

$$\left[E_1 \sin\frac{\pi y}{r}\,[\exp(ik^e_{11}x) - \exp(ik^e_{21}x)] + \sum_{n=2}^{N} E_n \sin\left(\frac{n\pi y}{r}\right)\cdot \exp(ik^e_{1n}x)\right] \quad (25)$$

$$\psi^s = S_1 e^{-i\sigma t}e^{cx/2}\sin\frac{\pi(x-\ell)}{s}\,[\exp(ik^s_{11}(y-r)) - \exp(ik^s_{21}(y-r))]$$

which must satisfy Equation 24. The system Equation 25 contains the N + 1 unknown complex coeffieients $E_i$, (i = 1,..., N) and $S_1$. These are usually determined by evaluating Equation 24 at M points along the curve Γ. This procedure is called method of collocation or point matching. From Equation 24 follows

$$\psi^e - \psi^s = 0, \qquad \text{on} \qquad \Gamma_1$$

$$\frac{\partial\psi^e}{\partial y} - \frac{\partial\psi^s}{\partial y} = 0, \qquad \text{on} \qquad \Gamma_1$$

$$\frac{\partial\psi^e}{\partial y} - \frac{\partial\psi^s}{\partial y} = 0, \qquad \text{on} \qquad \Gamma_2$$

which at the M points $\xi_i$, i = 1,..., M on Γ takes the matrix form (assume $E_{N+1} \equiv S_1$)

$$A_{ij}E_j = 0 \quad (26)$$

Therefore, N = 3M − 1 estuarine modes must be superposed and a nontrivial solution can be constructed, provided that

$$\det A_{ij} = 0 \quad (27)$$

This equation selects the distinct frequencies σ at which the shelf wave pattern resonates with the wave motion in the estuary. It is of importance to study the influence of the number of collocation points M (or the truncation order N) on these eigenfrequencies. Table 4 lists a convergence test of the first four eigenfrequencies. Only as few as eight collocation points on Γ yield already a satisfactory value of the eigenfrequency.

**TABLE 4**
**Convergence of Eigenfrequencies Obtained by Equation 27**

| M | $\ell = 4, r = 1, b = 1, s = 2, c = 1$ | | | |
|---|---|---|---|---|
| 8 | 0.15284 | 0.14170 | 0.12755 | 0.11338 |
| 10 | 0.15275 | 0.14143 | 0.12711 | 0.11284 |
| 12 | 0.15268 | 0.14116 | 0.12680 | 0.11243 |
| 14 | 0.15261 | 0.14095 | 0.12644 | 0.11197 |
| 16 | 0.15254 | 0.14082 | 0.12601 | 0.11173 |

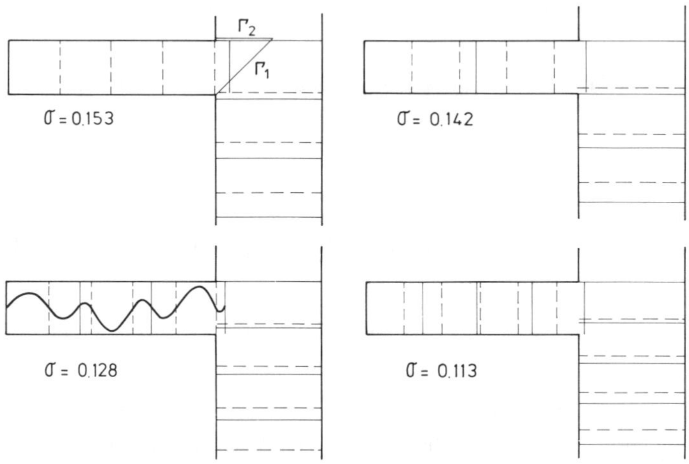


FIGURE 27. Distribution of nodal lines for the first four eigenfrequencies of the double trench shown in Figure 26 having the parameters $r = 1$, $b = 1$, $s = 2$, $\ell = 4$, $c = 1$.

It is instructive to investigate the influence of the five geometry parameters $\ell$, r, b, (length, width, and slope of the estuary) and s, c (width and slope of the shelf). For the following, the parameters are selected as follows: $M = 8$, $\ell = 4$, $r = 1$, $b = 1$, $s = 2$, $c = 1$ unless other values are stated explicitly. Figure 27 shows for the first four lowest eigenmodes how the nodal lines are distributed in the infinite and the finite estuarine trench. In one case we also show the distribution of the stream function. Amplitudes are large between nodal lines with the same signature (dashed-dashed) and small between those having different signature (solid-dashed). With this rule qualitative pictures of the distribution of the stream function can be constructed in the other cases.

Figure 28 shows how the eigenfrequencies of the first nine modes of the double trench vary with the length of the estuarine trench. For each mode $\sigma$ is monotonically increasing with the length of the estuary and for large $\ell$ seems to approach the cutoff frequency. Figure 29 shows how the eigenfrequencies of the four first modes vary with the width of the estuary trench (left) and with the strength of the topography, b (right). For all modes, the wider the trench is, the larger the eigenfrequency will be; equivalently, the stronger the trench topography is, the larger the eigenfrequency will become. This effect is particularly pronounced. Finally, in Figure 30 we show the dependence of $\sigma$ on the width and the topographic strength of the infinite trench. These dependencies are virtually absent or very weak. It follows that the resonating conditions of the double trench are critically influenced by the topographic features of the estuarine trench but only mildly by those of the shelf (which functions as the generator of the resonating effects). This indicates that local resolution of the topography

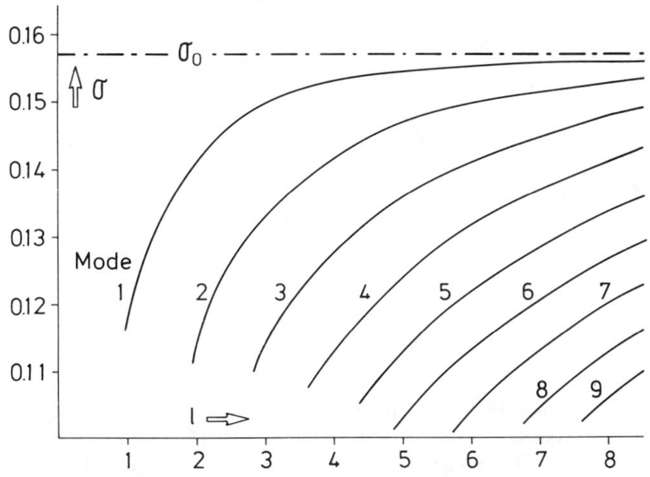


FIGURE 28. The first nine eigenfrequencies of the double trench shown in Figure 26 with parameters r = 1, b = 1, s = 2, c = 1, and $1 < \ell < 8$. Computations were done with M = 8 collocation points. For each mode $\sigma \rightarrow \sigma_0$ as $\ell \rightarrow \infty$, where $\sigma_0$ is the cutoff frequency.

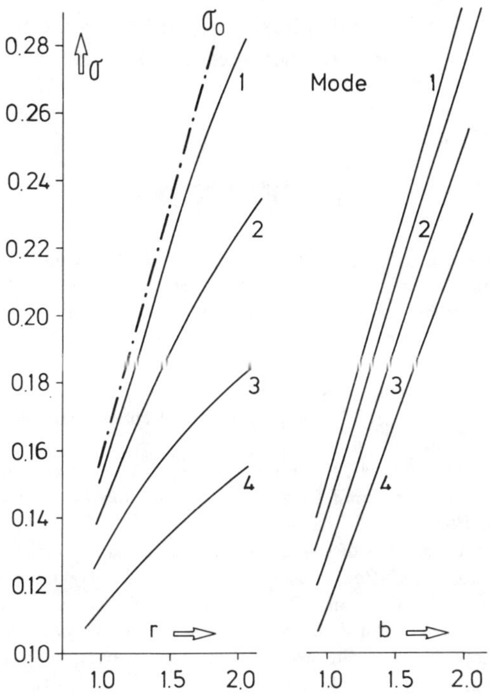


FIGURE 29. The first four eigenfrequencies of the double trench shown in Figure 26 with parameters b = 1 (left) and r = 1 (right); s = 2 and $\ell = 4$. Computations were done with M = 8 collocation points.

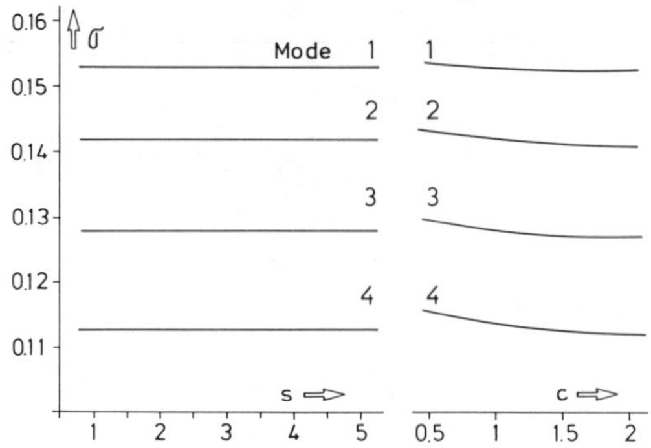


FIGURE 30. Dependence of the first four eigenfrequencies of the double trench shown in Figure 26 on the width s and topography parameter c of the infinite trench. Parameters are c = 1 (left) and s = 2 (right); and r = 1, b = 1, ℓ = 4. The number of collocation points was M = 8.

for topographic waves needs to be accurate, while far field conditions that generate the local activity may well only roughly have to be known.

The recent study by Stocker and Johnson[20] investigates the topographic wave problem in geometries consisting of a shelf channel interrupted by an estuary or a headland.

# REFERENCES


1. **Stocker, T. and Hutter, K.,** *Topographic Waves in Channels and Lakes on the f-Plane,* Lecture Notes on Coastal and Estuarine Studies, Vol. 21, Springer-Verlag, New York, 1987.
2. **Johnson, E. R.,** Topographic waves in elliptical basins, *Geophys. Astrophys. Fluid Dyn.,* 37, 279, 1987.
3. **Johnson, E. R.,** A conformal mapping technique for topographic wave problems: semi-infinite channels and elongated basins, *J. Fluid Mech.,* 177, 395, 1987.
4. **Mysak, L. A.,** Elliptical topographic waves, *Geophys. Astrophys. Fluid Dyn.,* 31, 93, 1985.
5. **Buchwald, V. T. and Adams, J. T.,** The propagation of continental shelf waves, *Proc. R. Soc. London Ser. A,* 305, 235, 1968.
6. **Lamb, H.,** *Hydrodynamics,* 6th ed., Cambridge University Press, Cambridge, 1932.
7. **Saylor, J. H., Huang, J. S. K., and Reid, R. O.,** Vortex modes in Southern Lake Michigan, *J. Phys. Oceanogr.,* 10, 1814, 1980.
8. **Ball, T. K.,** Second class motions of a shallow liquid, *J. Fluid Mech.,* 23, 545, 1965.
9. **Mysak, L. A., Salvade, G., Hutter, K., and Scheiwiller, T.,** Topographic waves in an elliptical basin, with applications to the Lake of Lugano, *Philos. Trans. R. Soc. London Ser. A,* 316, 1, 1985.
10. **Mysak, L. A.,** Recent advances in shelf wave dynamics, *Rev. Geophys. Space Phys.,* 18, 211, 1980.
11. **Stocker, T. and Hutter, K.,** One-dimensional models for topographic Rossby waves in elongated basins on the f-plane, *J. Fluid Mech.,* 170, 435, 1986.
12. **Stocker, T. and Hutter, K.,** Topographic waves in rectangular basins, *J. Fluid Mech.,* 185, 107, 1987.
13. **Trösch, J.,** Finite element calculation of topographic waves in lakes, in Proc. 4th Int. Conf. Applied Numerical Modelling, Tainan, Taiwan.
14. **Stocker, T.,** Topographic Waves. Eigenmodes and Reflections in Lakes and Semi-Infinite Channels, Mitteilung No. 93 der Versuchsanstalt fur Wasserbäu, Hydrologie und Glaziologie, ETH Zurich, 1987, 170.
15. **Stocker, T.,** A numerical study of reflections of topographic waves in semi-infinite channels, *J. Phys. Oceanogr.,* 18, 609, 1988.
16. **Ursell, F.,** Mathematical aspects of trapping modes in the theory of surface waves, *J. Fluid Mech.,* 183, 421, 1987.
17. **Stocker, T. and Johnson, E. R.,** Topographic waves in open domains. Part 2. Bay modes and resonances, *J. Fluid Mech.,* 200, 77, 1989.

18. **Gonella, J.,** A rotary-component methods for analysing meteorological and oceanographic vector time series, *Deep-Sea Res.,* 19, 833, 1972.
19. **Huthnance, J. M.,** On trapped waves over a continental shelf, *J. Fluid Mech.,* 69, 689, 1975.
20. **Stocker, T. F. and Johnson, E. R.,** Transmission and reflection of shelf waves by estuaries and headlands, *J. Fluid Mech.,* submitted.

# INDEX

## A



## B



## C

## D



## E

## F



## G



## H



## I



## J



## K



## L

## Q



## R



## S

## T



## U



## V